D1255685

LI

40.00
80U

The Preservation and Valuation of Biological Resources

The
Preservation
and Valuation
of Biological Resources

EDITED BY

Gordon H. Orians

Gardner M. Brown, Jr.

William E. Kunin

Joseph E. Swierzbinski

UNIVERSITY OF WASHINGTON PRESS

Seattle and London

QH
75
.P73
1989

Proceedings of an Interdisciplinary Workshop
held at Lake Wilderness, King County, Washington
June 12–16, 1985

Copyright © 1990 by the University of Washington Press
Printed in the United States of America

All rights reserved. No part of this publication may be reproduced or transmitted in any
form or by any means, electronic or mechanical, including photocopy, recording, or any
information storage or retrieval system, without permission in writing from the publisher.

Library of Congress Cataloging-in-Publication Data

The Preservation and valuation of biological resources / edited by
    Gordon H. Orians . . . [et al.].
    p.   cm.
    ISBN 0–295–97029–4
    1. Germplasm resources.   2. Biological diversity conservation.
    I. Orians, Gordon H.
    QH75.P73   1990   90–12118
    333.9516—dc20   CIP

The paper used in this publication meets the minimum requirements of American National Standard for Information Sciences — Permanence of Paper for Printed Library
Materials, ANSI Z39.48-1984.   ∞

Dedicated to the memory of
Deborah Rabinowitz,
one of the pioneers of conservation biology.
Her energy and insights have been valuable resources
that will be sorely missed.

9/30/92   Book House $21.00 buy

Contents

Preface

Many important environmental policy issues concern management of living resources. A real threat to human welfare is posed by the high rates of extinction of species expected during the next few decades, especially in the tropics. Predicting the rate of loss of species is a difficult task because so many species are yet to be described, the rate of conversion of tropical forests to other vegetation types is known only within limits, and future trends may deviate substantially from current predictions. Nonetheless, it is highly likely that loss of upwards of a million species of living organisms may occur by the end of the second decade of the twenty-first century.

It is widely recognized that a reduction of this magnitude in global genetic resources would seriously affect human welfare in many ways, including loss of genes to incorporate into genomes of existing domestic plants and animals, loss of new crops and plants that are potential sources of medicines and industrial chemicals, loss of species useful in biological and medical research, loss of potential biological control agents, and a reduction in the aesthetic richness of the earth.

A critical exigency is the need to develop approaches that will reduce both the number of losses and the impacts on human society of the losses that are inevitable despite our best efforts to avert them. However, as long as societies believe that preservation of natural habitats, the only way to preserve many species, is a waste of natural resources and a loss of potential income that could better improve the welfare of people, genetic resources are not likely to be high priority items on the political agenda of any country. This means that efforts must be expended to determine more accurately the real worth of genetic resources and to communicate those values to the general public and decision makers.

As a contribution to this important investigation, the Institute for Environmental Studies of the University of Washington, with the support of the William and Flora Hewlett Foundation, organized a Workshop on the Preservation and Valuation of Biological Resources, which was held at the Lake Wilderness Conference Center near Seattle, June 12–16, 1985. The workshop brought together experts from a number of disciplines, including biological sciences, economics, and management, for intensive discussions. To provide a focus for discussions, six major papers were commissioned, written in advance, and distributed to all participants prior to the workshop, together with the written commentaries of two discussants for each paper. Since all participants had read the papers before attending the workshop, valuable time was not expended in reading papers at the workshop but was devoted almost exclusively to in-depth discussions of the issues raised by the authors and discussants. The topics addressed were (1) laboratory based (*ex situ*) preservation technologies, (2) field based (*in situ*) preservation technologies, (3) genetic measures of uniqueness, (4) ecological measures of uniqueness, (5) the economic valuation of genetic resources, and (6) incentives for the conservation of genetic resources.

A half day was devoted to discussing each paper. A rapporteur assigned to each session prepared a written summary of the discussions, highlighting the main points raised, especially those not contained in the papers and commentaries. The final morning was devoted to meetings of three working groups whose objective was to synthesize the workshop proceedings with respect to needs for future research. The results of these deliberations are presented in the final chapter.

We believe that the materials contained in the papers and commentaries, supplemented by discussions in which people of varied backgrounds developed a basis for sharing different perspectives, represent a valuable contribution to conceptualizing the nature of problems surrounding biological resource preservation, how preservation can be accomplished, how decisions can best be made for deciding what to preserve, and how such decisions can be put into practice.

GORDON H. ORIANS
GARDNER M. BROWN, JR.
WILLIAM E. KUNIN
JOSEPH E. SWIERZBINSKI

The Preservation and Valuation of Biological Resources

1

Technologies
for Germ Plasm Preservation *Ex Situ*

Arthur K. Weissinger

Natural or "wild" organisms ideally exist as one or more populations in a dynamic state of continuous evolution. They change perpetually to remain adapted, shaped by interactions within the populations as well as with other organisms and forces in their environments. Change in the genetic constitution or "germ plasm" of these populations is, of course, integral to this dynamic process. Because of this, the gene pool of a species is best described as a continuum—a thing in process—rather than a static entity.

Germ plasm conservation techniques, in order to represent the natural condition perfectly, must retain the genes and genotypes of an organism and also their dynamic character. This requires protection not only of the organism but also of the environment responsible for its continuing evolution. In essence, this is the goal of nature reserves and other *in situ* methods of germ plasm conservation.

At times, however, *in situ* methods of germ plasm conservation fail, or their use is inappropriate. Such a situation might occur when no suitable natural area is available for protection as a reserve, or reserve areas are in danger of disruption and cannot be protected. Also the organism may be in imminent danger of extinction, and its numbers or genetic variability might be inadequate to found new colonies *in situ*. When these types of situations occur, *ex situ* or "laboratory" techniques of germ plasm preservation can be employed.

Ex situ technologies, while offering important alternatives to *in situ* conservation methods, are not replacements for them, nor are they entirely exempt from evolutionary forces. The purpose of this discussion is to examine some of the major *ex situ* preservation technologies in terms of how they are affected by and contend with the biological realities that dictate their design and application.

First, *ex situ* techniques cannot conserve the whole range of an organism's diversity; rather they preserve a sample of it. Since the gene pool as a whole is dynamic and large, this sample is necessarily incomplete. It represents only a portion of the population at the moment of its extraction.

Sampling germ plasm produces an artificial founder or "bottleneck" effect, since only a part of the gene pool is included (Franklin, 1980; Marshall and Brown, 1975). Specific individual alleles are thus likely to be lost, and certain genotypes with favorable or unusual gene associations may also be lost. Such linkage blocks are generally thought to be of major importance, for example, in crop improvement. Loss of these blocks affects not only strictly Mendelian or additive genetic effects, but also tends to reduce the variance of polygenic traits. Because these blocks are the accumulated product of many recombinational events, they are unlikely to be recovered.

Although gene losses due to sampling limitations may be small and largely neutral under normal conditions, such losses could be crucial to survival in a time of environmental crisis. For this reason, the amount and type of germ plasm that is to be preserved is probably the single most important decision in determining preservation strategies. The extent to which a preservation technology limits the scope of this decision is an important measure of its usefulness.

The way in which a preservation technology controls continuing evolutionary change is another major factor. The sample can be allowed to evolve freely or its evolution can be retarded. Allowing selection to occur is to some extent antithetical to germ plasm preservation. Suspending evolution entirely, however, is not without its own problems. New gene combinations cannot arise, and the sample will therefore become increasingly less representative of extant (natural) populations, which continue to evolve.

Another major biological phenomenon with important implications for preservation technologies is gene flow, the movement of new genes into the sample gene pool. This is of virtually no consequence if only a single sample of a natural population is acquired for preservation. In that case, there is no influx of new genes, and the sample composition is determined completely by the dynamics inherent to the preservation technology.

If wild populations remain extant, however, continued sampling is one way in which the preserved material can be made more representative of the natural condition. This applies equally to all *ex situ* techniques. It does not, however, guarantee that the sample gene pool will be enlarged in absolute terms. If the natural population is in a state of rapid decline, continued sampling may actually bias the sample unfavorably, particularly if the preservation technology allows uncontrolled selection within the sample population. An important feature of repeated sampling is that while new alleles may not be added to the sample population, new genotypes (gene blocks or groupings) may be acquired. If this is a major priority, the preservation technology must be designed to allow such genetic influx while at the same time limiting its unfavorable aspects.

Evaluation of preservation schemes must take into account mutation, genetic rearrangements, aneuploidy, and other genetic alterations associated with the technique. Cell and tissue culture, for example, may produce anomalies at relatively high rates, while technologies employing cryogenic storage may effectively eliminate genetic alterations altogether.

Recent information about molecular mechanisms is highly pertinent to the way in which we evaluate germ plasm preservation techniques. Work on transposable genetic elements in maize (Saedler and Nevers, 1985) suggests that the transposition of these elements produces specific kinds of changes at the sites from which they are excised. Such modifications may provide a large amount of genetic variability compared with that produced by mutation. Further, it has been suggested (McClintock, 1984; Michael Freeling, University of California, Berkeley, pers. comm.) that the movement of transposable elements may occur in response to "genomic shock" from such trauma as virus infection. If these theories are correct, direct genomic modification in response to environmental stimuli may produce genetic variability. This implies, perhaps, that populations can become more variable more rapidly than would be possible by point mutation alone.

These theories, if they are substantiated, will require us to modify our estimates of ideal sample sizes, at least for some organisms. Perhaps it will cause reevaluation of the need for germ plasm preservation in those organisms that, to some extent, may be capable of regenerating lost variability. Techniques may become available that can regenerate variability and thereby restore "lost" genetic diversity. These possibilities, speculative at present, may have a strong impact on future preservation strategies.

The major methods of germ plasm maintenance *ex situ* are addressed in a descending order of the extent to which they represent the natural condition of organisms and in an increasing order of technical complexity. The listing of organisms is not meant to be complete but provides examples that help to demonstrate the advantages and limitations of technologies. The boundaries between some of these technologies, such as tissue culture and gene libraries, are necessarily arbitrary, since the two methods are often used in conjunction with one another.

BOTANICAL GARDENS, ARBORETA, AND NURSERIES

Botanical gardens and arboreta are areas in which plants can be protected, propagated, and studied under conditions that in many respects mimic natural environments. Plants are maintained as mature individuals, and may also be stored as seed for long-term maintenance, propagation, exchange with other gardens, and reintroduction into natural habitats. The primary goals of botanical gardens are to educate and edify the public, and to provide an environment for studying and propagating plants. They can also serve important functions both directly and indirectly in germ plasm preservation (Simmons et al., 1976; Frankel and Soulé, 1981).

The use of gardens and arboreta for preservation of genetic stocks is especially important for the maintenance of species endangered through disruption of their natural habitats. Because of space limitations in botanical gardens and the large amount of space required for the growth and propagation of living plants, the number of species that can be preserved is strictly limited.

Rare plants characteristic of species-poor communities are well suited to preservation in botanical gardens, because of their small number and paucity of companion species. Botanical gardens are not very useful, however, for the preservation of large plant associations characteristic of major ecosystems.

Nurseries, as discussed here, refer to stands of plants either "wild" or domesticated that are grown for purposes of propagation, breeding, and study. They are usually pure stands of a single species or a few related species or varieties. Their chief feature is that they can be subjected to an intensive level of management and manipulation impossible, or very difficult, with most other strategies for the preservation of living plants.

The major contribution of nurseries to germ plasm preservation is as an adjunct to other technologies. They may be used for propagation, for evaluation and study of germ plasm resources, and for breeding and artificial selection. They also provide source materials for the reestablishment of plantings in natural habitats, for tissue culture, and for gene isolation and cloning.

Genetic Effects of Preservation in Botanical Gardens, Arboreta, and Nurseries

The extent to which the genetic base of the preserved species is reduced is largely dependent on the size of the sample and variability in the natural population. Very rare species with little genetic variability are affected less than more variable species by the "bottleneck" effect of sampling.

The reproductive strategy of plants must also be considered when estimating the genetic effects of sampling. Self-pollinated organisms usually exist as heterogeneous populations of highly homozygous individuals (Allard, 1960). Alleles may be carried by only a relatively small proportion of the population, but the genotypes of completely homozygous individuals can be replicated to high numbers within the population. Self-pollination of homozygous organisms results in progeny that are essentially duplicates of the parental genotype. Conversely, populations of cross-pollinated species tend to be homogeneous blends of largely heterozygous, essentially unique, individuals. Alleles, except for those of rare or deleterious genes, are expected to be represented within many members of the population.

Differences in the distribution of alleles and genotypes within a population require different sampling strategies in order to capture a representative gene base. Preservation objectives should also be considered, because it may be the goal of the project to retain certain favorable genotypes or specific alleles, or perhaps to recover as many alleles as possible.

Because of the small number of individuals that can be maintained in botanical gardens, arboreta, and nurseries, the choice of what to save, always an important decision, becomes an overriding consideration. However, other factors such as mutation, selection, genetic drift, and gene flow might affect the genetic diversity of preserved materials and must be monitored.

Mutation, including somatic mutation, is most important either in large popu-

6

lations of annual plants or, to a smaller degree, in rapidly growing and very long-lived perennial plants. Since mutations are relatively rare, they are unlikely to modify the genetic base of preserved materials unless populations are reproduced repeatedly over long periods, allowing rare mutations to arise and accumulate. The same can be said for perennials that grow rapidly over long periods, where extremely rare somatic mutations might become important. Somatic mutant stocks may arise and, even if unfavorable, could be maintained by contact with the normal portion of the plant. Many examples exist where such mutant scions have been selected and have provided useful or unusual varieties.

In the case of annuals, mutations are likely to be deleterious and would be selected against by natural forces or by artificial selection.

It is virtually impossible to eliminate the effects of intensive artificial selection on materials within botanical gardens, arboreta, or nurseries. The effects of artificial selection can be reduced, however, by dissemination of samples to more than one location. In this way, populations of the preserved organism could be exposed to an array of natural environments. Selection might then actually tend to diversify the genetic base of the organism. Artificial selection can also be limited by propagation of randomly chosen plants.

Genetic drift—that is, random sampling errors—may also tend to reduce the variability of preserved genetic stocks, or make them less representative of the natural population. In the case of annual plants, drift is difficult to control in small populations randomly sampled for propagation. As in the case of natural selection, a useful strategy may be to disperse subsamples to a number of remote locations. Pooled progeny from these isolated populations might tend to be more representative of the original population than would any single sample.

Genetic drift can also occur in perennial materials but normally at a slower pace than in annuals. To some extent the "error" in the original sample probably represents the most important form of genetic drift in perennials, especially long-lived trees that may be propagated vegetatively. Drift becomes an important factor, however, even in these perennial species, if they are used to produce sexual progeny.

For plants maintained in arboreta, botanical gardens, and nurseries, gene flow into and out of preserved populations can present unique problems. Populations of a species, or even members of different species, may be put into novel reproductive associations, and genetic combinations unknown under natural conditions can occur. This difficulty can easily be overcome, however, by the maintenance of progeny produced by controlled pollination.

Safety, Accessibility, and Flexibility
of Botanical Gardens, Arboreta, and Nurseries

A major limitation to the usefulness of botanical gardens, arboreta, and nurseries in germ plasm preservation is the time scale of preservation for which they are appropriate techniques. The effects of mutation, selection, and genetic drift could combine over long periods to alter significantly the genetic composition of preserved materials. A more important time constraint, however, is the resource-intensive nature of these preservation techniques and the way in which resources are allocated.

Large amounts of land, labor, management, and capital resources are needed, seriously limiting the number of species and population sizes that can be preserved. As priorities change, allocation of resources may change. Collections are sometimes lost or discarded, especially those maintained by one person. When that person changes research interests or leaves an institution, the continuity of the collections is at risk. Materials evolved over geological periods must be preserved in time spans measured in human lifetimes or careers. This serious constraint can be overcome only by establishing long-term priorities. Because of their resource-intensive nature and population dynamics, these techniques are of little use for very long-term maintenance, and may best be applied to the discovery or development of horticultural or agricultural uses for endangered plants, thereby ensuring the preservation of at least part of the genetic resources of a species. While materials treated in this way may not be truly representative of the wild condition, it is perhaps preferable to save a part of their inherent variability rather than lose all (Raven, 1976).

Evaluation and Documentation in Botanical Gardens and Nurseries

Botanical gardens, arboreta, and nurseries provide unique opportunities for the description and evaluation of materials. Information gathering should include inventories of collections and seeds to allow efficient collection with a minimum of redundancy. Data should also be collected regarding origins, ecology, reproductive physiology, flowering and seeding dates, diseases and pests, and distribution. Information storage in a computer data base provides efficient access to and distribution of data (Frankel and Soulé, 1981). Botanical gardens can also assume some responsibility for ascertaining the status and location of threatened species. This effort would be particularly helpful in the tropics, where the number of botanical gardens is unfortunately very small and their distribution uneven.

Availability, Costs, and Alternatives

While many excellent botanical gardens exist, they are, unfortunately, mostly located in temperate countries, not in tropical countries where the need is most acute because of the large variation in organisms and the high incidence of environmental threats, particularly deforestation. The cost of maintaining plants by these methods is variable, depending on size, location, staff, and so forth. Because cost is dependent primarily on the value of land and labor, it probably will increase in real terms over time. Therefore, the maintenance of plants in seed banks or in tissue culture is an attractive alternative. At this time, however, since nearly every storage strategy requires at least periodic rejuvenation, some reliance on botanical gardens is probably unavoidable.

ZOOS, FARMS, AQUARIA, AND CAPTIVE FISHERIES

These facilities are analogous to botanical gardens, arboreta, and nurseries in that they preserve organisms in the living state, and thus their effects on the genetic constitution of preserved materials are similar. While each of them serves a number of functions, their primary contribution to germ plasm preservation is in captive propagation programs. They also can provide a great deal of information about the animals they protect. This information-gathering function is critical, particularly for the reintroduction of animals into natural habitats.

There is a somewhat arbitrary but important distinction made here between zoos and farms, as well as between aquaria and captive fisheries. While zoos primarily house wild animals, farms are facilities for breeding and propagating domesticated animals. Similarly, aquaria have a major commitment to the propagation and study of wild marine animals, whereas fish hatcheries usually deal with the reproduction of game fish species for introduction into artificial habitats or natural rivers and streams.

The primary objective of captive propagation programs is the maintenance and reproduction of rare or endangered species, often with the goal of release in natural areas. Propagation can occur either by random mating within preserved populations or, more efficiently, by forced pairing. A major benefit of controlled matings is the minimization of inbreeding and the retention of maximum fitness within progeny that are to be released in the wild (see Chapter 3).

Genetic Considerations in Captive Propagation of Animals

Wild animals are propagated in captivity for a number of purposes. Of these, two—the development of new breeds of domestic animals or improvement of existing breeds, and the breeding of endangered species for preservation and release—contribute significantly to genetic preservation. Because of the resource-intensive nature of such enterprises, they often are affected by economic and political considerations that govern allocation of resources. Breeding and propagation programs appear only occasionally to proceed from ideal genetic considerations, particularly in the case of wild animals.

The domestication of wild animals is analogous in many respects to the domestication of wild plants for economic purposes. Such domestication can provide not only useful genetic material for improving agricultural systems but also a means for ensuring the survival of species through economic incentives (see

9

Chapter 6). In fact, however, only 16 of 4,500 species of mammals have been domesticated to meet basic needs (Frankel and Soulé, 1981; Spillett et al., 1975) even though many other animals are amenable to domestication (Coe, 1980).

Genetic and phenotypic changes in captive populations—a kind of "domestication" that must be avoided—have been divided into four major categories by Frankel and Soulé (1981). First, organisms might be selected for increased fecundity. Although this could be useful while maintaining the group in captivity, it is a departure from the (evolved) natural condition which may result in a reduction of fitness in progeny produced for release.

Second, selection for a particular phenotype may reduce genetic variability of the preserved population. The selected phenotype is likely to differ from those evolved under natural conditions, reducing its fitness when it is returned to the wild.

Third, behavioral changes, such as selection against aggressive tendencies in captive organisms, can be quite subtle. Nevertheless, those behaviors that make animals easier to manipulate in captivity may have strong deleterious effects in nature.

Fourth, the failure to exercise selection in captivity can also reduce overall fitness. Under natural conditions, many factors apply selection pressure against unfit individuals. If these less fit types are protected in captivity, their deleterious traits will be present in the population in an artificially high proportion. This would tend to weaken the population that is released in nature.

Captive populations offer the possibility of crossing varietal and other boundaries that limit cross-fertilization in the wild. This requires a decision about whether the preserved gene pool should be maintained as a series of distinct subpopulations or be allowed to intermate freely to produce a more variable "superpopulation" that could be used for colonizing diverse environments. In the case of wild marine organisms, the evidence for distinct levels of genetic variability and very different varieties of organisms associated with various environments suggests that colonization of these environments would require large, extremely varied superpopulations or smaller adapted populations (Bulnheim and Scholl, 1981; Ferguson and Mason, 1981; Gooch and Schopf, 1972). The much larger number of individuals required for release of superpopulations (because of selection after release) would unnecessarily burden limited preservation facilities. It would also be very difficult to eliminate the effects of selection and drift in such large populations in captivity.

The preservation of domesticated stocks is subject to all the forces that limit the variability of captive wild stocks. This condition is exacerbated by breeding and selection to achieve progeny with the very uniform phenotypes desired for efficient modern agriculture. Uniformity is also increased by the high levels of inbreeding typical of domestic stock (Mason, 1974; Maijala, 1974). In most instances, this reduces the ability of domesticates to respond to environmental changes.

There is an increasing awareness of the problems of genetic uniformity in farm livestock. Activity appears to be increasing to maintain rare breeds of livestock to improve this condition (see, e.g., Ryder, 1979). In some instances such breeds are being maintained in zoos or related farms. There is also much concern about the amount of variability maintained in these small preserved populations, as well as the accumulation of the effects of genetic drift, selection, and gene flow from other populations (Smith, 1976).

While maintenance of rare domesticated breeds in farms is still possible, concerns about their genetic integrity and the high resource requirement for their maintenance are serious problems. Since preservation of breeding stock in the form of frozen semen or embryos is now possible for most farm species, these methods offer excellent hope for maintaining a stable level of genetic diversity (Smith, 1984).

Safety, Accessibility, and Flexibility
of Zoos, Farms, Aquaria, and Captive Fisheries

Small populations maintained *ex situ* usually suffer less risk from environmental pressures than do similar size populations in the wild. They are often under similar or worse pressure, however, from genetic factors produced by the dynamics of their populations. For all these methods of preservation, genetic disruption appears to be the greatest single danger.

The flexibility of all these methods is primarily limited by available resources. Captive breeding programs are extremely expensive, and not all institutions will be able to preserve adequate breeding populations to ensure the diversity required to produce widely adapted animals for release. For institutions unable to maintain such populations, several options have been suggested that would allow them to contribute to the overall conservation effort (Frankel and Soulé, 1981). Animals or frozen semen or embryos can be loaned or exchanged with other institutions to enlarge their effective breeding gene pools. Other institutions can also offer temporary housing of "surplus" animals. Finally, all institutions can share results of research and experience.

Availability, Costs, and Alternatives to Ex Situ Maintenance of Living Animals

The availability of adequate facilities for captive breeding programs directed toward release in the wild is primarily dependent on resources. Costs of maintaining individual large animals in collections may range from hundreds to thousands of dollars annually (Frankel and Soulé, 1981). The costs of land, labor, and energy—all major contributors to this sum—can be expected to increase over time.

11

Two alternative options exist that can take the place of captive animal programs. The first (and best) solution is to maintain large habitats, such as nature reserves, where animals can be preserved as part of evolving ecosystems. Based on current trends, this option seems to be limited, for the foreseeable future, by competition with human needs for living space and arable land.

Preservation of frozen embryos and semen offers a reasonable and far less expensive alternative. The cost of this technology is small compared to the cost of maintaining even small breeding populations (Brem et al., 1982). If managed properly, these methods may also provide greater protection against genetic erosion and environmental challenge than is possible in most facilities where whole living animals are maintained.

SEED BANKS

Seed banks are facilities for the maintenance of living plants in the form of seeds (Roberts, 1975; FAO, 1975). They can vary in size and scope from small seed reserves maintained by the curators of botanical gardens, individual plant breeders, or germ plasm collectors, to extremely large seed repositories such as the National Seed Storage Laboratory at Fort Collins, Colorado. They may be involved primarily with one species, such as the seed storage facility at the International Rice Research Institute (IRRI) in Los Baños, the Philippines, or they may maintain seeds of many diverse species.

Genetic Stability of Seed Collections

As with any other germ plasm maintenance technology, plants stored as seeds can be subject to genetic erosion, thereby making the collections less representative of the original natural population. The first and perhaps most critical point of seed storage is the entry of seed into storage. Seed must be of a suitable germination quality for storage, ideally greater than 80 percent, the recommended standard of the FAO Panel of Experts on Plant Exploration and Introduction (FAO, 1975). Seed with low germination percentage must first be multiplied by growing plants. Either during the sampling of seed for storage or during this multiplication process, selection could eliminate a significant portion of the material to be preserved, and thereby lose part of the original genetic diversity. The goals of seed regeneration and multiplication are to increase seed stocks without altering their genetic composition. This requires that pollinations be tightly controlled. It is also extremely important that the effects of natural and artificial selection be minimized. This requires the regeneration or multiplication of seeds in a geographical location providing optimum survival. In reality, the practice is to idealize conditions to achieve maximum survival, so that most or all of the population reproduces. This is followed by random sampling during harvest and storage.

12

Once a seed accession has been tested or multiplied to achieve the desired level of viability, it must next be dried (in the case of "orthodox" or dry seeds) to a moisture level of 5 to 7 percent, a critical step in preservation. A high temperature during drying (greater than 75°C) causes a dramatic loss in viability, and overly fast drying has been found to reduce viability in some forage species. However, temperatures of approximately 60°C and drying over silica gel, at 26 to 28°C for approximately seven days, provides rapid drying of even large seeds without damage.

Seed must then be packaged in containers that will prevent moisture changes within the seeds, or else they must be deposited in humidity-controlled stores. Seeds are typically stored at reduced temperature. The FAO panel in 1975 recommended storage at less than or equal to 5°C ("acceptable") or, better, at −18°C ("preferred"). Storage under either of these temperature regimes should provide adequate protection against a decline in seed viability for periods of years to centuries.

There is, however, a growing body of information regarding the cryogenic storage of seeds at the temperature of liquid nitrogen, −196°C. Stanwood and Bass (1981) tested more than 120 species of vegetable, flower, shrub, and tree crop plants at that temperature without substantial reduction in viability. They also examined the effects of cooling and warming rates on seed viability. Towill (1982b), studying seed from *Solanum* species, found that most orthodox seeds have extremely low moisture contents and thus would be protected against freeze-thaw damage during storage in liquid nitrogen.

The interval at which seed should be tested for viability is controversial. Roberts et al. (1967) and Abdalla and Roberts (1969) have shown that mutational changes and chromosomal damage occur with increasing frequency as viability decreases. Roos (1984b) has shown that in seed lots of mixed genetic background there tends to be differential survival of genotypes. Thus, if viability is allowed to drop to low levels (ca. 50 percent), a sharp change in the genetic composition of the sample may occur. These findings argue for frequent germination testing, although other factors must also be considered.

First, testing requires the use of stored seed, thus reducing the quantity retained. While this may seem a minor loss, the sample size of any one group of seeds (place of origin, last regeneration, year of collection, etc.) must be large enough to be representative. If sampling occurs at relatively frequent intervals, then the cost, in terms of stored seed, could become substantial.

A greater concern is the exposure of genetic stocks to evolutionary change and other risks associated with growing plants. Selection, outcrossing, human error, and genetic drift are potential problems. Genetic drift reduces the diversity of the germ plasm and also increases the homozygosity of individual plants through inbreeding. An inevitable process could then be initiated whereby the genetic diversity of the sample decreases with each successive regeneration. This

13

problem can be controlled to some extent by taking large samples for the regeneration population. Seed from a minimum of 100 randomly chosen plants has been suggested as a reasonable sample size (Frankel and Soulé, 1981).

Decisions about the frequency of viability sampling and regeneration have to be made according to local conditions and considerations. If the seed storage laboratory has facilities for growing plants under optimal conditions, frequent regeneration may be preferable. Conversely, sampling and regeneration frequencies may be reduced at facilities where regeneration would be likely to alter the genetic makeup of the sample.

Various strategies have been attempted for the preservation of "recalcitrant" seeds—those that lose viability if dried to the low moisture contents typical of orthodox seed storage methods. Storage in carbon dioxide (Harrington, 1970) has been attempted with limited success for periods of less than one year. Storage of imbibed seeds has been more promising. Villiers (1975) found that seeds that have imbibed water retained viability and chromosomal integrity better than desiccated controls. There was even some indication that seeds damaged by low moisture conditions underwent chromosomal repair, since the viability of these samples actually increased after imbibition. However, more investigations are clearly required for recalcitrant seeds, since long-term storage is not possible and plants with recalcitrant seeds embrace a wide array of tropical genera, including a number of major food crops (Sykes, 1978).

Thompson (1976) listed criteria which can be helpful for evaluating the usefulness of seed banks in the maintenance of wild species. These include the availability of information regarding the taxonomy of the organism, information about its reproductive biology, the accessibility of the organism in the field, its response to normal seed storage conditions, and its amenability to cultivation. Preservation of wild species by seed storage can often be a useful adjunct to other preservation methods, allowing for the maintenance of a large number of individuals, thereby reducing the sampling error or "bottleneck effect" that can be associated with other techniques. This may be especially useful in large and long-lived species, which would otherwise be very difficult to maintain in representative population sizes (Frankel and Soulé, 1981). Many wild plant species produce dry seeds that can be handled as orthodox crop seeds. Others are recalcitrant, require dormancy, are hard seeded, or have other requirements that complicate their storage as seed.

Safety and Flexibility of Seed Storage Technologies

Stored under optimum conditions of temperature, humidity, and oxygen tension, seeds of most plants can be maintained over periods from years to decades. The extent of protection provided by some techniques will not be known for years. The most vulnerable stages occur during introduction and regeneration. Often, irreversible choices must be made about what is to be stored. If a collection is not representative or is in poor condition, its usefulness and safety are questionable.

Regeneration puts collections at risk to all the hazards of the external environment, including climatic fluctuations, pests, and diseases. Genetic changes, as described previously, can also occur during regeneration. All these factors could change collections, thereby making them less representative of the original populations. Fire, flood, and other catastrophes such as war would, of course, pose grave threats to the safety of seed banks. Against these contingencies, the best possible protection is the maintenance of duplicate accessions at two or more locations remote from one another.

Seed banks pose no threat to the surrounding environment, with the possible exception of the escape of potentially weedy species from regeneration nurseries. If appropriate measures are taken to prevent the dissemination of such plants, this threat should be insignificant.

Evaluation and Documentation

A broad array of technologies is available for the evaluation of stored germ plasm. These include evaluation of disease and pest resistance, multivariate analysis of morphological characteristics, and various electrophoretic and chromatographic techniques. The use of restriction fragment length polymorphisms (RFLP) for assignment to taxonomic group based on DNA structure is a powerful tool that is becoming applicable to an increasing number of species (David Grant, Pioneer Hi-Bred International, Inc., pers. comm.). RFLP analysis can discern between closely related taxonomic groups, and it may be useful for the identification of specific alleles or linkage groups. Thus RFLP analysis might allow the simultaneous documentation and evaluation for at least some traits of agronomic importance.

The evaluation of germ plasm resources and the dissemination of information is especially valuable in germ plasm that could be used for the improvement of economic species. Untested or uncatalogued materials are inaccessible and effectively useless to the plant breeder. The use of computer data bases and other such management tools, and the sharing of information, have been facilitated by a number of international collaborations (Hawkes, 1983) and should be encouraged wherever possible.

Cost, Availability, and Alternatives to Seed Storage

Some plant species produce seeds that are short-lived or otherwise unfit for seed storage, while others produce no seed, reproducing vegetatively. Few alternatives now exist for the safe preservation of such organisms in conditions analogous to seed storage. Maintenance of vegetative propagules is limited to short-term storage. A more promising alternative, however, is the use of tissue culture, particularly culture of organized tissues such as shoot tips and meristems. Cryo-

15

genic storage of recalcitrant seeds, vegetative propagules, and tissue cultures is still in the development stage but offers considerable promise.

Unfortunately, while there are several major seed storage facilities worldwide, most are found in developed countries in the Northern Hemisphere. Although not every country has the resources needed to support a large centralized seed storage facility, the general conditions of safe storage can be applied on a much smaller scale. Thus collection and short-term protection of some organisms can be carried out with later transfer to a central facility.

Labor for storage, rejuvenation, and evaluation are among the major expenses for seed banks, along with the costs of packaging materials and energy. All can be expected to increase over time. Advances in the efficiency of temperature and humidity control equipment may be expected to stabilize the cost of energy somewhat, but this probably represents only marginal savings.

Increase in cost of preservation could potentially alter the allocation of seed storage facilities in the future. At present, there is a clear bias toward the preservation of crop species and their relatives. This is quite understandable in light of their economic importance. As costs increase over time, this bias might well increase; thus storage of species perceived to be of no present economic importance is likely to decline from its already low level.

MICROBIAL CULTURES

The preservation of germ plasm resources of bacteria, fungi, viruses, and other microorganisms is carried out at several levels with varying degrees of sophistication. Investigators may maintain a few hundred working and reference strains, many of which require only short-term storage. Reference collections and large centralized culture collections may maintain many thousands of strains requiring very long-term storage. Many specialized research collections contain large numbers of genetic variants within a narrow range of species which may be invaluable in studies of genetics and taxonomy. As is the case with other collections maintained for use in special situations, however, these are often among the most valuable and least protected.

Microorganisms can be maintained by a broad array of techniques, which can conveniently be divided into cultures maintained in the active living state and those in which metabolic activity has been reduced or stopped. In the living condition, cultures can be maintained on minimal or rich media, with or without selective agents such as antibiotics. Life expectancy of such cultures is usually limited to periods from a few weeks to several years, depending on the stability of the strain, media components, and storage temperature. This limitation requires that cultures be transferred or propagated periodically.

Transfer of microbial cultures is not only labor intensive but also exposes cultures to the possibility of loss due to environmental hazards. Further, in the case

16

of pathogenic organisms, the transfer process exposes workers and the environ-ment to the danger of contamination.

A better long-term strategy is storage in a more or less inert form. In large-scale national repositories such as the American Type Culture Collection, most bacterial and fungal collections are maintained in a freeze-dried condition. This serves to preserve genetic stability, allows very long-term storage, and makes shipping safer and easier. Several types of freeze-drying are possible to increase life expectancy and to allow more convenient manipulation. These include freeze-drying of liquid cultures, drying on paper or other supports saturated with culture, or having organisms dispersed over a surface, and dried and stored in desiccant materials such as silica gel. Cryopreservation is also an effective technique for long-term storage. Organisms are usually preserved with cryopro-tectants such as dimethylsulfoxide or glycerol in liquid nitrogen or low tempera-ture freezers.

Genetic Stability of Preserved Microbial Stocks

The intrinsic rate of increase (i.e., the rate at which populations of an or-ganism can grow) is approximately inversely proportional to body size (Harrin, 1972). For microorganisms this means that generation times are usually short and growth rates very high. Population sizes can easily reach into the billions under even moderately favorable conditions. Such populations are likely to pos-sess variant individuals adapted to an extremely wide range of environments. Maintenance of actively growing microbial cultures can be complicated by this fact.

Organisms with highly adapted phenotypes will tend to survive preferentially so that over short periods (with high intrinsic rates of increase) progeny of these more fit individuals will predominate. Since the culture conditions may not rep-resent the environment from which an organism was collected, the predominant form may not accurately reflect the natural population.

Microorganisms tend to be more susceptible to some mutagenic agents in their environments than are larger organisms. This may translate into abnor-mally high mutation frequencies in microbial populations. Whereas most mutant types are likely to be less fit than average, hyper-fit individuals may arise at sig-nificant frequencies. Selection may then allow these newly arisen genotypes to predominate.

Genetic drift or sampling error may also cause successive transfers or sub-populations to represent the original population less and less over time. This is especially true if the microbial culture is large and contains diverse genotypes.

Preservation strategies that minimize the likelihood of mutation, selection, and genetic drift are best. Mutation and selection are likely to be minimized by storage under conditions that suspend normal metabolic functions. Where there is an option of either drying or cryogenic storage, the nature and require-

17

ments of the individual organism and available resources will determine which should be applied. Genetic drift can be minimized by the simultaneous storage of large numbers of samples from a single original culture, thus avoiding repeated sampling over time.

While "inert" storage schemes appear to offer the greatest possible protection against mutation and selection, there does not appear to be a large literature addressing this concern. Storage itself may be selective. In addition, neither desiccation nor cryopreservation prevents exposure to mutagenic phenomena such as irradiation. Selection or genetic drift can also occur during recovery of cultures from long-term storage.

A major use of bacterial strains is for the maintenance, replication, and selection of DNA molecules used in molecular biology and the study of genetic systems in bacteria and other organisms (see the section below on gene libraries). Bacterial plasmids, relatively small self-replicating circular molecules, usually carry one or more genes conferring resistance to antibiotics. Some bacterial strains also carry potentially unstable transposable elements, which may be lost or which can accumulate to high copy numbers. Long-term storage of these bacteria in the living condition often permits the loss of unstable plasmids and the mobilization of some transposable elements (Maniatis et al., 1982).

Loss of plasmids can often be prevented by growth of host strains in the presence of appropriate antibiotics. The stabilizing selection of antibiotics will usually retain the plasmids in the population. Where a plasmid contains genes for resistance to more than one antibiotic, multiple selective agents can help to retain the plasmid and prevent contamination by mutant strains, since double resistance mutants are extremely rare.

The method used to store transposon stocks will depend on the transposon in question. Tn10, an E. coli transposon, is stable enough so that it can be maintained safely in living culture. Tn5 insertion mutants, much less stable than Tn10, accumulate high copy numbers of the transposon in living cultures. For these and other transposon stocks, cryogenic storage ($-70°C$) is recommended (Davis et al., 1980).

Safety and Flexibility of Microbial Preservation Techniques

The major hazards to preserved microbial cultures are loss of viability, contamination, mutation, selection, and genetic drift. Most of these hazards can be circumvented to some extent by cryogenic preservation or freeze-drying. The main long-term preservation techniques, drying and cryogenic storage, are broadly applicable. Individual strains must usually be tested, however, for susceptibility to storage damage. Cryoprotectant and rate-of-cooling protocols may require adjustment to meet the storage characteristics of various species and strains.

A question of great concern is the danger posed to the environment by preserved microorganisms. This consideration has received even more attention in recent years because of a perceived danger from strains carrying recombinant DNA molecules. Except in the case of highly virulent pathogens and organisms modified by recombinant DNA techniques to produce toxins and other dangerous compounds, normal microbiological laboratory techniques constitute adequate protection. Pathogenic or other hazardous organisms can also be handled safely, but strict containment measures may be required, since even low-level contamination with these organisms could pose a danger to workers and the environment.

Evaluation and Documentation of Microorganisms

Although there are estimated to be hundreds of thousands of different microbial species, only about 40,000 are collected and exchanged (Frederick and Keyser, 1981). Data bases on strain information and availability are limited. Although there were an estimated 1,200 culture collections in the United States in 1981, only 24 of these offered catalogues for general distribution.

A 1981 panel (Frederick and Keyser, 1981) recommended a broad program for the improvement of documentation and accessibility of microbial genetic resources. They recommended that a national network of collections be created and that a national data base on microbial genetic resources be established. They also recommended that microbial systematists should be members of teams engaged in broad ecological surveys. Collaborations between the United States and other countries should be established for the exchange of cultures and resource data. The federal government should encourage the development of microbial collections and data bases. Scientific societies and/or government agencies should oversee the identification, collection, and preservation of microbial genetic resources. They should also establish criteria for the standardization of microorganisms used in research. Institution of these programs would greatly increase the accessibility and usefulness of microbial cultures in the United States and abroad.

Availability and Cost of Microbial Preservation Technologies

Whereas large, centralized microbiological culture collections are an invaluable resource, they represent a small minority of culture collections worldwide (Martin, 1973). In the United States, for example, there is only a single major national repository, the American Type Culture Collection (Williams, 1983); there are, however, a number of other important but less well known special purpose collections such as the *Phytophthora* collections at the University of California, Riverside, and the International Collection of Phytopathogenic Bacteria at the University of California, Davis. These working research collections are an extremely valuable genetic resource and are difficult to replace.

Although it is unlikely that every country will have a major microbial fa-

19

cility in the foreseeable future, this does not prevent work in the collection and evaluation of microbial strains in developing countries. Except for the most sophisticated preservation techniques, even the least developed countries should be able to maintain useful microbial collections. The advantages and limitations of preservation strategies discussed above apply equally to large collections and individual strains. Indeed, it is most often the developing countries, particularly of the tropics, that have the most to gain from such collections, since their microfloras are diverse and environmental disturbance is an increasing threat.

While the cost of some facets of microbial preservation, such as skilled labor, can be expected to increase worldwide, the cost of preserving an individual culture will probably remain relatively stable and may actually decrease in relative terms because of increased efficiency in evaluation and documentation, as well as in the preservation techniques themselves.

TISSUE CULTURE

Tissue culture comprises a range of systems for the growth *in vitro* of protoplasts, cells, calli, tissues, organs, and embryos. All these methods have great potential for use in germ plasm preservation. In both plant and animal systems, these methods can be used to augment or replace the maintenance of whole living organisms. A broad array of culture techniques have been developed and are now coming into limited application in the preservation of genetic stocks.

Techniques for *in vitro* or tissue culture growth of plants may be broadly divided into organ culture, callus culture, suspension cultures, cultures of single cells and protoplasts, and anther or pollen culture (Wilkins and Dodds, 1983). Some of these techniques are better suited to the special needs of gene preservation than others. Some are currently limited by shortcomings such as inability to regenerate whole plants, genetic changes in the culture materials, difficulty in establishing cultures, and so forth. The current state of the art often requires major revisions of techniques for a species, a time-consuming and resource-intensive process.

The current applications of tissue culture in germ plasm preservation center on those species that are reproduced vegetatively or produce short-lived recalcitrant seeds. Culturing these organisms is seen as a preferred storage method primarily because of high multiplication rates, small space requirements, and the potential for long-term storage with low rates of gene loss (Wilkins and Dodds, 1983; Withers, 1984). As a method for the vegetative reproduction of large numbers of genetically identical progeny, tissue culture is unparalleled. There is also great potential for extremely long-term storage through cryopreservation or other techniques by which metabolic activity of cultures is sharply reduced.

Animal tissue culture, while in many respects more advanced than that of plants, is currently limited for use as a germ plasm preservation strategy because

20

of a universal difficulty in the regeneration of whole organisms from disorganized cell masses. There has been substantial progress, however, in the use of *in vitro* fertilization techniques and cryopreservation of isolated sperm, ova, and embryos (Miller, 1977).

The preservation of important genetic stocks of cattle by the cryopreservation of semen has been a reality for a number of years (Miller, 1977). The first calf produced by *in vitro* development of an embryo to the blastocyst stage followed by artificial implantation occurred in 1973 (Wilmut and Rowson, 1973). While these techniques are increasingly used in domestic animal systems, their use in preservation of wild animal gene pools appears to have received only limited support.

Genetic Effects of In Vitro Culture

While there appears to be little information regarding the long-term genetic effects of *in vitro* culture in animal cell lines, genetic effects in plant cultures are now being studied intensively. For tissue culture to be a useful alternative for germ plasm preservation, the processes that cause genetic changes, and the extent to which they occur, must be fully understood and controlled.

Plant tissue cultures tend to be chromosomally unstable, particularly if kept in culture for long periods (Henshaw, 1975; Wilkins and Dodds, 1983; D. T. Tomes, Pioneer Hi-Bred International, Inc., pers. comm.). Thus the combination of culture and cryopreservation may be necessary to reduce metabolic activity and cell division, both of which appear to be responsible for these changes. Cryopreservation at liquid nitrogen temperatures has been effective for a number of species (see, for example, Bajaj, 1983a; Kartha et al., 1982), and storage at reduced temperatures (ca. 4°C) also has proved useful for the long-term preservation of calli and shoot tip cultures.

The high degree of isolation made possible by culture *in vitro* eliminates such problems as contamination due to outcrossing, selection by local environments, and genetic drift. However, it is important to consider the cultured material as a population evolving within a microenvironment *in vitro*. Cultures, especially protoplasts, cells, and calli, are subject to many of the evolutionary forces at work in larger environments. The fact that these forces are generated by a wholly alien environment can exacerbate potential genetic shifts. Mutations and other changes at the DNA level may be generated by the culture procedures, and selection can occur at several levels within the system. Selective properties of the cell-media interaction (including carbohydrate, nitrogen, and other nutrients, hormonal balance, and subtle cross-feeding properties of the cultures themselves) can potentially produce a subpopulation of cells capable of maximum exploitation of the medium. But such a subpopulation may be only partly representative of the whole organism, and less representative of the population represented by that organism. Continued transfers, which are de facto selection of cells that are able to grow well in culture, may likewise give rise to unrepresentative subpopulations by selection and genetic drift. Conditions of temperature and light might produce similar selection pressures.

The extent to which these evolutionary considerations affect the utility of tissue culture for genetic preservation appears to depend primarily on the species in culture, the medium and other culture conditions, the degree of organization of the cultured material (i.e., the extent to which it mimics the whole organism), the conditions and period of culture and storage, and the proposed uses of the cultured material. In general, minimum disruption seems to accrue in relatively organized tissues (e.g., meristems, shoot tips, or embryos) in which infrequent transfer is made possible by control of metabolic rate through cooling or cryopreservation. The effect of variation in these materials can be disruptive, but excellent results have been reported in some systems (Morel, 1975; Jones et al., 1979). Successful and stable reproduction of orchids and grapes (Morel, 1975) and apples (Jones et al., 1979) has been reported for meristem and shoot tip cultures respectively. These results are encouraging, and suggest that the technology can be expanded.

Genetic variation seen under culture conditions must be balanced against that resulting from other forms of storage. For species where growth is limited to nurseries, botanical gardens, arboreta, zoos, or farms, exposure to the large-scale environment may well pose the greater risk of loss or change in genetic reserves.

Safety, Accessibility, and Flexibility of Tissue Culture

The period for which tissue culture maintenance of genetic stocks is appropriate depends on the state of the art of culture and storage conditions for a particular species. Meristem cultures of grapes, for example, can probably be maintained indefinitely with little genetic disruption (Morel, 1975). The durability of cultures stabilized by cryopreservation or cooling is variable but would seem to be quite good. Shoot tip cultures of potato have been successfully stored at liquid nitrogen temperatures for three years (Towill, 1982a). Thus cryopreservation of cultures may be suitable for storage of other species for intervals of at least several decades (Grout and Henshaw, 1978; Bajaj, 1983a and b).

Genetic resources for domestic animals can probably be maintained in the form of cryopreserved semen and cultured fertilized ova for periods in excess of their useful lifetime (Miller, 1977). It may also be more desirable to produce diploid progeny from single fertilized ova. Work is progressing in this area (Miller, 1977), and, together with cryopreservation, it may provide long-term storage of valuable genotypes.

At present, a major limitation of the flexibility or applicability of tissue culture techniques is the requirement of specific culture protocols for each species. Often this requires slight modification and further testing for each set of related

22

genotypes or varieties within a species (Tomes, Pioneer Hi-Bred International, Inc., pers. comm.). Once these conditions are defined, however, it should be possible to culture large numbers of individuals without further amendment of protocols. Gene banks already with limited resources are unlikely to be able to conduct such research.

A serious constraint placed on cultures for germ plasm maintenance is the need for regeneration of whole organisms. In mammals, this can be accomplished by the use of surrogate mothers in which cultured or preserved embryos are gestated. Implantation and insemination can even be accomplished surgically where required (Polge, 1981). For plants, the ability to regenerate depends on the species, degree of organization of cultured material, and length of time in culture. This is especially true in the grasses, the family from which a large proportion of important crop species are derived.

In maize, the regenerability of cultures provides an excellent example of regeneration requirements (Tomes, pers. comm.). Callus cultures maintained for periods of less than one year appear to regenerate readily with relatively simple media modifications. Beyond one year, however, regeneration of normal plants becomes increasingly difficult.

Tissue culture can provide relatively secure storage of genetic materials. Cultures can be protected from external environments by controlled-temperature chambers, or through cryopreservation. The equipment for the maintenance of these conditions is readily available and reliable. Cultured materials are also unlikely to threaten the environment.

While tissue culture is a relatively secure technology for germ plasm preservation, it is perhaps more susceptible to disruption of the infrastructure (system of support facilities) than other techniques. For example, liquid nitrogen and other cryopreservation materials are dependent on an advanced and complex infrastructure of controlled laboratory conditions, which could be easily disrupted by a natural disaster or political upheaval. Stored seeds, on the other hand, could probably survive even relatively protracted periods of disruption so long as the seeds were maintained at cool temperatures and low humidity.

Evaluation of Cultured Materials

Cultured materials become useful when they are evaluated and catalogued. For traits expressed only in whole organisms, such as milk production in cows or yield of grain in crop plants, such evaluation must be done in vivo prior to or following culture. There is as yet no in vitro substitute for evaluating most such traits. For some traits, however, cultured materials represent unique systems for evaluation or selection. Traits such as antibiotic or toxin resistance in plants, for example, can be modeled with great success in vitro (see, for example, Gegenbach and Connelley, 1981).

As cultures undergo genetic changes during protracted culture periods, it may

23

be necessary to reevaluate them at intervals analogous to the regeneration intervals used in seed storage banks. Such periodic regeneration and reevaluation may be one way in which the cumulative effects of long-term culture can be circumvented. Such periodic regeneration would also introduce periodic selection and the risk of genetic drift.

Availability, Economics, and Alternatives to Tissue Culture

As discussed previously, tissue culture can require considerable research effort to establish cultures and storage conditions for a species. This, together with the physical infrastructure which it requires, limits the application of tissue culture generally to technologically developed countries and to well-known organisms. Its application is also somewhat limited by the costs (particularly labor costs) of maintaining cultures for long periods.

Against these considerations of cost and infrastructure, however, one must weigh the land and labor costs of maintaining even small populations of animals and plants, as well as the environment to which these populations must be exposed with potentially damaging effect. It has been noted, for example, that in grapes, 800 cultivars can be maintained in six replications in a space of two square meters using meristem cultures (Morel, 1975). These cultures require transfer only yearly. When one considers the savings of land and labor of tissue culture in this example over conventional maintenance strategies, the economic benefits become apparent.

As countries develop, the cost of land, labor, and other resources tends to increase. Under these conditions, a shift from conventional means of germ plasm maintenance to tissue culture may gain considerable favor.

GENE LIBRARIES

The discovery of restriction endonucleases (enzymes that cleave double-stranded DNA only at specific recognition sites) set the stage for a wholly new method of germ plasm preservation. DNA sequences of diverse lengths can be incorporated into bacterial plasmids or bacteriophage genomes. These "recombinant" molecules can then be replicated or "cloned" in bacteria to produce an essentially limitless number of copies.

If a suspension of double-stranded DNA from the nucleus or cytoplasmic organelles of an organism is partly cleaved with a restriction enzyme, an array of partly digested DNA fragments of varying sizes are produced, each having identical terminal sequences. Many restriction enzymes leave a short, single-stranded DNA "tail" protruding from the molecule's terminus. Plasmid or phage DNA can be cleaved in an analogous manner to produce single-stranded tails which can be ligated to those of the genomic DNA fragments. In this way the DNA fragments derived from organisms can be incorporated into the plasmid or phage molecules, which also contain bacterial genetic markers and origins of replication. The "recombinant" molecules can then be transformed into bacteria or packaged into phage particles for replication.

Randomly cut DNA incorporated into plasmid or phage DNA and subsequently replicated produces a population of DNA molecules which together represent all the DNA sequences in the genome of the original organism, usually referred to as a "genomic library."

DNA derived from virtually any source, including synthetic DNA and DNA from extinct organisms, can be incorporated into a DNA library. Cloned genes or other DNA sequences can, in many instances, be incorporated into DNA "vectors" which allow their expression in whole organisms. If a vector is successfully incorporated into the cells of an organism or tissue culture, and if the gene it carries is integrated into the genome of the host, the genetic information it contains can be expressed and may also be stably inherited. In this way a gene recovered from one source can be integrated into the germ plasm of another, and can be expressed to modify the phenotype of the recipient organism.

It is possible to store cloned DNA for indefinite periods by simply freezing it at $-20°C$ in an appropriate buffer solution. It is also possible to store cloned DNA as plasmid or phage DNA resident in bacteria used for its propagation, provided that appropriate measures are taken to prevent the genetic modification or loss of the cloning vehicle from the bacterium, due to selection or genetic drift in the bacterial population. Finally, genomic DNA can be kept indefinitely as an ethanol precipitate or in freeze-dried form.

Present efforts to use cloned DNA center on the genetic manipulation of extant organisms, including economically important plants and animals, as well as humans. As genes are cloned and characterized, they become valuable genetic stocks in their own right. The vectors (plasmids and bacteriophage genomes) used for gene cloning are also valuable and potentially labile genetic resources. They must be readily available and must not be modified by unfavorable conditions during their replication and storage.

The discussion below provides a brief overview of the preparation, characterization, and maintenance of genomic libraries with special emphasis on the protection of their genetic integrity. For the purpose of this discussion, libraries constructed in both bacteriophage and bacterial vectors are addressed without the usual distinctions that are applied in practice. It is sufficient for our purposes here to realize that plasmids and bacteriophage libraries are used for different purposes, that bacteriophage libraries require passage through two organisms (the phage and the bacterium) for their propagation, and that the two types of libraries are typically used for the cloning of different size sequence fragments, usually less than 20 kilobase pairs (kb) in plasmid libraries, and up to approximately 50 kb in bacteriophage libraries. The general concepts of cloning, and the population dynamics and other factors that contribute to their potential instability in storage, are important considerations in both systems.

Preparation and Maintenance of Genomic Libraries

For a genomic library to be truly representative of the genome of the organism from which it is constructed, it must contain all of the sequences present in the genome, or a desired subpopulation of sequences. It should contain all sequences in copy numbers that are representative of their relative frequencies in the genome. It is also desirable that any gene be found completely intact in a single cloned fragment. For greatest efficiency in establishing linkage relationships and DNA homology, closed sequences should be as long as possible.

Maniatis et al. (1982) have provided a step-by-step description of the cloning process used to construct genomic libraries in bacteriophage. They describe and compare two ways in which libraries of eukaryotic DNA can be constructed (the reader is referred to this excellent laboratory manual for a more detailed description).

In the first protocol, eukaryotic DNA is digested to completion with restriction enzymes, and the resulting fragments are inserted into an appropriate vector and packaged for replication in phage. However, a sequence of interest may not be contained within a single clone, and numerous small DNA fragments are produced, thus making evaluation laborious and expensive.

The second protocol for production of libraries of eukaryotic DNA employs partial digestion with restriction enzymes cleaving randomly at recognition sequences. This results in a random array of much larger fragments on the order of 20 kb or greater, allowing easier evaluation and ensuring that a gene of interest is likely to be contained in a single fragment.

After digestion, DNA sequences are inserted into vectors that are introduced *in vitro* into phage particles. Recombinant phage are then allowed to infect a bacterial host, where they are amplified by phage replication in the bacteria.

DNA sequences can also be cloned by insertion into bacterial or hybrid bacterial-eukaryotic plasmids. Experience dictates, however, that these are best used for small DNA fragments, usually of 10 kb or less (Maniatis et al., 1982). Plasmids can be replicated by insertion into bacterial hosts via transformation.

Genetic Stability of Genomic Libraries

A number of factors can erode genomic libraries. First, plasmids, especially large plasmids, are sometimes eliminated spontaneously from the bacterial host. Should this occur in a significant proportion of the bacterial population, a part of the library will be lost. If the population continues to grow, this phenomenon could seriously erode the library. Loss of cloned DNA might be exacerbated by even slight selective pressure against bacterial cells containing recombinant plasmids. Such selection might reduce the effective "allele frequency" of certain clones, thus making their loss by genetic drift more likely. This effect is minimized by growth of host bacteria under selective conditions or maintenance of bacteria in an inert form.

Similar genetic changes can occur in bacteriophage libraries during amplification. Amplification is usually carried out by infecting E. coli cultures with bacteriophage containing recombinant molecules. The phage particles are then replicated to high copy numbers to obtain a substantial increase in total library DNA. If sequences are represented only very infrequently in the original library, this passage through bacteria may reduce the frequency of those rare sequences through genetic drift.

Maximum stability is achieved by storing the library in an inert state. This could be done by isolating and pooling DNA from the host bacteria or bacteriophage. The DNA could then be dried and might remain stable for very long periods. In practice, however, libraries are usually maintained by reducing the metabolic activity of host bacteria (for plasmid libraries), to a nearly inert state by cryogenic storage or freeze-drying. Phage particles can be retained in an inert state by storing at low temperatures in chloroform.

Another potential source of genetic erosion in libraries is the deletion of duplicate segments of target DNA by rearrangements that occur during propagation of libraries in E. coli. This can be particularly damaging in the case of plant DNAs, which may contain as much as 98 percent repeated DNA. This difficulty appears to have been overcome by the construction of new families of vectors (Loenen and Blattner, 1983).

Availability and Cost of Gene Libraries and Associated Technology

Recombinant DNA technology is now widely available. At present there does not appear to be any scheme for centralized maintenance of DNA libraries. There are, however, a few repositories, such as the American Type Culture Collection, which maintain collections of organisms bearing DNA vectors. Typically, DNA libraries are made in the laboratories where they are used and are often exchanged among laboratories. This situation may change as use of libraries increases.

Recovery of Genetic Material from Extinct Organisms

Germ plasm maintenance has always had as its main concern the preservation of extant organisms. The use of recombinant DNA techniques to recover genetic material from preserved remains now offers the exciting possibility of retrieving sequences for comparison from both extinct and extant organisms. Recently, Higuchi and coworkers (Higuchi et al., 1984) recovered short gene sequences from the quagga (Equus quagga), which became extinct in 1883. More recently Paabo (1985) obtained clonable DNA from a mummy preserved 2,400 years ago.

27

This result showed that it is possible to recover relatively large (3.4 kb) DNA fragments from preserved tissue. It also demonstrated that little or no change had occurred postmortem.

This technology is new and limited. It may be extremely useful, however, for the recovery of genes from extinct organisms for comparison with those of extant organisms. This could provide useful insights into the evolutionary dynamics and movements of organisms that gave rise to our modern flora and fauna. For the present, results are limited to archaeological time scales. In future, however, it may be possible to recover DNA from the paleontological past (Allan C. Wilson, University of California, Berkeley, pers. comm.).

REFERENCES

Abdalla, F. H., and E. H. Roberts. 1969. Effects of temperature and moisture on induction of genetic changes in seeds of barley, broad beans and peas during storage. *Ann. Bot.* 33:153.

Allard, R. W. 1960. *Principles of Plant Breeding.* New York: John Wiley and Sons.

Bajaj, Y. P. S. 1983a. Casava plants from meristem cultures freeze-preserved for three years. *Field Crops Res.* 7:161–167.

———. 1983b. Production of normal seeds from plants regenerated from the meristems of *Arachis hypogaea* and *Cicer arietinum* cryopreserved for 20 months. *Euphytica* 31: 425–430.

Brem, V. G., F. Graf, and H. Kraublich. 1982. Möglichkeiten der Anlag von Genreserven—genetische Probleme und Kosten. *Eingang des Manuskripts* 8(4):380–383.

Bulnheim, H. P., and A. Scholl. 1981. Genetic variation between geographic populations of the amphipods *Gammarus zaddachi* and *G. salinus. Mar. Biol.* 64:105–115.

Burton, G. W. 1976. Gene loss in pearl millet germ plasm pools. *Crop Sci.* 16:251–255.

Chang, T. T. 1984. Conservation of rice genetic resources: Luxury or necessity? *Science* 224:251–256.

Coe, M. 1980. African wildlife resources. In *Conservation Biology: An Evolutionary-Ecological Perspective,* ed. M. E. Soulé and B. A. Wilcox, 273–302. Sunderland, Mass.: Sinauer.

Cushing, D. H. 1981. *Fisheries Biology.* 2d ed. Madison: University of Wisconsin Press.

Damania, A. B., E. Porceddu, and M. T. Jackson. 1983. A rapid method for the evaluation of variation in germ plasm collections of cereals using polyacrylamide gel electrophoresis. *Euphytica* 32:883–977.

Davis, R. W., D. Botstein, and J. R. Roth. 1980. *A Manual for Genetic Engineering: Advanced Bacterial Genetics.* Cold Spring Harbor, N.Y.: Cold Spring Harbor Laboratory.

Durzan, D. J., and R. A. Campbell. 1973. Prospects for the introduction of traits in forest trees by cell and tissue culture. *New Zealand J. For. Sci.* 4(2):261–266.

Ellis, R. H., and E. H. Roberts. 1980. Improved equations for the prediction of seed longevity. *Ann. Bot.* 45:13–30.

FAO 1975. Report of the Sixth Session of the FAO Panel of Experts on Plant Exploration and Introduction. Rome: United Nations Food and Agriculture Organization.

Ferguson, A., and F. M. Mason. 1981. Allozyme evidence for reproductively isolated sympatric populations of brown trout *Salmo trutta*, L. M. Lough Melvin. *Ireland J. Fish. Biol.* 18:629–642.

Frankel, O. H., and M. E. Soulé. 1981. *Conservation and Evolution*. Cambridge: Cambridge University Press.

Franklin, I. R. 1980. Evolutionary change in small populations. In *Conservation Biology: An Evolutionary-Ecological Perspective*, ed. M. E. Soulé and B. A. Wilcox, 135–149. Sunderland, Mass.: Sinauer.

Frederick, L., and H. Keyser. 1981. Report of the Microbial Resources Panel. In *Proceedings of the U.S. Strategy Conference on Biological Diversity*, 86–92. U.S. Department of State, Agency for International Development, Washington, D.C.

Gegenbach, B. G., and J. A. Connelley. 1981. Mitochondrial DNA variation in maize plants regenerated during tissue culture selection. *Theor. Appl. Genet.* 59:161–167.

Gooch, J. L., and T. J. M. Schopf. 1972. Genetic variability in the deep sea: Relation to environmental variability. *Evolution* 26:545–552.

Grout, B. W. W., and G. G. Henshaw. 1978. Freeze preservation of potato shoot-tip cultures. *Ann. Bot.* 42:1227–1229.

Grout, B. W. W., K. Shelton, and H. W. Pritchard. 1983. Orthodox behaviour of oil palm seed and cryopreservation of the excised embryo for genetic conservation. *Ann. Bot.* 52:381–384.

Harrin, A. C. 1972. Population ecology of a colonizing species: The pelagic tunicate *Thalia democratica*. I. Individual growth rate and generation time. *Oecologia* 10:269–293.

Harrington, J. F. 1970. Seed and pollen storage for conservation of plant gene resources. In *Genetic Resources in Plants: Their Exploration and Conservation*, ed. O. H. Frankel and E. Bennett, 501–521. IBP Handbook 11. Oxford: Blackwell Scientific Publications.

Hawkes, J. G. 1983. *The Diversity of Crop Plants*. Cambridge, Mass.: Harvard University Press.

Henshaw, G. G. 1975. Technical aspects of tissue culture storage for genetic conservation. In *Crop Genetic Resources for Today and Tomorrow*, ed. O. H. Frankel and J. G. Hawkes, 349–357. IBP 2. Cambridge: Cambridge University Press.

Higuchi, R., B. Bowman, M. Frieberger, O. A. Ryder, and A. C. Wilson. 1984. DNA sequences from the quagga, an extinct member of the horse family. *Nature* 312:282–284.

Johnson, R. S. 1983. Changing germ plasm storage technology may brighten future of rare crop genes. *Genetic Engineering News*, January–February, p. 12.

Jones, O. P., C. A. Pontikis, and M. E. Hopgood. 1979. Propagation *in vitro* of five apple scion cultivars. *J. P. Hort. Soc.* 54:155–158.

Justice, O. L., and L. N. Bass. 1978. *Principles and Practices of Seed Storage*. Agriculture Handbook 506. Washington, D.C.: U.S. Department of Agriculture.

Kartha, K. K. 1982. Cryopreservation of germ plasm using meristem and tissue culture. In *Application of Plant Cell and Tissue Culture to Agriculture and Industry*, ed. D. T. Tomes, 139-161. Guelph, Ontario: Plant Cell Culture Center, University of Guelph.

Kartha, K. K., N. L. Leung, and L. A. Mroginski. 1982. *In vitro* growth responses and plant regeneration from cyopreserved meristems of cassava (*Manihot esculenta* Crantz). *Z. Pflanzenphysiol.* 107:133–140.

Loenen, W. A. M., and F. R. Blattner. 1983. Lambda charon vectors (Ch. 32, 33, 34 and 35) adapted for DNA cloning in recombination-deficient hosts. *Gene* 26:171–179.

Maijala, K. 1974. Conservation of animal breeds in general. In *Round Table: The Conservation of Animal Genetic Resources*. Rome: United Nations Food and Agriculture Organization.

Mak, C., and B. L. Harvey. 1982. Exploitable genetic variation in a composite bulk population of barley. *Euphytica* 31:85–92.

Maniatis, T., E. F. Fritsch, and J. Sambrook. 1982. *Molecular Cloning: A Laboratory Manual*. Cold Spring Harbor, N.Y.: Cold Spring Harbor Laboratory.

Marshall, D. R., and A. H. D. Brown. 1975. Optimum sampling strategies in genetic conservation. In *Crop Genetic Resources for Today and Tomorrow*, ed. O. H. Frankel and J. G. Hawkes, 53–80. IBP 2. Cambridge: Cambridge University Press.

Martin, S. M. 1973. Culture collections and the preservation of genetic resources. *Impactos Globais Da Microbiologia Aplicada* 1:41–49.

Mason, I. L. 1974. The conservation of animal genetic resources: Introduction to round table. *First World Congr. Genet. Appl. Livest. Prod.* 2:13–21.

McClintock, B. 1984. The significance of responses of the genome to challenge. *Science* 226:792–801.

Miller, R. H. 1977. The need for and potential application of germ plasm preservation in cattle. *J. Hered.* 68:365–374.

Morel, G. 1975. Meristem culture techniques for the long-term storage of cultivated plants. In *Crop Genetic Resources for Today and Tomorrow*, ed. O. H. Frankel and J. G. Hawkes, 327–332. IBP 2. Cambridge: Cambridge University Press.

Orshinsky, B. R., and D. T. Tomes. 1985. Effects of long term culture and low temperature incubation on plant regeneration from a callus line of birds foot trefoil (*Lotus corniculatus L.*) *J. Plant Physiol.* 119:389–397.

Paabo, S. 1985. Molecular cloning of ancient Egyptian mummy DNA. *Nature* 314:644–645.

Polge, C. 1981. New biological techniques for the conservation of animal resources. In *Animal Genetic Resources Conservation and Management: Proceedings of the FAO/UNEP Technical Consultation*. Rome: United Nations Food and Agriculture Organization.

Polunin, N. V. C. 1983. Marine "genetic resources" and the potential role of protected areas in conserving them. *Environ. Conserv.* 10(1):31–39.

Raven, P. R. 1976. Ethics and attitudes. In *Conservation of Threatened Plants*, ed. J. B. Simmons, R. I. Beyer, P. E. Brandham, G. L. Lucas, and V. T. H. Parry, 155–179. New York: Plenum Press.

Roberts, E. H. 1975. Problems of long-term storage of seed and pollen for genetic resources conservation. In *Crop Genetic Resources for Today and Tomorrow*, ed. O. H. Frankel and J. G. Hawkes, 269–295. IBP 2. Cambridge: Cambridge University Press.

Roberts, E. H., F. H. Abdalla, and R. J. Owen. 1967. Nuclear damage and the ageing of seeds, with a model for seed survival curves. *Symp. Soc. Exp. Biol.* 21:65–100.

Rognoni, G., and A. Finzi. 1984. Aspects of conservation of animal genetic resources: Italian experiences. *Livestock Prod. Sci.* 11:61–64.

Roos, E. E. 1984a. Characterization and maintenance of genetic variability in seed germ plasm: A case study using *Phaseolus vulgaris*. *Hortscience* 19(3):599.

——. 1984b. Genetic shifts in mixed bean populations. I. Storage effects. *Crop Sci.* 24:240–244.

Ryder, M. L. 1979. Preserving rare breeds of sheep and goats. *Span* 22:11–13.

Saedler, H., and P. Nevers. 1985. Transposition in plants: A molecular model. *European Molecular Biol. Org. J.* 4:585–590.

Senner, J. W. 1980. Inbreeding depression and the survival of zoo populations. In *Conservation Biology: An Evolutionary-Ecological Perspective*, ed. M. E. Soulé and B. A. Wilcox, 209–224. Sunderland, Mass.: Sinauer.

Simmons, J. B., R. I. Beyer, P. E. Brandham, G. L. Lucas, and V. T. H. Parry, eds. 1976. *Conservation of Threatened Plants*. New York: Plenum Press.

Slack, S. A. 1980. Meristem-tip culture. *Plant Disease* 64(1):15–17.

Smith, C. 1976. Alternative forms of genetic controls. *Anim. Prod.* 23:403–412.

——. 1984. Genetic aspects of conservation in farm livestock. *Livestock Prod. Sci.* 11: 37–48.

Spillett, J. J., T. D. Bunch, and W. C. Foote. 1975. The use of wild and domestic animals and the development of new genotypes. *J. Anim. Sci.* 40:1009–1015.

Stanwood, P. C., and L. N. Bass. 1981. Seed germ plasm preservation using liquid nitrogen. *Seed Sci. Technol.* 9:423–437.

Suzuki, S., and K. Kumagai. 1981. Present status of genetic resource information management in Japan. *JARQ* 15(2):75–84.

Sykes, J. T. 1978. The conservation of crop genetic resources: International actions in long-term seed storage. *Seed Sci. Technol.* 6:1053–1058.

Thompson, P. A. 1976. Factors involved in the selection of plant resources for conservation as seed in gene banks. *Biol. Conserv.* 10:159–167.

Towill, L. E. 1982a. Freezing potato shoot tips. *Agric. Res.* 31(½):6.

——. 1982b. Low temperature (−196°C) storage of true seed from the tuber-bearing *Solanum* species. *Amer. Potato J.* 59:141–147.

Villiers, T. A. 1975. Genetic maintenance of seeds in imbibed storage. In *Crop Genetic Resources for Today and Tomorrow*, ed. O. H. Frankel and J. G. Hawkes, 297–315. IBP 2. Cambridge: Cambridge University Press.

Wilkins, C. P., and J. H. Dodds. 1983. The application of tissue culture techniques to plant genetic conservation. *Sci. Prog.* 68:259–284.

Williams, P. H. 1983. Conservation of plant symbiont germ plasm. In *Challenging Problems in Plant Health*, ed. T. Kommedahl and P. H. Williams, 131–134. St. Paul, Minn.: American Phytopathological Society.

Wilmut, I., and L. E. I. A. Rowson. 1973. The successful low temperature preservation of mouse and cow embryos. *J. Reprod. Fertil.* 33:352.

Withers, L. A. 1984. *In vitro* techniques for germ plasm storage. In *Efficiency in Plant Breeding: Proceedings of the 10th Congress of the European Association for Research on Plant Breeding*, ed. W. Lange, A. C. Zeven, and N. G. Hogenbloom, 182–193. Wageningen, Netherlands: EUCARPIA.

Wochok, Z. S. 1981. The role of tissue culture in preserving threatened and endangered plant species. *Biol. Conserv.* 20:83–89.

Zobel, B. 1978. Gene conservation—as viewed by a forest tree breeder. *Forest Ecol. and Mgmt.* 1:339–344.

Commentary

Brian J. Harvey

RESEARCH IN FISH CULTURE

When discussing *in situ* preservation of germ plasm, it is important to remember that technologies developed for one species may not be readily adaptable to other species even if they are close relatives. A notable example is the failure of the techniques developed for the preservation of bovine semen to work when they are applied to swine. Consequently, separate research on germ plasm preservation techniques will have to be conducted on many different taxa. Therefore, as a complement to Dr. Weissinger's excellent review, in this paper I describe the state of preservation technologies in fish, and discuss their use in aquaculture.

Aquaculture, though a human activity dating from antiquity, remains in many ways unsophisticated; one need only compare present fish farming with agricultural techniques. Whether one believes that the future development of aquaculture can steer clear of mistakes already committed in agriculture depends on one's interpretation of our record of learning from past mistakes. One of the aims of this paper is to show how some of the genetic consequences of extensive breeding in crop plants and domestic animals can be forestalled in aquaculture. The arguments that follow relate mainly to fish culture, the area of my own expertise. It is also in this area that long-term genetic conservation in the form of a gamete bank is feasible.

Since genetics and selective breeding of fishes are disciplines demanding long-term study with many animals, advances come slowly (Stickney, 1979). There is a danger that genetic improvement of fish is regarded as a sort of "fine-tuning" to be done once all the outstanding problems with culture have been solved. This argument is fallacious simply because, as we have been forced to learn in agriculture, any sort of husbandry results in a narrowing of the species' genetic base (Harlan, 1975; Wilkes, 1977). If we neglect research on genetics and selective breeding, this narrowing will take its own course.

Current topics in fish genetics include hybridization, selective breeding, genotype-environment interactions, and specialized techniques like gynogenesis, sterilization, induced polyploidy, and introgression of genes. The potential for some of the latter is high indeed, because of the enormous fecundity of many species of cultured fish; this essential difference between fish and farm animals should not be forgotten when we compare sperm banks for aquaculture and existing artificial insemination facilities for livestock.

Until recently, methods for routine indefinite frozen storage of fish spermatozoa were adequate only for salmonids (Scott and Baynes, 1980). Application of these techniques to spermatozoa from warm-water fishes had met with only modest success (Harvey and Hoar, 1979). Reliably high fertility in frozen sperm from a warm-water species was first demonstrated for the tilapia *Oreochromis mossambicus* (Harvey, 1983a). Further development of this basic method in my laboratory has yielded a simple technique that uses apparatus standard in artificial insemination facilities throughout the world. Preliminary results show the method to be applicable to cyprinid fishes, including carps. As a fundamental technique used in combination with improved means of transport of specimens frozen in liquid nitrogen, it will allow geneticists to do artificial fertilizations at their own convenience and with the knowledge that at least one of the gametes is from an identified source. This development also means that fish spermatozoa can be collected from a multitude of sources, including research facilities and farms, banked indefinitely, and distributed widely.

CURRENT TOPICS IN APPLIED GENETICS
AND THE POTENTIAL UTILITY OF FROZEN SPERMATOZOA

Hybridization

Most species of cultured fishes are highly fecund, and there is an understandable tendency among breeders to derive their stock from a small number of parents. Continual selection of broodstock from closely related individuals leads inevitably to generation after generation of inbreeding, and the resulting homozygosity for unfavorable genes results in inbreeding depression (FAO/UNEP, 1981).

The symptoms of inbreeding are especially evident in traits associated with reproduction, but they can also include decreased survival of eggs and larvae, reduced growth rate, and an increase in deformities (Moav, 1979; Stickney, 1979). There is ample evidence for the harmful effects of inbreeding in managed fish populations (Kincaid, 1983) and the attendant economic losses are probably extensive.

The antidote to inbreeding depression is outcrossing, or the mating of distantly related individuals. The resulting increase in heterozygosity is termed "hybrid vigor," and would improve growth rate and viability in many fish farms. Because hybrid vigor usually disappears by the second generation, a culturist must combat inbreeding either by maintaining two or more inbred lines of broodstock to cross and produce vigorous offspring or by continually obtaining new strains. Both tactics are costly, for they demand extra holding facilities (rotational line mating, for example, involves maintenance of three lines of brood animals); yet roughly the same effect can be obtained by handling not new breeders but new gametes. This application of a sperm bank for fish culture has already been recommended by an international consultative group (FAO/UNEP, 1981). Effective broodstock populations are kept large, and inbreeding is countered by providing frozen spermatozoa from a large number of males from all possible strains so that a high level of genetic variability is maintained.

The ability of a sperm bank to act as a repository dedicated to the long-term maintenance of hybrid vigor will rest squarely on its ability to collect and exchange gametes. Strains can be collected from fish farms and hatcheries, from laboratories, and from nature. Naturally, the more known about the donor, the greater the benefits to the breeder using its milt. Ideally, milt likely to yield superior crossbreeds should be identified by testing crosses in a uniform environment, in the manner of a plant seed bank equipped with experimental plots. In cases where particular strains merit widespread promotion, this function of a sperm bank would resemble an artificial insemination facility for animal husbandry.

There are other kinds of hybridization for which a sperm bank would be useful. The ability to perform fertilizations *in vitro* removes behavioral barriers to interspecific and even intergeneric crosses, and the potential practical benefits of such crosses are great. For example, hardy and fast-growing sterile hybrids are known in most groups of cultured fishes and hold great potential because they do not expend energy on reproduction. A bank of frozen sperm from males known to produce sterile offspring in specific crosses would permit easier and wider dissemination of these breeds. In general, there is an enormous reservoir of behavioral, physiological, and territorial adaptations in unexploited species that could have an important role in future programs of genetic improvement. Artificial propagation techniques, aided by storage of sperm, will help tap this resource (FAO/UNEP, 1981).

Domestication and Conservation

The only fish bred long enough to have domestic races are the common carp and trout (Moav, 1979). Several other species, including Chinese carps, Indian carps, tilapias, catfishes, and Atlantic and Pacific salmon, are in the process of becoming genetically accustomed to living under artificial conditions throughout their life cycle. Bitter experience in crop and animal breeding has shown the folly of allowing genetic impoverishment to accompany domestication (Wilkes, 1977), and fish breeders still have the opportunity to preserve variability as domestication proceeds. As the genetic base narrows with domestication, genetic determinants likely to be lost include those for disease resistance and fitness in artificial environments, and it is vital that a reservoir of variability for such characters be maintained.

Many indigenous fishes in Africa, Asia, and South America remain undescribed. Many more are described yet untested for local or widespread use in aquaculture. These unexploited species represent a vast genetic repertoire of adaptations of potential use in aquaculture, whether through domestication or as sources of genetic material for interspecific hybridizations. One of the major recommendations of the FAO Consultation on the Genetic Resources of Fish in 1980 was to gather taxonomic and other information concerning these fishes so that a better idea could be had of their importance in scientific, economic, and ecological terms (FAO/UNEP, 1981).

Two steps in the domestication process will be served by the establishment of a sperm bank. For many of the fishes grown for food today, the process has already advanced to the second. Ideally, domestication should begin with exploration for, and testing of, wild stocks with potential in aquaculture. Once a species is settled on, founder stocks should then be collected from as wide a distribution as possible to ensure the broadest possible genetic base. One important advantage of a sperm bank would be to allow the deposit of genetic material from these varied stocks as they are collected and tested, thus establishing a collection for future reference during domestication.

In cases where domestication is already advanced, a sperm bank will be important in maintaining a reservoir of variability so that loss of fitness in captivity is not debilitating and so that progressively genetically impoverished populations are not at the mercy of unexpected environmental changes. Of all the functions of a sperm bank, this one is closest to that of an agricultural seed bank, for it involves the collection of genetic material from wild relatives of cultivated species and its maintenance as a hedge against an unforeseen breakdown in the ability of managed stocks to survive in an artificial environment.

Selective Breeding

Mass selection based on the performance of individuals has been widely if not rigorously practiced in aquaculture in order to improve such characteristics as growth rate, disease resistance, and docility. Fish geneticists agree that additive variation is the greatest genetic resource available for stock improvement and that such improvement can be achieved through selection. Nevertheless, progress in this area is still limited because, with the exception of common carp, diverse, genetically improved stocks from which to build a selection program are not yet available (Gall, 1983). There is clearly a role for a sperm bank in the buildup and maintenance of at least the male complement of such a genetic resource.

In the long term, designing efficient selection programs and predicting gains requires genetic analysis of the relevant characters, particularly because most "production-related" characters are controlled by several genes and are strongly influenced by environmental variation (Moav, 1979). The design of selection experiments and programs is the subject of some debate among fish geneticists. A sperm bank would facilitate the process of genetic analysis of production

35

characters by providing a stable control for evaluating the results of selection experiments when year-to-year environmental variation is large.

Study of Genotype-Environment Interaction

A strain that grows well in one environment may do poorly in another. Growth rate rankings of the Chinese and European races of the common carp, for example, are reversed when pond conditions are changed (Wohlfarth, 1983). In general, expression of genotype can be affected not only by pond conditions (water quality, climate, etc.) but by season, age, and husbandry system (Moav, 1979). There is an obvious role for a sperm bank in the study of these interactions, for it would allow researchers and farmers in different locations to inseminate female broodstock with milt from a single strain or even with aliquots of frozen milt from the same male. The success of such an approach would, of course, depend on the male's genetic contribution to the expression of traits likely to be environmentally sensitive.

Artificial Gynogenesis

Artificial gynogenesis is an example of genetic engineering for which a sperm bank will be an invaluable resource. Artificial gynogenesis makes use of artificial fertilization techniques and employs inactivation of the male genetic complement by irradiation or chemical treatment of sperm, followed by induction of diploidy in the "fertilized" embryo using temperature or pressure shock. In most cases, gynogenetic offspring are exclusively female (Donaldson and Hunter, 1982; Purdom, 1983).

The great appeal of artificial gynogenesis as a tool in selective breeding is that it provides highly inbred lines virtually instantaneously. Useful homozygous genotypes can thereby be created and preserved without the attendant disadvantages of inbreeding depression. If several such inbred lines are maintained, genetic variability is preserved even though each line is highly inbred (FAO/UNEP, 1981). When these lines are crossed, the results are new heterotic combinations with the added bonus of hybrid vigor.

There are two specific applications of a sperm bank in artificial gynogenesis. In the first, irradiated sperm from a number of species and strains is stored for distribution to breeders wishing to produce gynogenetic stocks but lacking irridiation facilities. A high degree of cooperation with donors is implied. A second application of a sperm bank in artificial gynogenesis relies on the technique of sex reversal, in which diploid gynogenetic female embryos are treated with a low dose of androgen to produce highly homozygous phenotypic males that remain genotypically female (Donaldson and Hunter, 1982). The procedure yields "female sperm," and if such sperm from one gynogenetic stock is used to fertilize ova from another, control over the resulting degree of heterozygosity is even greater than if gynogenetic females are fertilized with conventional sperm (as they must be in the simplest application of gynogenesis). Here a sperm bank would be in a position to supply highly homozygous spermatozoa guaranteed to produce all-female offspring. The attractiveness of all-female populations for enhanced growth and suppressed reproduction alone make this function an important one.

ORGANIZATION OF A SPERM BANK FOR AQUACULTURE

Establishment of a sperm bank for cultured fishes is implied in the recommendations of several recent symposia dealing with genetic resources in fish in general (FAO/UNEP, 1981; Moav, 1979), and with research on the biology and culture of tilapias in particular (Fishelson and Yaron, 1983; Pullin and Lowe-McConnell, 1982). Although research continues on the long-term preservation of fish ova and embryos (Harvey, 1983b; Harvey et al., 1983; Harvey and Ashwood-Smith, 1982), success is not likely to come soon. When, and if, it does, the results may be in the form of cryopreservation of embryonic cells or primordia rather than of intact embryos. For the time being, ova fertilized with frozen-thawed spermatozoa will have to be obtained fresh from the female breeder. Another topic for discussion should be identification of donors; electrophoretic techniques exist for some fishes (Pullin and Lowe-McConnell, 1982), but their application will need to be closely considered.

A central storage facility could interact with a number of regional centers whose function is to identify, collect, transport, and utilize spermatozoa. Regional centers might be set up at existing aquaculture facilities. Functions of the main facility would include cataloguing and viability testing in addition to storage, as well as training personnel from international network members in the collection, processing, and use of sperm. Decisions will have to be made on the biological and physical scope of a sperm bank, as well as on the legality of its operation. Tilapias are an obvious starting point by virtue of their global importance and the state of our genetic knowledge; carps, catfish, and many brackish water species are also excellent candidates. The growing importance of aquaculture in Asia, Africa, and Latin America suggests that most of the regional centers will be in these areas; arguments for and against establishing a central storage facility in the developing world are complex (Ford-Lloyd and Jackson, 1984) and will not be dealt with here.

ACKNOWLEDGMENTS

I am grateful to the International Development Research Centre of Canada for financial support during the development of these arguments. Thanks are due to Dr. R. S. V. Pullin, Dr. R. Doyle, and Dr. E. M. Donaldson for critical reading of the manuscript.

REFERENCES

Donaldson, E. M., and G. A. Hunter. 1982. Sex control in fish with particular reference to salmonids. *Can. J. Fish. Aquat. Sci.* 39:99–110.

FAO/UNEP. 1981. Conservation of the genetic resources of fish: Problems and recommendations. Report of the Expert Consultation on the Genetic Resources of Fish. FAO Fish. Tech. Pap. 217. Rome, June 9–13, 1980.

Fishelson, L., and Z. Yaron, eds. 1983. International Symposium on Tilapias in Aquaculture. Nazareth, Israel, May 8–13, 1983.

Ford-Lloyd, B., and M. Jackson. 1984. Plant gene banks at risk. *Nature* 308:683.

Gall, G. A. 1983. Genetics of fish: A summary of discussion. In *Genetics in Aquaculture*, ed. N. P. Wilkins and E. M. Gosling, 383–395. Amsterdam: Elsevier.

Harlan, J. R. 1975. Our vanishing genetic resources. *Science* 188:618–621.

Harvey, B. 1983a. Cryopreservation of *Sarotherodon mossambicus* spermatozoa. *Aquaculture* 32:313–320.

———. 1983b. Cooling of embryonic cells, isolated blastoderms, and intact embryos of the zebra fish *Brachydanio rerio* to −196°C. *Cryobiology* 20:440–447.

Harvey, B., and M. J. Ashwood-Smith. 1982. Cryoprotectant penetration and supercooling in the eggs of salmonid fishes. *Cryobiology* 19:29–40.

Harvey, B., and W. S. Hoar. 1979. The theory and practice of induced breeding in fish. International Development Research Centre, Ottawa.

Harvey, B., N. Kelley, and M. J. Ashwood-Smith. 1982. Cryopreservation of zebra fish spermatozoa using methanol. *Can. J. Zool.* 60:1867–1870.

———. 1983. Permeability of intact and dechorionated zebra fish embryos to glycerol and dimethyl sulfoxide. *Cryobiology* 20:432–439.

Kincaid, H. L. 1983. Inbreeding in fish populations used for aquaculture. In *Genetics in Aquaculture*, ed. N. P. Wilkins and E. M. Gosling, 215–229. Amsterdam: Elsevier.

Lam, T. J. 1982. Applications of endocrinology to fish culture. *Can. J. Fish. Aquat. Sci.* 39:111–137.

Moav, R. 1979. Genetic improvement in aquaculture industry. In *Advances in Aquaculture*, ed. T. V. R. Pillay and W. A. Dill, 610–622. Farnham, Surrey: Fishing News Books Ltd., for FAO.

Pullin, R. S. V., and R. H. Lowe-McConnell. 1982. The biology and culture of tilapias. ICLARM Conference Proceedings 7. International Centre for Living Aquatic Resources Management, Manila, Philippines.

Purdom, C. E. 1983. Genetic engineering by the manipulation of chromosomes. In *Genetics in Aquaculture*, ed. N. P. Wilkins and E. M. Gosling, 287–301. Amsterdam: Elsevier.

Scott, A. P., and S. M. Baynes. 1980. A review of the biology, handling and storage of salmonid spermatozoa. *J. Fish. Biol.* 17:707–739.

Stickney, R. R. 1979. Principles of warmwater aquaculture. New York: John Wiley.

Wilkes, G. 1977. The world's cropplant germplasm: An endangered resource. *Bull. Atom. Sci.* 33:8–16.

Wohlfarth, G. W. 1983. Genetics of fish: Application to warmwater fishes. In *Genetics in Aquaculture*, ed. N. P. Wilkins and E. M. Gosling, 373–383. Amsterdam: Elsevier.

Commentary
Mark S. Hafner

Dr. Weissinger's contribution to this conference is an overview of five methods currently used to maintain (in an artificial environment) tissues containing native DNA. The methods discussed include pseudonatural environments (such as zoos and botanical gardens), seed banks, microbial cultures, tissue cultures, and genomic libraries. Because Dr. Weissinger has focused on the technical aspects of each of these preservation methods, I concentrate here on the interface between present germ plasm preservation technology and biological conservation.

From the outset, it is important to distinguish between those technologies designed solely for germ plasm *preservation* (maintenance in a steady state for long periods) versus those that show promise in the area of germ plasm *conservation* (maintenance in the natural, dynamic state for long periods). Although the title of Dr. Weissinger's paper contains the word "preservation," it is clear that his emphasis is placed (correctly, I believe) on genetic *conservation*. Most of the technologies discussed in this paper are not really new. We have long had the ability to maintain populations of plants and animals in captivity, and techniques for microbial and tissue culturing and long-term preservation of undenatured macromolecules are not especially new. The critical issue is: how can we translate these abilities and techniques into something that actually *aids* the organism, rather than *uses* it?

As Weissinger is undoubtedly aware, most *ex situ* germ plasm preservation today (excluding zoos, botanical gardens, and the like) is done strictly in the interest of basic and applied research. It is, therefore, incorrect to imply that *ex situ* methods are a backup or safety net protecting organisms from extinction in instances where *in situ* methods (such as nature reserves) have failed. Although this is a noble goal, it is not a proper characterization of the state of *ex situ* technology today. This technology was developed largely through basic research, and it is used most heavily today in basic research. We can only *hope* that it will be applied more rigorously to biological conservation problems in the future. If material collected for use in basic research proves fortuitously to be of value in the conservation effort, so much the better, but it is important to remember that most existing samples of viable germ plasm were originally collected (and are maintained today) largely for reasons unrelated to biological conservation.

To understand, and perhaps strengthen, the link between present *ex situ* technology and biological conservation, we must face the current state of affairs squarely. For example, most zoos do not keep animals in the interest of conservation or preservation. For economic reasons, only a few can afford this relative luxury; and at that, only a small aspect of their operations can be truly conservation oriented. The same can be said for botanical gardens, arboreta, nurseries, farms, fisheries, and aquaria. Likewise, most plant seeds and animal semen are *not* preserved today in the interest of biological conservation, and microbial cultures and genomic libraries are rarely established to protect wild organisms from extinction. Understandably, these complex technologies are driven, in large part, by economic needs and goals. Dr. Weissinger has outlined for us the uses and limitations of each of these techniques; and even assuming that current *ex situ* preservation methods *could* help in the area of biological conservation, where are the driving economic needs and goals? They do not exist (or do not *yet* exist), and as a consequence, state-of-the-art *ex situ* germ plasm preservation technology is a long way from becoming *conservation* technology.

As we work toward the development of true *ex situ* germ plasm conservation technology, it would be helpful to decide exactly what we want this technology to do. The few zoos and botanical gardens with restoration-reintroduction programs for endangered species have a clear mission, but the carrying capacity of these institutions is simply too small to manage well-designed breeding programs for the many thousands of threatened or endangered organisms on earth today (see Myers, 1979). In fact, Conway (1980) estimated that all the zoos in the world have the combined resources to maintain adequate levels of genetic diversity in only 100 species of animals.

Perhaps, then, we will want to use laboratory-based technology to supplement the conservation efforts of zoos and botanical gardens. If so, we should focus our efforts on the development of protocols for long-term (cryo?)preservation of seeds, vegetative propagules, semen, ova, and/or embryos to maintain the genetic integrity of a species during periods of great risk or prolonged restoration efforts. Although current *ex situ* technology is helpful, it is far from ideal. As Dr. Weissinger points out, long-term seed preservation is limited to species that produce dry seeds, and vegetative propagules are limited to short-term storage. As for animals, semen cryopreservation is limited to fewer than fifty mammalian species plus a few fish, reptile, and bird species (e.g., see the contributions by Harvey, this volume, Sexton and Gee, 1978; Graham et al., 1978; Seager et al., 1978). At present, researchers are only able to preserve viable embryos from dairy cattle and a few species of laboratory mammals (Seidel, 1981), and the widespread use of embryo transfer, perhaps the most promising of all the above techniques, must await the solution of many practical problems (Ryder and Benirschke, 1984).

In time, it is possible that the restoration-reintroduction programs of zoos and botanical gardens, augmented by improved *ex situ* technologies designed to preserve gametes and embryos needed to maintain the genetic integrity of captive populations, will constitute an effective *ex situ* conservation program for higher plants and animals. As Dr. Weissinger points out, the technology for long-term preservation of organisms at the other end of the biological spectrum (bacteria, fungi, viruses, and other microorganisms) is already available. It is important to note, however, that large-scale microbial collections (such as the American Type Culture Collection) are designed to serve the research community. Thus their orientation toward the conservation of free-living microbes may be similar to the white-rat breeder's orientation toward the conservation of wild *Rattus* species. This orientation must change to some extent if microbial culturing is to become an effective *ex situ* conservation tool for naturally occurring microbes.

Unfortunately, there appears to be no large-scale germ plasm conservation effort focusing on those organisms midway between higher plants or animals and microbes (i.e., the lower plants and animals). Again, we see what happens in the absence of incentive, or "economic drive."

Thus far, I have commented on possible future roles in the conservation effort for zoos, botanical gardens, seed banks (to which I add semen and embryo banks), and microbial cultures. What roles will tissue culture technology and genomic libraries play in our future *ex situ* germ plasm conservation program? Certainly, these are extremely valuable research tools, and basic knowledge of plant and animal biology is necessary for well-designed conservation programs; necessary, but *not* sufficient. If we agree that *ex situ* germ plasm conservation is a means to an end, the end being reintroduction of the species into the wild, then these technologies offer little that is not already available through cryogenic gamete and embryo storage. If, on the other hand, we regard *ex situ* germ plasm conservation as the genetic equivalent of traditional systematic collections (e.g., animal skins and skeletons; dried plants and seeds) designed to serve as a historical record of the genetic diversity of the earth, then long-term DNA preservation through tissue cultures and genomic libraries may be the ultimate answer. Although Dr. Weissinger points out that we are as yet unable to regenerate a complete animal from a tissue culture (and only a few plants regenerate successfully from culture), this technology should develop rapidly over the next few decades. Even so, what conservation goal will be served when we finally have the ability to "grow a rhinoceros in a petri dish" if its natural habitat has long since been destroyed?

Finally, if we view *ex situ* germ plasm preservation technology as a potential means of placing living organisms in long-term suspended animation—a modern-day "Noah's ark"—until such time as the "flood" of ignorance concerning the need for habitat preservation has subsided, we are hiding our heads in the sand. To serve in the conservation effort, this technology must be an adjunct to programs designed to eventually reintroduce organisms into the wild. Otherwise, the germ plasm storehouse of the future will be no more than high tech "genetic menageries," and the practice of resurrecting (from culture) extinct organisms whose natural habitats no longer exist will be, at best, good for the economy (e.g., improved crops and livestock through genetic engineering) and, at worst, a sad curiosity for future generations to ponder.

41

REFERENCES

Conway, W. G. 1980. An overview of captive propagation. In *Conservation Biology: An Evolutionary-Ecological Perspective*, ed. M. F. Soulé and B. A. Wilcox, 199–208. Sunderland, Mass.: Sinauer.

Graham, E. F., M. K. L. Schmehl, B. K. Evensen, and D. S. Nelson. 1978. Semen preservation in non-domestic animals. In *Symposium on Artificial Breeding of Non-domestic Animals*, ed. P. F. Watson. *Zool. Soc. London* 43:153–173.

Myers, N. 1979. *The Sinking Ark*. Oxford and New York: Pergamon Press.

Ryder, O. A., and K. Benirschke. 1984. The value of frozen tissue collections for zoological parks. In *Collections of Frozen Tissues*, ed. H. C. Dessauer and M. S. Hafner, 6–9. Lawrence, Kans.: American Association of Systematics Collections.

Seager, S., D. Wildt, and C. Platz. 1978. Artificial breeding on non-primates. In *Symposium on Artificial Breeding of Non-domestic Animals*, ed. P. F. Watson. *Zool. Soc. London* 43:207–218.

Seidel, G. E., Jr. 1981. Superovulation and embryo transfer in cattle. *Science* 211:351–358.

Sexton, T. J., and G. F. Gee. 1978. A comparative study on the cryogenic preservation of semen from the sandhill crane and the domestic fowl. In *Symposium on Artificial Breeding of Non-domestic Animals*, ed. P. F. Watson. *Zool. Soc. London* 43:89–95.

Summary of the Discussion

Ruth Shaw

The paper by Dr. Weissinger provides a comprehensive review of the diverse methods available for the preservation of genetic material, including maintenance of whole organisms in seminatural environments, seed banks, and microbial cultures, of cell lines in tissue cultures, and of DNA segments in cloning vectors. Although these methods offer a wide array of possibilities for preserving genetic entities, from single genes to individual genomes to populations, their inherent deficiencies in the context of long-term conservation of evolving populations were emphasized in both the paper and ensuing discussion. As noted by Dr. Hafner, the methods were developed using species that are of economic interest (crop plants, livestock) or are the subject of intensive medically related research (pathogenic bacteria and viruses, mouse, human). During the course of the discussion, two consequences of the historical focus on this circumscribed set of species became clear. First, despite the diversity of techniques, these are applicable, at present, to rather few species. Taxonomic barriers to general implementation of particular techniques remain, despite intensive research efforts.

For example, cryogenic preservation of swine semen has been unsuccessful to date, although the method is standard practice for the preservation of cattle semen. Dr. Harvey's contribution notes a similar failure to develop methods for long-term storage of fish embryos. More generally, there is no broadly applicable method for storing plant species that are soft-seeded or that propagate exclusively vegetatively. Thus the goal of conserving genetic material of broad taxonomic diversity demands development of techniques of preservation that specifically apply to species that remain intractable, especially those that are endangered in the wild. Experimentalists involved in developing such technologies warned against naive belief that the necessary technologies are available or will become available simply as spin-offs from research on a few widely studied, commercially important, species.

A second consequence of the historical emphasis in the development of these methods is that there are few repositories for genetic materials that are without direct economic value or medical research interest. Zoos and botanical gardens serve this function, but they are severely limited in the number of species and the number of individuals of each species that they can accommodate. A seed storage facility for wild plant species would be of great value and is technically feasible. Such a collection could provide a short-term "safety net" for endangered species and could serve as a source of useful genes for transfer within and between species, capabilities that undoubtedly will be available in the near future. Likewise, although the possibility of regenerating organisms from their DNA is quite remote, collections of full complements of DNA would be of great value as a store of genes that might be of use in both wild and domesticated populations in the future. Moreover, such a collection would contain a tremendous store of information relevant to inferring phylogenies. At present, DNA libraries generally contain only genome fragments bearing genes of current interest. Thus there is a pressing need for the establishment of facilities for the preservation of a wide variety of genetic materials.

Dr. Weissinger notes numerous liabilities of laboratory-based preservation technologies, and these are particularly relevant in the context of conservation of dynamic populations that are genetically representative of their counterparts in natural conditions. These liabilities include (1) limitations on population size, both at the time of sampling and in subsequent generations, with attendant inbreeding, drift, and loss of genetic variability, (2) genetic response to selection in the novel, rather than native, environment, and (3) genetic stasis of cryogenically preserved material. Breeding programs for species held in captivity are designed to minimize inbreeding, but further research, including application of pedigree analysis to increase effective population size, was called for. At present, however, because the above-named liabilities are virtually unavoidable and their effects on the genetic constitution are irrevocable, the goal of long-term conservation of natural populations dictates the use of *ex situ* techniques primarily as an adjunct to *in situ* methods whenever this is possible. The additional inappropriateness of *ex situ* techniques for the preservation of associated species (e.g., obligate pollinators) or whole ecosystems further reinforces this judgment.

Nevertheless, several discussants emphasized the value of laboratory-based preservation technologies in distributing the risk of extinction, given the unpredictability of political and economic conditions, hence the insecurity of wildlife preserves over centuries. Considering the increasing numbers of species nearing extinction, a great deal of reliance will continue to be placed on the methods for preserving genetic material out of the natural context.

Clearly, the issues of (1) what do we want to preserve and (2) in what state (or for what purpose) bear very strongly on the choice of conservation approaches. Moreover, decisions regarding priorities for funding depend on assessment of the urgency of direct conservation effort relative to the expected gain from research on improving methods. The session's discussion addressed these issues with no final resolution.

2

In Situ Conservation of Genetic Resources

Bruce A. Wilcox

In situ technologies for conserving genetic resources are methods of protecting such resources in the habitats in which they naturally occur, or in which they have become naturalized. This is contrasted with *ex situ* methods, which involve removal of genetic resources from their natural or naturalized habitats for storage in artificial environments (Frankel and Soulé, 1981; Oldfield, 1984).

The essential requirement for the *in situ* conservation of genetic resources is therefore implementation of methods of land and resource use that provide for the protection of natural habitats. Thus the most immediately critical problem is the prevention of habitat destruction. Unfortunately, this seems a hopeless task even for the most extensive and genetically diverse ecosystems on earth, tropical forests. Conversion of tropical forests to pastureland, plantations, and, in many cases, wasteland is taking place on a scale of millions of acres per year (Myers, 1980; FAO, 1981). As Raven (1984) points out, because exploitation is tied to population growth, which has sufficient built-in momentum to double human populations in the tropical countries by the year 2020, this depletion "cannot be reversed or even appreciably slowed down."

Because of this, strategies for the protection of genetic diversity *in situ* must be developed in the context of the expanding human population's needs for food, fuel, fiber, and space. There is little question that few of today's extensive tracts of relatively pristine natural habitat will remain a generation from now. What will remain, if we are lucky, is a network of relatively tiny habitat fragments serving as conservation areas for a variety of purposes, including *in situ* conservation of genetic resources.

The technical challenge for resource planners and managers as the architects for this system of natural enclaves is to determine the location, size, shape, configuration, and management regime that maximizes their capacity to maintain their natural integrity despite varying degrees of exploitation and modification. A basic premise of this paper is that all tracts of land, regardless of how natural and undisturbed they appear at present, are (or soon will be) subject to significant human influences, potentially threatening their role as repositories of genetic diversity. In short, the laissez-faire view of ecosystem or habitat protection is no longer an option; even reserves exempt from human interference are subject to a syndrome of indirect effects as ecological islands (Diamond and May, 1976; Soulé and Wilcox, 1980; Frankel and Soulé, 1981).

This situation calls for an expanded view of habitat or wildland conservation,

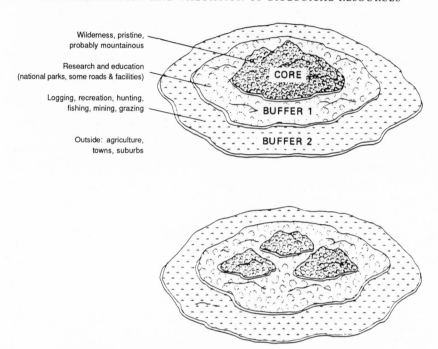

Wilderness, pristine,
probably mountainous

Research and education
(national parks, some roads & facilities)

Logging, recreation, hunting,
fishing, mining, grazing

Outside: agriculture,
towns, suburbs

FIGURE 1. *In situ* conservation area based on the biosphere reserve concept. A conservation area consists of several contiguous regions differing in the degree of human modification. Zones surrounding the remnant natural areas at the core double as disturbance areas and buffer zones for adjacent (inward) zones. The zones represent, to varying degrees, human-created ecological islands within islands.

a redefinition of a "natural preserve." A convenient model exists in the form of the "biosphere reserve" as conceived by UNESCO's Man and the Biosphere Program (Project 8: The Conservation of Natural Areas and the Genetic Diversity They Contain; UNESCO, 1973, 1974). This is the only program that includes *in situ* conservation of both *natural* and *human-modified* ecosystems (Oldfield, 1984). It does not view reserves as pristine, sacrosanct areas but as conservation planning regions, consisting of zones within which varying degrees of modification or manipulation are allowed. Biosphere reserves may consist, for example, of undisturbed "core" natural areas surrounded by buffer zones and light to heavily exploited ecosystems supporting agriculture, forestry, or other economic uses, all within the reserve (Figure 1).

So defined, *conservation areas* are the fundamental means of *in situ* conservation of genetic resources, as they represent protected regions with defined boundaries within which natural and seminatural systems and their components are maintained. Ideally, this includes maintenance of the composition and abundance of species, the genetic diversity they contain, the physical structure of the habitat, and all basic ecosystem processes. In reality, conservation areas are established and maintained for a variety of purposes which often conflict with these uses.

Existing conservation areas differ in their specific goals and in the range of legal and regulatory means for accomplishing them. They may consist, for example, of the biosphere reserves, typically a combination of administratively and legally separate conservation areas in a host country, internationally designated as representative samples of the entire biota of a specific biogeographic province (McNeely, 1982). At another extreme they may consist of parcels of critical habitat designated by state or federal agencies merely to protect a single, endangered species. Despite such a wide array of designations and widely different aims, few conservation areas exist for the express purpose of conserving genetic diversity. While this is a stated purpose of biosphere reserves, the constituent national parks, national forests, wildlife refuges, and so forth, are managed for a variety of purposes other than conserving genetic resources. In addition to a handful of areas only recently established explicitly for the protection of wild relatives of important crop plants (see Prescott-Allen and Prescott-Allen, 1984), the system of private and public lands on which wildland resources are ostensibly managed or protected can be described as a de facto system for the conservation of genetic resources (Thibodeau, 1981).

There is at least one form of conservation area in the United States created explicitly for conserving genetic resources. The U.S. Forest Service has established "genetic resource management units" where timber genetic resources are maintained, although logging and other forms of exploitation may take place (Krugman, 1985; see also NCGR, 1982).

For the purposes of this discussion, *in situ* conservation technology is defined as the application of technical information to three component problems related to conservation areas: (1) their *selection*, or siting; (2) their *design*, or physical configuration; and (3) their *management*, or long-term maintenance. Until recently, these components have been addressed with minimal scientific input and little consideration of genetic diversity in general, or genetic resources in particular. The purpose of this paper is to provide an overview of the state of technology applicable to each of these areas. It is not a comprehensive review but an attempt to tie together some still largely undeveloped concepts related to "genetic resources" and the relatively new field of conservation biology.

DEFINING THE PROBLEM

What Are Genetic Resources?

The terms "genetic resources" and "genetic diversity" have been used rather loosely and often synonymously with biological diversity. Wilcox (1984b) defines

47

biological diversity as "the variety of life forms, the ecological roles they perform and the genetic diversity they contain." This includes the variation that exists among individual organisms and within populations, and the geographical variation within and among species and among the assemblages of species that constitute rich ecological communities. The concept of biological diversity thus is applicable to different levels of biological organization, from genes to habitats or entire landscapes. Genetic diversity can be viewed as the key component of biological diversity at the molecular level of biological organization, and therefore as the source of variation at all other, "higher" levels of organization.

Miller et al. (1985) regard *biological diversity resources* as that proportion of biological diversity of actual or potential usefulness or importance to humans, although they limit their definition of biological diversity to only the "organismal level." They also urge that in view of human ignorance of the potential usefulness of biological diversity, virtually all biological diversity must be assumed to be a resource. Consistent with this definition of biological diversity, biological diversity resources include elements of biological diversity at all levels of organization (genes, populations, and ecosystem and community types) as well as the ecological processes they drive (nutrient cycling, pollination, watershed capacity, climatic amelioration) (Ehrlich and Mooney, 1983).

These definitions may pose difficulties for economists. "Free goods and services" are not economic resources, because they are not of limited quantity or exhaustible, and thus cannot be allocated among competing uses. Yet this clearly does not take into account the potential diminution of the provision of free goods and services due to ecosystem damage associated with the extraction of economic goods (e.g., timber, minerals, etc.). The problem is: at what point does the protection of biological diversity for its "nonresource" values represent a competing use? Our understanding and expression of the effects of resource development on the provision of free goods and services and of their values is extremely limited. As a consequence, economic analysis will tend to skew estimates of such values (Westman, 1977).

Leaving ecosystem-based goods and services aside, and limiting our consideration to genetic materials, Oldfield (1981, 1984) provides perhaps a more useful operational definition of genetic resources. She views genetic resources as "the socioeconomic value of the genetic diversity contained within (intraspecific) or among (interspecific) populations or species of organisms" (1981: 280). In this context, intraspecific variation, especially at the molecular level, is viewed as "germ plasm resources," since it is ultimately maintained by and derived from individual organisms (genotypes). In comparison, the term "gene pool resources" refers to interspecific or interpopulational genetic diversity (Oldfield, 1981). Oldfield (1984) points out that when new economically valuable species are sought *in situ* (e.g., biomedical research animal models, pharmaceuticals, industrial oils), workers search among taxa likely to exhibit the greatest potential for yielding the desired product. Once an organism is brought under cultivation or husbandry for the purpose of producing market quantities of the desired product, efforts at genetic improvement necessitate use of the germ plasm resources derived from its various gene pools. Thus the term "gene pool resources" applies to genetic resources conserved *in situ*, while the term "germ plasm resources" refers to *ex situ* conservation.

In Situ *versus* Ex Situ *Methods*

There has been an unfortunate tendency for "conservationists" to split into separate philosophical camps favoring either *in situ* or *ex situ* conservation methods. Heated debates have centered on whether to attempt to recover a dwindling population through *in situ* management or *ex situ* captive propagation (e.g., the California condor; see Phillips and Nash, 1981). Proponents of captive propagation point to *ex situ* maintenance as a last redoubt for a species going extinct in the wild, with the possibility of producing sufficient stock for eventual reintroduction and reestablishment of a "wild" population. Those critical of this rationale argue that the emphasis on captive breeding may undermine habitat protection efforts, since once a species is removed from the wild, it will be more difficult to protect its habitat from development pressures. Furthermore, reintroduction poses enormous technical difficulties. Nevertheless, the technology of captive propagation and reintroduction has made heroic recent advances (see Benirschke et al., 1980; Kleiman, 1980; and Benirschke, 1985). See Soulé and Wilcox (1980, chaps. 11–15) and also Foose (1983) for discussions of the pros and cons of captive propagation and reintroduction.

Similarly, breeders of agricultural plants and animals and *in situ* proponents are often at odds. The former view gene resource conservation in terms of resources of known, immediate value, and concentrate on problems related to inadequate breeding and storage facilities, often with little concern for the potential of *in situ* methods for conserving gene pool resources. The latter employ gene resource conservation in their repertoire of arguments for protecting natural habitat and tend to be less than enthusiastic about possible alternatives.

It is becoming clear, however, that *in situ* and *ex situ* conservation both have important roles to play as subdisciplines of a still very young science of conservation biology. The effort of Kleiman and colleagues (Kleiman et al., 1985) to save the highly endangered golden lion tamarin is an excellent example of the potential for the integration of *in situ* and *ex situ* methods. A large *ex situ* population has been established through an intensive, long-term captive breeding effort. The surplus individuals are being reintroduced in a careful and elegantly designed protocol based on extensive ecological and behavioral data from combined studies of wild and captive animals. The captive bred stock is not only helping to shore up a dangerously small wild population, but it is likely that the former has maintained genetic diversity already lost in the latter, and therefore represents a supplemental genetic reservoir.

If we define genetic resources in the broad sense, as the total pool of genetic diversity extant in nature, in situ preservation is the principal method for protecting genetic resources (Franklin, 1980; Oldfield, 1981, 1984). This is because the amount of genetic diversity is simply too vast and its distribution too poorly known (merely in economically important species) for field collection and ex situ protection to provide for more than a minuscule fraction of the genetic diversity extant on earth. For higher plants alone, with some 250,000 species described, and as many as three times this number undescribed (Raven, 1984), one might imagine an expenditure rivaling the cost of the U.S. space program to preserve the genetic diversity of but a small percentage of these species ex situ.

In the more limited context of economically important crop genetic resources, in situ preservation is an essential component of preservation technology (Frankel and Soulé, 1981; Oldfield, 1981, 1984; Prescott-Allen and Prescott-Allen, 1984; Ingram and Williams, 1984). Wild relatives of crop plants that have "recalcitrant" seeds, those that cannot be kept successfully in cold storage (Harrington, 1970; Roberts, 1975; Wang, 1975; Hawkes, 1982), and very long-lived perennials (e.g., forest trees) are best preserved in situ.

Conservation of gene pool resources is just one of many benefits of in situ conservation. The protection afforded natural ecosystems ensures the provision of economically and socioculturally significant products and services, already mentioned, and also makes possible a "genetic library" (see Ehrenfeld, 1976; Myers, 1979; Ehrlich and Ehrlich, 1981; Ehrlich and Mooney, 1983). Moreover, to the extent that ecosystem functioning is dependent on intra- and interpopulation genetic variation of constituent species, the sustained use of these products and services is ultimately dependent on the maintenance of genetic diversity.

Perhaps most important, in situ methods "preserve" the evolutionary dynamic characteristics of genetic resources, allowing gene pools to continue to generate new variants of potential value, as well as to facilitate the creative forces of diversifying and stabilizing selection. With the exception of a very few major crop plants in which the genetic basis of key traits (salinity and drought tolerance, resistance to disease, etc.) is sufficiently well understood to allow maintenance through artificial selection, potentially important germ plasm is undoubtedly lost with the cessation of natural selection associated with ex situ methods.

Genetic variation can be induced in vitro through breeding of mutants. This can be particularly useful in advanced or modern crop plants in which variation has been exhausted and where it has been found difficult to transfer useful genes from wild relatives or primitive germ plasm. However, artificially generated variation cannot substitute for natural genetic variation, because breeding of mutants is costly, and screening techniques are limited to a few traits (in species for which screening is feasible at all). Breeding of mutants is technically infeasible for many organisms, particularly long-lived plants and animals in general (Oldfield, 1984).

Moreover, while it is technically difficult enough to create and select economically useful single-gene (point) mutations from a welter of mostly neutral or deleterious mutations, it is effectively impossible to assemble multigene complexes forming the basis of polygenic traits and coadapted gene complexes. Coadapted gene complexes, which have been suggested as the basis of many economically important traits of major crop plants, are created and maintained via natural selection *in situ*. This aspect clearly differentiates gene pool resources from germ plasm resources, since the former can, for the foreseeable future, be maintained only by conserving natural populations as continually evolving units.

While such genetic systems can be transferred from wild relatives of advanced cultivars or livestock breeds via recombinant DNA techniques, they nonetheless cannot be created *in vitro*. Thus, to realize its full potential, genetic engineering depends on *in situ* gene pool resources. In sum, *in situ* reservoirs provide the necessary genetic resources for *ex situ* reservoirs.

The Problem: Level of Focus

The problem of *in situ* conservation is that of maintaining adequate intraspecific and interspecific genetic variation in wild populations (or managed populations in the wild). Conservation efforts may focus on a specific gene pool resource as just defined, or an entire ecosystem as genetic reservoir. Depending on the objective, the proper selection, design, and management strategies differ. If conservation of the gene pool of a particular taxon is the aim, concern for other species, and the genetic diversity they contain, are important only inasmuch as they influence the maintenance of the target resource.

A community type can be preserved independently of the genetic integrity of its component populations. In fact, because community types are classified on the basis of general features such as vegetation structure and dominant species, extirpations of populations and even species resulting in a loss of unique genes could take place without the apparent loss or degradation of the community. However, this may be a reflection of our ignorance of ecosystem and community taxonomy and functioning.

From an operational standpoint, three alternative foci of conservation efforts can be described: gene, species, and ecosystem and community conservation, any one or combination of which may dominate in a particular instance. Fortunately, there appear to be some universal protection criteria that apply to all three, because of the tendency for mechanisms involved in the maintenance and the loss of biological diversity to relate to processes primarily on the population level (see Wilcox 1984a, 1984b). For example, the loss of diversity at nearly all levels of biological organization involves the extinction of populations. While

51

a population can, of course, survive and still lose genetic variability, the same factors that threaten the survival of populations reduce their genetic variability.

THE NATURAL DISTRIBUTION OF GENETIC RESOURCES

As of 1982, four million square kilometers worldwide were under protection as national parks or equivalent protected areas (Harrison et al., 1982). Therefore, the diversity contained in about 3 percent of the world's natural habitat is "reasonably" protected. Given increasing population growth and the associated demands on land and natural resources, it seems unlikely that much more than this amount of land will be dedicated to habitat protection. Even nations with the most vigorous conservation programs generally have less than 10 percent in national parks and equivalent reserves. Placement of reserves in a manner that maximizes the amount of diversity protected thus becomes the critical problem. Furthermore, the rapid pace at which habitat conversion is occurring, thus closing reserve selection options, makes expedience a critical element of conservation strategy.

Biological Diversity

The first task in preserving biological diversity is to list the known components (ecosystem and community types, species, unique populations, and economically important gene pools) and their geographic occurrences. This is the justification for UNESCO/IUCN's development of a classification scheme of natural ecosystems (Dasmann, 1972, 1973; Udvardy, 1984), and for the establishment of biological inventory programs such as those pioneered by the British Nature Conservancy Council and the Nature Conservancy. This task poses the first major challenge to the state of our technical knowledge. Classification schemes and inventories attempt to define and list community types, species, and, when practical, subspecies, races, and varieties. However, since only a small fraction of the variation in ecosystems and organisms has been classified, our ability to inventory biological diversity is limited.

Reserve selection, design, and management decisions clearly cannot await completion of the task of describing the remaining several millions, perhaps tens of millions, of species and lower taxa that are undescribed and even undiscovered. Nor is the genetic characterization of even an infinitely small fraction of the known taxonomic diversity feasible. Instead, the existing information and whatever additional information is practical to obtain, and the theory derived from it, must be used as a basis for decision making. Research in biology of potential relevance to this problem has focused on understanding biological diversity at three levels. Variation in the diversity of known taxa (generally at the species level) has been investigated for (1) geographic patterns, (2) ecological patterns, with attempts at ecological and evolutionary explanation, and

(3) levels of genetic diversity maintained in different populations and species as they reveal relationships to various characteristics.

Geographic Patterns. Geographic patterns in species diversity have been a major concern of ecologists and biogeographers for some time (e.g., Pianka, 1966; MacArthur, 1972). Recent advances in computer technology, particularly management and graphics, and the continually expanding data base on species' distributions, suggest much untapped potential in this area. For example, sufficient data exist to examine distributional patterns of species richness over a continent-wide system of quadrats with much greater resolution and for a larger number of taxa than has previously been done for mammals (Simpson, 1964), birds (Cook, 1969), and reptiles and amphibians (Kiester, 1971) in North America. This approach should be particularly useful for identifying regions of high diversity as candidates for reserves, as well as for generating a more refined predictive theory of diversity. Areas of high endemism in certain taxa in South America are being considered for siting of reserves (Lovejoy, 1979). However, much more work is required to uncover less obvious geographic patterns of diversity.

Ecological Patterns. The study of diversity and its causes on the habitat level has been receiving wider attention (see Connell, 1978, for community studies; and Hubbell, 1979, for a population study). Also receiving attention has been the relevance of dynamics on both community and population levels to selection and design (e.g., Foster, 1980; L. E. Gilbert, 1980) and management of reserves (Green, 1972; Grime, 1973; Silverton, 1980).

Research on the relationships between the diversity of invertebrates and plants is of much potential importance (see Mound and Waloff, 1978, for a review). Opler (1978) has documented the consequences of the decline of the American chestnut in terms of the insects likely to disappear along with this tree. Clearly, a greater understanding of the relationships between plant and animal diversity is essential to the development of *in situ* preservation technologies. There is little evidence that even the meager knowledge in this area is so employed.

Genetic Patterns. Understanding the maintenance and distribution of genetic diversity in organisms has been the major preoccupation of population genetics for at least two decades (Dobzhansky, 1970; Lewontin, 1974). While major theoretical issues remain unresolved regarding the adaptive significance of genetic variation, the application of a variety of biochemical techniques, particularly electrophoresis, for genetically characterizing individual organisms, populations, and species has revolutionized the study of genetic diversity. Genetic variability has now been measured in hundreds of species. Broad patterns relating diversity to taxonomic, distributional, life history, and ecological characteristics are beginning to emerge (Nevo, 1978; Hamrick et al., 1979). This information should be helpful in maximizing protection of genetic diversity through protection strategies at the population and species level.

53

Genetic Diversity

Given the assumption that the amount of genetic diversity in a community is proportional to the diversity of species, genetic resources clearly are not distributed uniformly. Tropical forests, for example, support many times the number of species as temperate forests, grassland, and tundra biomes do. Sharp differences also exist between areas within biome types. Furthermore, on smaller geographic scales, community types and habitat types can differ greatly in species diversity. Certain types of organisms exhibit more genetic diversity than others (e.g., invertebrates versus vertebrates, sexual versus asexual reproducers, widespread versus restricted species, etc.) as to populations within species (e.g., central populations contain more allelic diversity than marginal populations).

This type of information is important for strategically siting reserves to maximize the amount of genetic diversity protected. Unfortunately, even general information about the distribution of diversity is underutilized.

Gene pool resources for the world's crop species exist in regions formerly identified as Vavilov centers (after the plant collector and breeder V. I. Vavilov, who first recognized them) but currently called "crop gene centers" (Oldfield, 1984). The areas contain the wild populations of ancestors of most of the economically important crops and the traditional agroecosystems in which they were domesticated. As such, they represent areas of exceptional importance from the standpoint of *in situ* conservation, although their initial identification was as key collecting areas for making accessions for *ex situ* preservation.

Crop gene centers are defined on large geographic scales, hence they do not provide sufficient resolution for siting individual reserves. They demonstrate by their existence, however, that the major genetic resources are not evenly distributed geographically, and give a general indication of where *in situ* preservation activities ought to be concentrated.

Not all important crop genetic resources are limited to crop gene centers. Some are widely distributed, presenting the additional problem of which populations are most important. A species of wild rice, *Oryza nivara*, which grows as a weed throughout South and Southeast Asia to Northern Australia, has been found to have valuable germ plasm at only one site in Central India, despite extensive screening throughout its range (Prescott-Allen and Prescott-Allen, 1984). This is the only known source of resistance for Asian rice (*O. sativa*) to the devastating grassy-stunt virus. For genetic resources in which screening has not been so thorough, as in most wild relatives of important crop plants, we may not know which populations are most important to protect. In fact, there are many examples of important tolerance or resistance factors that have arisen in the absence of the particular environmental condition or pathogen to which protection is conferred (Oldfield, 1984:25).

Lack of knowledge of the geographic distribution of genetic resources is thus

54

a major impediment to the development of *in situ* conservation strategies. It is known that populations in the center of a species' range tend to be genetically more variable than those at the margins. Thus, to increase the likelihood of conserving as yet unidentified genetic resources, central populations should receive priority. However, rare and potentially useful genetic properties may occur more frequently in isolated populations at the margins of a species range. A possible strategy might be to preserve populations along ecological gradients in the center of a species range.

Methods of Reserve Selection

There are apparently only four reserves in the world (three established and one in the planning stages) designated specifically for *in situ* protection of genetic resources (Prescott-Allen and Prescott-Allen, 1984). In these cases, site selection was simply based on the locations of wild relatives of important crop plants. This relatively straightforward approach to reserve selection obviously is not applicable to global genetic diversity as a whole. Two approaches are currently employed for this purpose.

The first of these is exemplified by the UNESCO/IUCN classification of biogeographic provinces mentioned above. Biogeographic provinces are large-scale divisions of terrestrial environments based on gross ecological and environmental features, including the types of plants and animals. As such, they are considered distinct ecological entities. An objective of Man and the Biosphere (MAB) is the establishment of a network of biosphere reserves across all biogeographic provinces.

A single biosphere reserve will encompass perhaps a small percentage of the land areas within a province, potentially excluding elements of diversity unique to it. Thus, on the scale of a province (roughly the area of a medium-size country), further categorization of a biotic region is required. There are ongoing efforts in biotic classification in many countries which, in theory, will be employed in the development of protected area networks at this geographic level.

Biotic classification, although not developed for such purposes, is a form of crude biotic inventory. Species lists, another form of inventory, are likewise poorly developed for most regions. In fact, few existing protected areas have adequate species lists for even the most conspicuous and well-known groups (e.g., vertebrates and vascular plants). Since only a small fraction of the species on earth have been scientifically described or "discovered" (probably well under 10 percent), reliance on species lists for protected area selection is problematical.

Biological Diversity Inventories. Despite such incomplete inventory data, many habitat types, species, and unique populations are known to be rare, threatened, or endangered. Since the establishment of the federal Endangered Species Act and similar legislation, volumes of technical information have been generated on determining the status and assisting in recovery efforts for hundreds of species and habitats. Protected area selection for these elements of diversity requires knowing their geographic occurrence, or at least refining the existing knowledge of their occurrence, and siting the protected areas accordingly.

The cataloguing of rare, threatened, and endangered biological elements (i.e., communities, species, and populations) as a means of assisting in the conservation of biological diversity is being used extensively throughout the United States in State Heritage Programs established with the assistance of the Nature Conservancy. Digitized for efficient storage, retrieval, and updating of information on sites that contain rare, threatened, or endangered elements, these inventories provide an efficient data management system for ranking occurrences according to importance and conservation potential, as well as allowing follow-up on the progress of their conservation.

A more comprehensive approach to "bioinventory" is the cataloguing and mapping of the occurrences of *all* species of the major plant and animal taxa (e.g., vascular plants, vertebrates, and major insect groups) known in an area. Such a system was established by the British Nature Conservancy and is maintained at the Biological Records Centre at Monks Wood Experimental Station, England. In this system, the presence or absence of each species is recorded in quadrats ten by ten kilometers square making up a network or grid covering the entire British Isles. The principal advantages of this type of system are that it allows a more complete assessment of biological diversity, instead of just known endangered elements, and it can produce maps showing species' distributions. The maps can also be compared with geographic patterns in land use and features of the natural environment. Among other things, this provides for ongoing monitoring and protection of biological elements before they become endangered. Further, land use and resource agencies have at their fingertips a diversity profile of their entire jurisdiction. Biodiversity inventory coupled with long-term monitoring is in fact a central feature of UNESCO's Biosphere Reserve Program.

Natural Diversity Data Base: California as an Example. The California Natural Diversity Data Base is one of the most advanced state inventory programs and provides a good example of the potential of this technology for *in situ* preservation. Two attributes contribute to the particular potential of this system: (1) a graphics or mapping capability and (2) the incorporation of species distributions for virtually the entire vertebrate fauna of the state (over 600 species of mammals, birds, reptiles, and amphibians). Superimposing a grid on the species distributions would provide the beginnings of a system similar to the British one, although there are currently no such plans. Aside from this potential, the system contains the digitized boundaries of several important land use features, including major hydrographic basins and watersheds, biogeographic provinces, and counties.

Securing habitat protection for rare, threatened, and endangered species and representative tracts of ecosystem and community types is a beginning. However, for every known element of biological diversity there are many more of similar status which remain unknown, particularly among plants and invertebrates. For example, there are an estimated 280,000 insect species in California. Only a fraction have been recorded, and for only a smaller fraction are distributions and conservation status known. Since conservation decisions cannot await the inventory of tens of thousands of poorly known plants and animals in the United States (and millions globally), indirect estimates of biological diversity must be employed. Such estimates could be based on indices of biological diversity developed from a biogeographic data base. For example, it might be found that for certain groups of organisms (e.g., mammals, birds, butterflies, vascular plants), species richness might be a good indicator of overall biological diversity. This has not been adequately tested.

As in most states in the United States, California has numerous reserve systems, established and operated virtually independently of one another. In California these include the reserve systems of the Bureau of Land Management, the National Park Service, the U.S. Forest Service, the California Department of Parks and Recreation, the Audubon Society, the California Fish and Game Ecological Reserves and Wildlife Areas, the Nature Conservancy, the California State Reserve System, the Research Natural Areas, the Wilderness Areas, the University of California Land and Water Reserves, and the National Wildlife Refuges, just to name the major ones. Few of these reserves have lists of even the common species, and there is no coordinated effort among these systems to attempt to maximize coverage for the state's biological diversity.

For these systems, which represent about 7 percent of the land area in California, Klubinikin (1979) found very irregular coverage of vegetation types, with some receiving little or no protection at all. Similar analyses could easily be carried out with more detailed vegetation classifications and distribution data on all vertebrate species, as well as rare, threatened, and endangered plant and animal species, all of which are already digitized.

As mentioned above, geographic variation in species density has been examined on a continental scale for North American mammals, birds, reptiles, and amphibians. Some informative patterns that emerged indicated that species richness was related to certain environmental factors. Similar studies on a smaller geographic scale would greatly enhance understanding of the influence of physical features on biological diversity, and of biotic features on various components of biological diversity, and could be useful in reserve selection.

THE SIZE, SHAPE, AND SURROUNDINGS OF CONSERVATION AREAS

Virtually every form of habitat disturbance—conversion to agriculture, timber harvesting, construction of roads, pipelines, power lines, and so forth—tends to isolate fragments of natural habitat in a matrix of disturbed habitat. As conversion outside the boundaries of reserves proceeds, the habitat area becomes effectively smaller and more distant from other such habitats (Wilcox and Murphy, 1985). A continuous natural landscape is thus transformed into an archipelago of habitat islands in a sea of developed land. Thus most reserves are, or inevitably will become, ecological islands. Furthermore, habitat fragmentation due to the construction of roads, power lines, and various support facilities, may occur in protected areas. From the standpoint of species supported by a habitat that becomes fragmented, habitat components necessary for their survival may be lost or significantly reduced and scattered among different fragments.

Organisms are not uniformly distributed throughout the natural landscape. They have minimum spatial requirements and limited capacity to disperse freely among isolated habitat fragments. Habitat fragmentation therefore poses a considerable threat to the survival of populations and, as a consequence, genetic diversity. From the standpoint of genetic resources, it is useful to assess the risk, measured in the loss of diversity, associated with varying degrees of habitat fragmentation, and on that basis to develop strategies to reduce the risk. Applicable technical information is unfortunately limited. The following relatively crude approaches, which assume that the loss of genetic diversity is proportional to population extinction, may be helpful.

Application of Island Biogeography

Recognition of the parallel between anthropogenically created habitat fragments, particularly nature reserves, and islands has led scientists to attempt to draw conclusions about the efficacy of existing protected areas and to develop criteria for designing future protected areas based on the scientific discipline known as island biogeography. The establishment of this discipline can be largely credited to Preston (1962) and, especially, MacArthur and Wilson (1963, 1967).

Two basic conclusions applicable to conservation can be drawn from contemporary island biogeographic theory.

1. *The number of species supported by an island is proportional to its size.* The relation between the number of species and island size, as determined by examining such data for species on archipelagoes (e.g., plants on the Galapagos, birds on the Hawaiian Islands, Antilles, West Indies, East Indies, etc.), reflects a virtually universal quantitative pattern. This "species-area relationship" as it is called, can be roughly simplified to this rule: division of island areas by ten reduces the number of species by a factor of two. This relation between the size of an area and the number of species it contains is referred to as the "area effect."

2. *The number of species supported on islands is proportional to their distance from the nearest land masses.* All else being equal (e.g., size), islands that are more isolated from other islands or continents support fewer species. No uni-

versal quantitative relationship exists, and the "distance effect" is not universal, although it is commonly found.

To explain these observations, which have been apparent to naturalists since Darwin but not developed theoretically until recently, Preston and MacArthur and Wilson independently proposed the so-called equilibrium theory. Briefly, this theory states that the number of species on an island reflects a balance between the loss of species through extinction and the gain of species via immigration. They pointed out that since the likelihood of extinction is inversely proportional to population size, and smaller islands support smaller populations, smaller islands will equilibrate at fewer species. Similarly, since new species can be added only by immigration from other terrestrial sources, and the likelihood of immigration is proportional to the distance from those sources, more distant islands equilibrate at fewer species.

Thus for example, a small, near island will have a number of species similar to that of a large, distant island. The existence of area and distance effects are empirically substantiated by scores of studies (Connor and McCoy, 1979), but there has been much debate about the causes of those relationships (Simberloff and Abele, 1976; F. S. Gilbert, 1980).

MacArthur and Wilson pointed to the significance of these observations and the potential of island biogeographic theory in preserving biological diversity in an increasingly fragmented natural landscape. By the mid-1970s specific recommendations about very general aspects of *in situ* preservation began to emerge (e.g., Diamond, 1972, 1974, 1975; Terborgh, 1974, 1975). These applications are based on the assumption that protected areas are islandlike or insular, and on the general tenets of island biogeographic theory (Figure 2). As a result of debates about these general recommendations (e.g., Simberloff and Abele, 1976; Diamond and May, 1976; Terborgh, 1976; Whitcomb et al., 1976; Gilpin and Diamond, 1980; Higgs, 1981; Cole, 1981; Simberloff and Abele, 1982; Wilcox and Murphy, 1985), refinements in procedures have been suggested (Margules et al., 1982).

Fragmented Ecosystems: Experimental Results

Research on the effects of habitat fragmentation, including the insularization of reserves, has increasingly moved away from extrapolating studies of true islands to hypothetical reserves and has shifted toward experimental research on actual mainland habitat fragments. Several workers (Moore and Hooper, 1975; Whitcomb et al., 1981; Galli et al., 1976; Robbins, 1979; Butcher et al., 1981; Willis, 1979; and others) have surveyed birds in forest fragments in various parts of the world in attempts to relate effects of fragment size and isolation to species composition and diversity. In general, the following conclusions can be drawn from these and other studies (see also Burgess and Sharpe, 1981): (1) In most cases,

PRINCIPLES FOR DESIGN OF FAUNAL PRESERVES

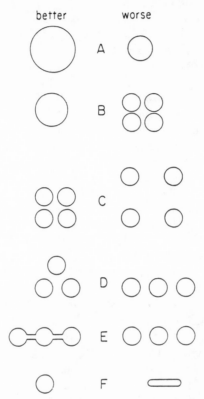

FIGURE 2. General design principles suggested by Diamond (1975) on the basis of island biogeographic theory: (A) a larger reserve will hold more species at equilibrium and have lower extinction rates; (B) given a certain area available, it should be fragmented into as few pieces as possible, because species incapable of dispersing readily among fragments may be confined to areas too small to avoid extinction; (C) if an area must be fragmented, the pieces should be as close as possible to minimize inhibition of dispersal; (D) a cluster of pieces in contrast to a linear arrangement will enhance dispersal among the fragments; (E) maintaining or creating corridors of natural habitat between fragments will enhance dispersal; and (F) a circular shaped area will minimize dispersal distance with a conservation area. See Margules et al. (1982) for exceptions to these principles.

there is a highly significant correlation between species diversity and fragment size, higher in fact than that found on average for true archipelagoes (cf. Connor and McCoy, 1979). (2) The greater number of species in larger fragments is in part the result of an increased likelihood of the occurrence of species with larger area requirements. The smallest fragment occupied varies greatly among species, and this "minimum area requirement" is often consistent with independent information on the ecological requirements of a species discussed elsewhere. Thus the strong species-area relationship is due in part to a cumulative effect of incorporating species with progressively larger minimum area requirements.

There are three nonexclusive explanations for these observations (Martin, 1981). The first is that larger fragments contain greater habitat diversity, and thus an increased likelihood the habitat components for any given species will be met. The second is that larger fragments support larger populations, which have less likelihood of extinction due to chance events. A third possibility is that larger fragments represent bigger targets for dispersing or migrating organisms, and will therefore accumulate more species. The relative importance of these mechanisms undoubtedly varies among species, and the rules useful for preservation technologies may be applicable only to specific types of organisms (e.g., flying versus flightless vertebrates, small versus large animals, "cold-blooded versus warm-blooded," etc.) (Wilcox, 1980).

Barro Colorado Island

Perhaps the best documented case of extinctions apparently related to the problem of minimum area requirements is that of birds on Barro Colorado Island (BCI). Barro Colorado was a hilltop of lowland tropical forest which became isolated in 1914 when the surrounding forest was flooded, creating Lake Gatun, during the construction of the Panama Canal. Shortly thereafter, BCI was established as a research reserve of the Smithsonian Institution, and since has been the subject of extensive biological study.

The decline in avian diversity on BCI was first brought to light by Willis (Willis, 1974; Wilson and Willis, 1975), whose extensive surveys and compilation of data during the decade 1961–70 showed that by 1970, 45 of the original 208 resident bird species known from surveys in the 1920s had disappeared. By the end of the following decade, six more extinctions had occurred (Willis and Eisenmann, 1979). Thus in total, 51 of the 209 resident species originally recorded on BCI have become extinct, without replacement by other species. However, this is a conservative estimate, since the original censuses were probably incomplete.

Detailed field studies (Karr, 1981, 1982) comparing contemporary avifaunas on BCI and at an ecologically matched site on the "mainland" push the number of avian extinctions on BCI to 107 or higher. About half appear to be a result of habitat changes due to forest succession following the establishment of Barro Colorado as a protected area. The other half are presumably due to the combined effects of habitat loss and insularization of BCI.

Minimum Critical Size Ecosystem Experiment

Except for the fragmentary information deduced from the decay of BCI's biota, what occurs in the early stages of an isolated ecosystem, and what the ultimate outcome will be, remain poorly documented. Only after decades of observations of habitat fragments will a reasonably complete picture emerge—from a relatively short-term ecological perspective. A historic experiment is under way in Brazil, however, to study the effects of ecosystem fragmentation in the tropical forests of Amazonia, as a joint effort of the Instituto Nacional de Pesquisas da Amazona and the World Wildlife Fund (Lovejoy and Oren, 1981; Lovejoy et al., 1984).

In this project, forest areas were selected and inventoried before the surrounding forest was cut down and burned for conversion to pasture. Although only a few of these reserves have been isolated for more than several years, the results are illuminating. So far, the reserves first isolated show marked and parallel changes in their wildlife communities. For birds, immediately following removal of the surrounding forest there is a dramatic increase in activity as individuals are apparently displaced from surrounding habitat and become concentrated in the reserve. Within months, however, the activity declines and the bird community begins to exhibit significant compositional change. Progressively more individuals must be caught in mist nets to achieve similar species totals as months go by. This is the expected pattern as certain species begin to decline and disappear. So far, the details seem to be in agreement with what is seen on Barro Colorado Island.

Another effect of fragmentation is to increase the proportion of the blocks of forest that are close to an edge and under the influence of microclimatic and vegetational changes associated with edges. Species that live near edges and those that normally live in gaps along streams or rivers, or in ones created by tree falls and other natural disturbances, find suitable habitats around the perimeters of the forest blocks. It is species that require forest interiors that are among the first to disappear in isolated fragments. Many forest-interior butterflies are strongly affected and have become rare or have disappeared entirely with reserve fragmentation.

In these studies, the effect of fragmentation on diversity is mostly inferential. While local extinctions are known on the basis of comparisons of past historical records with present-day observation—and, much less often, direct observation during the period of study—some extinctions are nonetheless expected naturally. A particular difficulty in this area of research is how to mask the normal background extinctions to determine the effect of fragmentation, particularly since most fragmentation has occurred either long before modern studies or very recently.

Practical Implications

Because the surroundings of a reserve are generally not as distinct as those of an island, and vary greatly even along the boundary of a single reserve, the practical value of the island paradigm is often questioned by resource managers. Obviously, surrounding habitat may range from utterly islandlike, due to urbanization, to barely distinct at all, such as only slightly modified by selective timber harvesting. Furthermore, what constitutes an insular system for some species may not for others. Thus the utility and the limitations of the insular paradigm and of the results of fragmentation research require further investigation. Nonetheless, there is a consensus among scientists and resource managers that the island perspective and information emerging from fragmentation research are important for the design and management of conservation areas.

Perhaps the most important consequence of island studies is that they have compelled consideration of the consequences of large-scale habitat conversion on the diversity of natural ecosystems. And they have helped define key aspects of this problem, particularly the relationship of habitat loss and insularization to extinction.

Two additional topics related to the surroundings of conservation areas have recently emerged that may be of considerable significance to reserve design and perhaps have been ignored because of the emphasis on comparing reserves with islands. Janzen (1983) argues that insular reserves may be more vulnerable to disturbance than true islands because they are often surrounded by earlier successional habitat, which is a good source of invasive, competing species. Brittingham and Temple (1983), Temple (1986), Wilcove (1985), and others have documented these so-called *edge effects* in terms of avian parasitism and predation associated with proximity to disturbed habitat. Edge effects may be subsumed under a syndrome of problems called *boundary effects* (Schonewald-Cox, 1983). These effects reveal the multiple interactions between conditions that define boundaries in anthropogenic and natural ecological terms as well as legal or political terms.

MAINTAINING VIABLE CONSERVATION AREAS

Scientists and resource managers are discovering that the maintenance of natural or desired biological features within conservation areas will require intensive intervention. The technical challenge of reserve management is most apparent from the nature of the threats to the species and habitats therein. Simply managing the natural disturbances in conservation areas that threaten the safety of visitors (e.g., fire, flood, tree falls, landslides), yet are essential to maintaining natural conditions, poses major difficulties for *in situ* conservation. Aside from this, a number of human activities pose a threat to the maintenance of biological diversity in conservation areas. Ziswiler (1967) and Ehrlich and Ehrlich (1981) distinguish between "direct" and "indirect" threats (or causes of extinction): the former involve the intentional taking of organisms (hunting, fishing, harvesting, collecting, trapping, poisoning, etc.), and the latter include all other inadvertant causes (habitat destruction and modification, pesticides and pollution, transport of exotics, etc.).

These threats are fairly easy to identify, and, for some, corrective measures can be designed. Yet, more frequently than not, they are not the immediate causes of extinction. Instead, random, undirected factors probably play the major role in pushing a population over the brink. Recognition of this has led to the development of the minimum viable population concept (MVP), which essentially attempts to address the problem of assigning risk in terms of the likelihood of extinction to activities that threaten a population (Shaffer, 1981; Ginzberg et al., 1982; Salwasser et al., 1985).

The Minimum Viable Population Concept

Direct and indirect threats all produce a common, predictable effect: they cause (or potentially cause) populations to decline in size. Because these threats, and their effects, typically cannot be entirely eliminated, but only managed, the problem reduces to that of how small and fragmented a population can be and still survive.

There is little doubt that smaller populations have greater likelihood of extinction, although the precise explanation for this and the implications to conservation have only recently begun to be explored. With the exception of a small handful of theoretical studies (e.g., Levins, 1970; Richter-Dyn and Goel, 1972; Leigh, 1981; Wright and Hubbell, 1983), the population biology of extinction has received almost no scientific attention. Its relevance to conservation strategy, however, appears to have spawned increasing interest (Shaffer, 1981; Simberloff and Abele, 1982; Wilcox 1984a, 1984b; Wilcox and Murphy, 1985).

The MVP problem has become a focus of attention for conservation biologists interested in developing the body of theory applicable to the fundamental problem of determining how many individuals, dispersed in what pattern, over which habitats, and throughout how extensive a region, are necessary to preclude a reasonable likelihood of total extinction due to environmental uncertainty and chance demographic and genetic events. MVP is not only becoming a major area of basic research in conservation biology, but also has become a central concern of wildlife management planning in the United States (Salwasser et al., 1985).

MVP and Preserving Genetic Variability

The application of genetics to determining population and habitat minima is technically complex and not yet fully developed. Some aspects can nonetheless be reduced to relatively simple criteria. Genetic variability plays an important role in the survival of populations. The production, maintenance, and loss of genetic variability in a species is determined to a significant degree by population size and spatial structure. Populations of small size and simple spatial structure tend to be more susceptible to the loss of genetic variability than large, geographically dispersed and subdivided populations.

The loss of genetic variability can diminish the chances of survival of a population in two important ways. First, the reduction in genetic variability in a population may lead to *inbreeding depression*. This phenomenon, well known to animal breeders probably for centuries, includes a wide variety of symptoms affecting the general viability, reproduction, and survival of individuals (see Soulé and Wilcox, 1980; Frankel and Soulé, 1981; Schonewald-Cox et al., 1983).

Second, the reduction in genetic variability within and among populations comprising a species amounts to a loss of *evolutionary potential* (Franklin, 1980; Soulé, 1980). This is because genetic variability represents the "raw material" for natural selection. When genes are lost, so are potentially adaptive traits and future evolutionary options. This includes not only long-term evolutionary change, and even speciation, but short-term evolutionary adaptation, sometimes called microevolution (Dobzhansky, 1970).

Genetic variability is rapidly lost in small populations as a result of genetic drift (random changes in gene frequencies) and inbreeding (breeding among close relatives). On the basis of experimental work and theoretical considerations, 500 is presently accepted as the size threshold for maintaining genetic variability in an "ideal" population. To be "ideal" a population must meet certain criteria. Among the most important are (1) the number of reproductive individuals is the same for both sexes (1:1 sex ratio) and (2) the numbers are the same each generation. These conditions are rarely (if ever) fulfilled by natural populations. For instance, many animals are polygynous, with just the dominant males breeding. Since only the individuals that breed successfully contribute to the gene pool of the next generation, the size of a population from a genetic standpoint may be significantly less than its total number of individuals. To account for this problem geneticists developed the concept of *effective population number* (see Soulé and Wilcox, 1980; Frankel and Soulé, 1981), the calculation of which is an essential component of MVP determination.

Aside from their significance for population survival, these considerations obviously bear importantly on the problem of maintaining maximum genetic diversity in protected gene pool resources. For instance, conditions adequate for population survival are likely to be inadequate for maintaining rare alleles in a population. Reasonably sound models exist to predict the loss of alleles given different population management criteria (see Soulé and Wilcox, 1980; Schonewald-Cox et al., 1983).

65

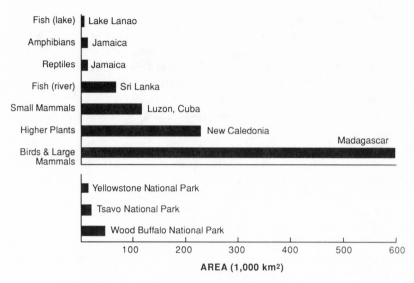

FIGURE 3. Smallest "islands" within which speciation of vertebrate taxa has occurred, compared with three large national parks (from Soulé, 1980).

Minimum Area Requirements for Speciation

Speciation, the process by which new species are generated, usually requires some form of geographic or spatial isolation of populations within a species. When populations are sufficiently isolated (temporally as well as spatially) by way of some geographic or ecological barrier, they will accrue distinct heritable characteristics. Once they are too distinct to interbreed, by definition they are considered separate species. The essential criterion for this "allopatric" speciation is isolation. Obviously, for large organisms in small reserves, geographic isolation is impossible. But how large an area is needed and for which types of species?

Soulé (1980) employed an ingenious method to attempt to answer this question. Compiling data on apparent autochthonous cases of speciation on islands of different size, he found that for most vertebrate taxa and higher plants, the areas required were generally larger than the biggest national parks (Figure 3). This suggests that for these groups, speciation will effectively come to a halt probably early in the next century.

Conversely, reserve isolation may well enhance speciation in certain groups of smaller organisms, particularly sedentary invertebrates. Thus anthropogenic fragmentation arguably may generate genetic diversity, but primarily in groups with relatively low conspicuousness and aesthetic appeal.

66

Minimum Area Requirements for Ecosystems

While it is possible to determine MVP requirements for a species, and indeed this is being attempted for several endangered species in the United States (e.g., grizzly bear [Shaffer, 1982] and spotted owl [Salwasser et al., 1985]), such requirements assume that the ecosystem providing those requirements will remain viable. In fact, the stability of the ecosystem will depend on the viability of numerous of its constituent species. Yet how can the "minimum area requirement" for an entire ecosystem be determined?

One approach is to estimate the amount of area needed to satisfy the MVP requirements for area-sensitive species, and then assume that adequate survival conditions are thereby ensured for many other species in a biota. Preserving these species thus will provide a "protective umbrella" for others. In addition to this rationale, species may be chosen because of their ecological significance.

Good candidates for target species are species with characteristics generally associated with low population density. Typical examples are species of large body size (e.g., some ungulates), high trophic level (e.g., mammalian carnivores), high metabolic requirements (i.e., mammals and birds as opposed to reptiles and amphibians) (see Eisenberg, 1980, and Wilcox, 1980), patchy distributions (see Diamond, 1980), and species dependent on successional, rare, or unpredictable habitats or resources (see L. E. Gilbert, 1980; Foster, 1980; Terborgh and Winter, 1980; Terborgh, 1974; Karr, 1982).

As a result of their central position in the food web of a community, some species may be of such enormous importance that their decline or extinction would cause a cascade of extinctions similar to the faunal collapse described for islands. Examples are "mobile links" and "keystone mutualists" described by Gilbert (1980). In many communities, but particularly in tropical forests, animals act as agents of pollen and seed dispersal for plants. Bats and birds, for example, are the major seed dispersers in neotropical forests. They act as mobile links between fruit-bearing trees and otherwise separate food webs based on the other plants. These other plants depend on the mobile links for pollination and seed dispersal. The decline in numbers or loss of one or more mobile links in an isolated protected area could result in the extinction of one or more plant species, and along with each of them, the numerous host-specific insect species.

More critical yet are keystone mutualists. These are usually plants, particularly trees, providing resources that support large numbers of mobile links. Single, large fruit-bearing tree species may depend on as few as one mobile link for effective seed dispersal, yet they can provide critical support for several other fruit-eating mobile links. The loss of a single keystone mutualist could conceivably cost the survival of hundreds of other species.

As conservation areas become increasingly insular, there undoubtedly will be a need for close observation to detect cases of incipient extinction. Both the

species and even the type of intervention required will in many cases be anticipated by a target species selection procedure. The long-term monitoring of such
species will provide managers with the opportunity for intervention before they
become seriously endangered. An additional benefit of equal importance will be
the acquisition of long-term population data. Research based on this information will enhance our understanding of biological diversity and how it can be
preserved.

Wilcox (1980) distinguished three broad phases occurring in a time sequence
to describe the loss of diversity due to the establishment and subsequent insularization of nature reserves. The description should provide a useful synopsis of
the risks inherent in *in situ* conservation. The following is generally applicable
to biological diversity in general, as well as genetic resources in particular.

Sample Effect

According to the species-area relationship, as a very general rule, for every
tenfold decrease in the size of a surveyed area, 30 percent fewer species are found.
This means that if a reserve is gazetted to cover a "representative" fraction of
general habitat type, the reserve will contain only a "sample" of all species found
throughout the range of that habitat. This may be true even if all of the variability within that general habitat type is included (see Diamond, 1980). The
principle can be extended to other components of biological diversity. For example, of the species included in the area, only a sample of the genetic diversity
contained among all the populations that comprise many of those species will
occur within the protected area.

This effect applies most directly to the problem of selecting reserves—how
many, how large, and where protected areas should be established. The effect,
of course, can be minimized by taking advantage of the fact that biological diversity is not uniformly distributed. In the absence of solid information on the
distribution of biological diversity, and without a strategy based on that information, even an extensive system of protected areas will fail to capture a significant
fraction of biological diversity.

Short-term Insularization Effects

The sampling principle describes how a protected area may fail to contain a
portion of the entire ensemble of species, and their full range of genetic variability. However, of those species included in the sample—that is, found to
occur within the boundaries of the protected area—some fraction will not be
self-sustaining as the area becomes increasingly surrounded by modified habitat.

In these cases, the protected area will either (1) fail to incorporate a sufficient number of individuals to represent a demographically stable unit or (2) fail to incorporate a sufficient quantity or range of resources to support such a unit. The ecosystem that supports these species in effect extends beyond the reserve boundaries. Once the surrounding habitat is converted, their numbers will begin the decline to extinction.

Long-term Insularization Effects

Many species in protected areas will consist of stable populations, at least over the short term. Yet all species, given enough time, go extinct. In fact, extinction of local populations appears to be relatively common in many organisms. Normally, the loss of local populations through extinction is counterbalanced by their eventual reestablishment by migrants from surviving populations of the same species. However, the reduction in habitat size that accompanies insularization will result in fewer populations to provide this dynamic balance, and in population sizes that are smaller than normal (Pickett and Thompson, 1978). The result will be the tendency for a process (extinction of a species) normally occurring on a geological time scale to condense to an ecological time scale.

The long-term effects of insularization could prove to be the most devastating for protected areas. Studies of islands created by the rising sea level in the past several thousand years indicate large losses of species as a result of insularization (Soulé et al., 1979). Based on these extinction rates, it has been estimated that even the largest protected areas will suffer an attrition of species amounting to as much or more than half their large mammal species in little more than a thousand years.

The extent to which protected areas will actually exhibit this "faunal collapse" is uncertain. For one thing, the above estimates assume no human intervention, positive or negative. Moreover, five thousand years far exceeds the time scale of concern for most governments and other institutions involved in land use planning. However, the best evidence suggests an exponential collapse such that most of the extinctions would occur in less than 1,000 years in larger areas (approximately 10,000 square kilometers or greater) and less than one or two hundred years in smaller areas (approximately 5,000 square kilometers and smaller).

Simberloff (1984) carried out a similar analysis for vascular plants and birds in the New World tropics and the Amazon basin, respectively. Assuming habitat was constricted to only that area currently protected (less than 1 percent), his figures showed a loss of 69.2 percent of some 740 birds unique to the Amazon basin, and 65.6 percent of some 92,000 plants unique to tropical Latin America.

Extinctions in Reserves

The risks of extinction of species in reserves have been empirically substantiated by documented extinctions of mammals in regions subject to recent frag-

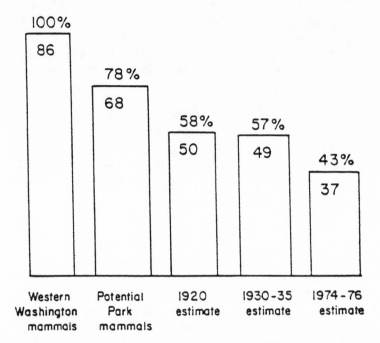

FIGURE 4. Number of mammal species occurring in western Washington and in Mount Rainier National Park at various times in recent history (from Weisbrod, 1976).

mentation. Harris (1984) provides a useful discussion of the evidence. For example, Mount Rainier National Park has already lost nearly half of the mammal species that probably occurred there originally (Figure 4).

Further empirical evidence that reserves will suffer significant losses of at least large mammals is apparent from examination of data on population densities in national parks and equivalent protected areas. Schonewald-Cox (1983) found that of 21 carnivores in 7 reserves in Africa, 13 of their populations number less than 100. In another study, East (1983) found that of 55 populations of large mammals (20 different species) in 27 reserves, 17 were smaller than 100.

ACKNOWLEDGMENTS

I am grateful to Dennis Murphy for his comments on an earlier version of this paper and Margery Oldfield for numerous helpful suggestions.

REFERENCES

Benirschke, K. 1985. *Towards a Self Sustaining Population of Captive Primate*. San Diego: Zoological Society of San Diego.

Benirschke, K., B. Lasley, and O. Ryder. 1980. The technology of captive propagation. In *Conservation Biology: An Evolutionary-Ecological Perspective*, ed. M. E. Soulé and B. A. Wilcox, 225–242. Sunderland, Mass.: Sinauer.

Brittingham, M. C., and S. A. Temple. 1983. Have cowbirds caused forest songbirds to decline? *BioScience* 33:31–35.

Burgess, R. L., and D. M. Sharpe, eds. 1981. *Forest Island Dynamics in Man-Dominated Landscapes*. New York: Springer-Verlag.

Butcher, G. S., W. A. Niering, W. J. Barry, and R. H. Goodwin. 1981. Equilibrium biogeography and the size of nature preserves: An avian case study. *Oecologia* 49:29–37.

Cole, B. J. 1981. Colonizing abilities, island size, and the number of species on archipelagos. *Amer. Nat.* 117:629–638.

Connell, J. H. 1978. Diversity in tropical rain forests and coral reefs. *Science* 199:1302–1310.

Connor, E. F., and E. D. McCoy. 1979. The statistics and biology of the species-area relationship. *Amer. Nat.* 113:791–833.

Cook, R. E. 1969. Variation in species density of North American birds. *System. Zool.* 18:63–84.

Cooley, J. L., and J. H. Cooley, eds. 1984. *Natural Diversity in Forest Ecosystems: Proceedings of the Workshop*. Athens, Ga.: Institute of Ecology.

Dasmann, R. F. 1972. Towards a system for classifying natural regions of the world and their representation by national parks and reserves. *Biol. Conserv.* 4:247–255.

———. 1973. A system for defining and classifying natural regions for purposes of conservation. *IUCN Occasional Paper* 7:1–18.

Diamond, J. M. 1972. Biogeographic kinetics: Estimation of relaxation times for avifaunas of southwest Pacific islands. *Proc. Nat. Acad. Sci. (USA)* 69:3199–3203.

———. 1974. Relaxation and differential extinction on land-bridge island: Applications to natural preserves. In *Proceedings of the 16th International Ornithological Congress*. Australian Academy of Sciences, Canberra City, A.C.T. 2601.

———. 1975. The island dilemma: Lessons of modern biogeographic studies for the design of natural reserves. *Biol. Conserv.* 7:129–146.

———. 1980. Patchy distributions of tropical birds. In *Conservation Biology: An Evolutionary-Ecological Perspective*, ed. M. E. Soulé and B. A. Wilcox, 57–74. Sunderland, Mass.: Sinauer.

———. 1984. "Normal" extinctions of isolated populations. In *Extinctions*, ed. M. H. Nitecki, 191–246. Chicago: University of Chicago Press.

Diamond, J. M., and R. M. May. 1976. Island biogeography and the design of nature reserves. In *Theoretical Ecology: Principles and Applications*, ed. R. M. May, 163–183. Sunderland, Mass.: Sinauer.

Dobzhansky, T. 1970. *Genetics of the Evolutionary Process*. New York and London: Columbia University Press.

East, R. 1983. Application of species-area curves to African savannah reserves. *African J. Ecol.* 21:123–128.

Ehrenfeld, D. W. 1976. The conservation of non-resources. *Amer. Sci.* 64:648–656.

Ehrlich, P. R., and A. H. Ehrlich. 1981. *Extinction: The Causes and Consequences of the Disappearance of Species.* New York: Random House.

Ehrlich, P. R., and H. A. Mooney. 1983. Extinction, substitution, and ecosystem services. *BioScience* 33:248–254.

Eisenberg, J. F. 1980. The density and biomass of tropical mammals. In *Conservation Biology: An Evolutionary-Ecological Perspective*, ed. M. E. Soulé and B. A. Wilcox, 35–55. Sunderland, Mass.: Sinauer.

FAO. 1981. Report on tropical moist forests. Rome: United Nations Food and Agriculture Organization.

Feinsinger, P., J. A. Wolf, and L. A. Swarm. 1982. Island ecology: Reduced hummingbird diversity and the pollination biology of plants, Trinidad and Tobago, West Indies. *Ecology* 63:494–506.

Foose, T. J. 1983. The relevance of captive populations to the conservation of biotic diversity. In *Genetics and Conservation: A Reference for Managing Wild Animal and Plant Populations*, ed. C. M. Schonewald-Cox, S. M. Chambers, B. MacBryde, and W. L. Thomas, 374–401. Menlo Park, Calif.: Benjamin/Cummings.

Foster, R. B. 1980. Heterogeneity and disturbance in tropical vegetation. In *Conservation Biology: An Evolutionary-Ecological Perspective*, ed. M. E. Soulé and B. A. Wilcox, 75–92. Sunderland, Mass.: Sinauer.

Frankel, O. H., and M. E. Soulé. 1981. *Conservation and Evolution.* Cambridge: Cambridge University Press.

Franklin, I. R. 1980. Evolutionary change in small populations. In *Conservation Biology: An Evolutionary-Ecological Perspective*, ed. M. E. Soulé and B. A. Wilcox, 135–149. Sunderland, Mass.: Sinauer.

Galli, A. E., C. F. Leck, and R. T. Forman. 1976. Avian distribution patterns in forest islands of different sizes in central New Jersey. *Auk* 93:356–364.

Gilbert, L. E. 1980. Food web organization and the conservation of neotropical diversity. In *Conservation Biology: An Evolutionary-Ecological Perspective*, ed. M. E. Soulé and B. A. Wilcox, 11–33. Sunderland, Mass.: Sinauer.

Gilbert, F. S. 1980. The equilibrium theory of island biogeography: Fact or fiction? *J. Biogeogr.* 7:209–235.

Gilpin, M. E., and J. M. Diamond. 1980. Subdivision of nature reserves and the maintenance of species diversity. *Nature* 285:567–568.

Ginzburg, L. R., L. R. Slobodkin, K. Johnson, and A. G. Bindman. 1982. Quasiextinction probabilities as a measure of impact on population growth. *Risk Analysis* 2:171–181.

Green, B. H. 1972. The relevance of serial eutrophication and plant competition to the management of successional communities. *Biol. Conserv.* 4:378–384.

Grime, J. P. 1973. Competitive exclusion in herbaceous vegetation. *Nature* 242:344–347.

Hamrick, J. L., Y. B. Linhart, and J. B. Mitton. 1979. Relationships between life history characteristics and electrophoretically detectable variation in plants. *Ann. Rev. Ecol. Syst.* 10:173–200.

Harrington, J. F. 1970. Seed and pollen storage for conservation of plant gene resources. In *Genetic Resources in Plants: Their Exploration and Conservation*, ed. O. H. Frankel and

E. Bennett, 501–521. IBP Handbook 11. Oxford: Blackwell Scientific Publications.

Harris, L. D. 1984. *The Fragmented Forest*. Chicago: University of Chicago Press.

Harrison, J., K. Miller, and J. McNeely. 1982. The world coverage of protected areas: Development goals and environmental needs. *Ambio. XI* 5:28–245.

Hawkes, J. G. 1982. Genetic conservation of "recalcitrant species": An overview. In *Crop Genetic Resources and Conservation of Difficult Material*, ed. L. A. Withers and J. T. Williams, 83–92. Proceedings of International Workshop, University of Reading, U.K. Paris: International Union of Biological Sciences, Series B42.

Higgs, A. J. 1981. Island biogeography theory and nature reserve design. *J. Biogeogr.* 8: 117–124.

Higgs, A. J., and M. B. Usher. 1980. Should nature reserves be large or small? *Nature* 285:568–569.

Hubbell, S. P. 1979. Tree dispersion, abundance, and diversity in a tropical dry forest. *Science* 203:1299–1309.

Ingram, C. B., and J. T. Williams. 1984. *In situ* conservation of wild relatives of crops. In *Crop Genetic Resources: Conservation and Evaluation*, ed. J. H. W. Holden and J. T. Williams, 163–179. London: Allen and Unwin.

Janzen, D. H. 1983. No park is an island: increase in interference from outside as park size decreases. *Oikos* 41:402–410.

Karr, J. 1981. Population variability and extinction in the avifauna of a tropical land bridge island. *Ecology* 63:1975–1978.

———. 1982. Avian extinction on Barro Colorado Island, Panama: A reassessment. *Amer. Nat.* 119:220–239.

Kiester, A. R. 1971. Species density of North American amphibians and reptiles. *System. Zool.* 220:127–137.

King, M. W., and E. H. Roberts. 1979. *The Storage of Recalcitrant Seeds: Achievements and Possible Approaches*. Rome: IBPGR.

Kleiman, D. G. 1980. The sociobiology of captive propagation. In *Conservation Biology: An Evolutionary-Ecological Perspective*, ed. M. E. Soulé and B. A. Wilcox, 243–261. Sunderland, Mass.: Sinauer.

Kleiman, D. G., B. Beck, B. Rettberg, and J. Dietz. 1985. *Preparation of Golden Lion Tamarins for Reintroduction to the Wild*. San Diego: Zoological Society of San Diego.

Klubinikin, K. 1979. An analysis of the distribution of park and reserve systems relative to vegetation types in California. M.S. thesis, California State University, Fullerton.

Krugman, S. L. 1985. Policies, strategies, and means for genetic conservation in forestry. In *Plant Genetic Resources*, ed. C. W. Yeatman, D. Kafton, and G. Wilkes, 71–78. Boulder, Colo.: Westview Press.

Leigh, E. G., Jr. 1981. The average lifetime of a population in a varying environment. *J. Theoret. Biol.* 90:213–239.

Levins, R. 1970. Extinction. In *Some Mathematical Questions in Biology*, vol. 2, ed. M. Gerstenhaber, 77–107. Providence, R.I.: American Mathematical Society.

Lewontin, R. C. 1974. *The Genetic Basis of Evolutionary Change*. New York: Columbia University Press.

Lovejoy, T. E. 1979. Designing refugia for tomorrow. In *Proceedings, Fifth Symposium Association Tropical Biology*. New York: Columbia University Press.

Lovejoy, T. E., R. O. Bierregaard, J. M. Rankin, and H. O. R. Shubard. 1983. Ecological dynamics of forest fragments. In *Tropical Rainforests*, ed. S. C. Sutton, T. C. Whitmore, and A. C. Chadwick, 377–384. Oxford: Blackwell Scientific Publications.

Lovejoy, T. E., and D. C. Oren. 1981. The minimum critical size of ecosystems. In *Forest Island Dynamics in Man-Dominated Landscapes*, ed. R. L. Burgess and D. M. Sharpe, 8–12. New York: Springer-Verlag.

Lovejoy, T. E., J. M. Rankin, R. O. Bierregaard, Jr., K. S. Brown, Jr., L. H. Emmons, and M. E. Van der Voort. 1984. Ecosystem decay of Amazon forest fragments. In *Extinctions*, ed. M. H. Nitecki, 296–325. Chicago: University of Chicago Press.

MacArthur, R. H. 1972. *Geographical Ecology*. New York: Harper and Row.

MacArthur, R. H., and E. O. Wilson. 1963. An equilibrium theory of insular zoogeography. *Evolution* 17:373–387.

———. 1967. *The Theory of Island Biogeography*. Princeton, N.J.: Princeton University Press.

Margules, C., A. J. Higgs, and R. W. Rafe. 1982. Modern biogeographic theory: Are there any lessons for nature reserve design? *Biol. Conserv.* 24:115–128.

Martin, T. E. 1981. Species-area slopes and coefficients: A caution on their interpretation. *Amer. Nat.* 118:823–837.

McNeely, J. A. 1982. The world coverage of protected areas: Development goals and environmental needs. In *National Parks, Conservation, and Development: The Role of Protected Areas in Sustaining Society*, ed. J. A. McNeely and K. R. Miller, 24–33. Washington, D.C.: Smithsonian Institution Press.

Miller, K. R., J. Furtado, C. de Klemm, J. A. McNeely, N. Myers, M. E. Soulé, and M. C. Trexter. 1985. *Maintaining Biological Diversity: The Key Factor for a Sustainable Society*. Gland, Switzerland: IUCN.

Moore, N. W., and M. D. Hooper. 1975. On the number of bird species in British woods. *Biol. Conserv.* 8:239–250.

Mound, L. A., and N. Waloff, eds. 1978. *Diversity of Insect Faunas*. New York: Halsted (Wiley).

Myers, N. 1979. *The Sinking Ark: A New Look at the Problem of Disappearing Species*. Oxford and New York: Pergamon Press.

———. 1980. *Conversion of Tropical Moist Forests*. Washington D.C.: National Academy of Sciences.

National Council on Genetic Resources (NCGR). 1982. Douglas-fir genetic resources: An assessment and plan for California. Sacramento: Department of Food and Agriculture, State of California.

Nevo, E. 1978. Genetic variation in natural populations: Patterns and theory. *Theoret. Pop. Biol.* 13:121–177.

Oldfield, M. L. 1981. Tropical deforestation and genetic resources conservation. *Stud. Third World Soc.* 14:277–345.

———. 1984. *The Value of Conserving Genetic Resources*. Washington, D.C.: U.S. Department of the Interior, National Park Service.

Opler, P. A. 1978. Insects of American chestnut: Possible importance and conservation concern. Proceedings of the American Chestnut Symposium.

Phillips, D., and H. Nash. 1981. *Captive or Forever Free? The Condor Question*. San Francisco: Friends of the Earth.

Pianka, E. R. 1966. Latitudinal gradients in species diversity: A review of concepts. *Amer. Nat.* 100:33–46.

Pickett, S. T. A., and J. N. Thompson. 1978. Patch dynamics and the design of nature reserves. *Biol. Conserv.* 13:27–37.

Picton, H. D. 1979. The application of insular biogeographic theory to the conservation of large mammals in the northern Rocky Mountains. *Biol. Conserv.* 15:73–79.

Prescott-Allen, R., and C. Prescott-Allen. 1984. Park your genes: Protected areas as *in situ* genebanks for the maintenance of wild genetic resources. In *National Parks, Conservation, and Development: The Role of Protected Areas in Sustaining Society*, ed. J. A. McNeely and K. R. Miller, 634–638. Washington, D.C.: Smithsonian Institution Press.

Preston, F. W. 1962. The canonical distribution of commonness and rarity: Part I. *Ecology* 43:185–215.

Raven, P. 1984. Knock down drag out on global futures. Speech at the Global Futures Conference Forum. AAAS, New York.

Richter-Dyn, N., and N. S. Goel. 1972. On the extinction of a colonizing species. *Theoret. Pop. Biol.* 3:406–433.

Robbins, C. S. 1979. Effect of forest fragmentation on bird population. In *Management of North Central and Northeastern Forests for Nongame Birds*, 198–213. USDA For. Serv. Gen. Tech. Rep. NC-51. North Central For. Exp. Stn., St. Paul, Minn.

Roberts, E. H. 1975. Problems of long-term storage of seed and pollen for genetic resources conservation. In *Crop Genetic Resources for Today and Tomorrow*, ed. O. H. Frankel and J. G. Hawkes, 269–295. Cambridge: Cambridge University Press.

Salwasser, H., S. P. Mealey, and K. Johnson. 1985. Wildlife population viability: A question of risk. *Transactions, North American Wildlife and Natural Resources Conference* 49: 421–439.

Schoener, T. W. 1983. Rate of species turnover decreases from lower to higher organisms: A review of the data. *Oikos* 41:372–377.

Schonewald-Cox, C. M. 1983. Conclusions: Guidelines to management: A beginning attempt. In *Genetics and Conservation: A Reference for Managing Wild Animal and Plant Populations*, ed. C. M. Schonewald-Cox, S. M. Chambers, B. MacBryde, and W. L. Thomas, 414–445. Menlo Park, Calif.: Benjamin/Cummings.

Schonewald-Cox, C. M., S. M. Chambers, B. MacBryde, and W. L. Thomas, eds. 1983. *Genetics and Conservation: A Reference for Managing Wild Animal and Plant Populations.* Menlo Park, Calif.: Benjamin/Cummings.

Shaffer, M. L. 1981. Minimum population sizes for species conservation. *BioScience* 31: 131–134.

———. 1982. Determining minimum viable population sizes for the grizzly bear. Conference on Bear Resources and Management 5:133–139.

Silverton, J. 1980. The dynamics of a grass land ecosystem: botanical equilibrium in the park grass experiment. *J. Applied Ecol.* 17:491–504.

Simberloff, D. S. 1984. Tropical forests in danger. *Garden* 8(2):2–8, 32.

Simberloff, D. S., and L. G. Abele. 1976. Island biogeography theory and conservation practice. *Science* 191:285–286.

———. 1982. Refuge design and island biogeographic theory: Effects of fragmentation. *Amer. Nat.* 120:41–50.

Simpson, G. G. 1964. Species density of North American recent mammals. *System. Zool.* 13:57–73.

Soulé, M. E. 1980. Thresholds for survival: Maintaining fitness and evolutionary potential. In *Conservation Biology: An Evolutionary-Ecological Perspective*, ed. M. E. Soulé and B. A. Wilcox, 151–169. Sunderland, Mass.: Sinauer.

Soulé, M. E., and B. A. Wilcox, eds. 1980. *Conservation Biology: An Evolutionary-Ecological Perspective*. Sunderland, Mass.: Sinauer.

Soulé, M. E., B. A. Wilcox, and C. Holtby. 1979. Benign neglect: A model of faunal collapse in the game reserves of East Africa. *Biol. Conserv.* 15:259–272.

Temple, S. A. 1986. Predicting avian response to forest fragmentation. In *Wildlife 2000: Modelling Wildlife Habitat Relationships*. Madison: University of Wisconsin Press.

Terborgh, J. 1974. Preservation of natural diversity: The problem of extinction prone species. *BioScience* 24:715–722.

———. 1975. Faunal equilibria and the design of wildlife preserves. In *Tropical Ecological Systems: Trends in Terrestrial and Aquatic Research*, ed. F. B. Golley and E. Medina, 369–380. New York: Springer-Verlag.

———. 1976. Island biogeography and conservation: Strategy and limitations. *Science* 193:1029–1930.

Terborgh, J., and B. Winter. 1980. Some causes of extinction. In *Conservation Biology: An Evolutionary-Ecological Perspective*, ed. M. E. Soulé and B. A. Wilcox, 119–133. Sunderland, Mass.: Sinauer.

Thibodeau, F. 1981. The preservation of genetic diversity in America: The adequacy of present ecological reserves and a biological foundation for further action. Ph.D. diss., Tufts University, Medford, Massachusetts.

Udvardy, M. D. F. 1984. A biogeographical classification system for terrestrial environments. In *National Parks, Conservation, and Development: The Role of Protected Areas in Sustaining Society*, ed. J. A. McNeely and K. R. Miller, 34–38. Washington, D.C.: Smithsonian Institution Press.

UNESCO. 1973. Conservation of natural areas and the genetic material they contain. Report Series 12. Paris: MAB.

———. 1974. Task force on criteria and guidelines for the choice and establishment of biosphere reserves. Report Series 22. Paris: MAB.

Wang, B. S. P. 1975. Tree seed and pollen storage for hybrid genetic conservation: Possibilities and limitations. In *Report on a Pilot Study on the Methodology of Conservation of Forest Genetic Resources*. FAO/UNEP. Rome: FAO.

Westman, W. E. 1977. How much are nature's services worth? *Science* 197:960–964.

Whitcomb, R. F., J. F. Lynch, P. A. Opler, and C. S. Robbins. 1976. Island biogeography and conservation: Strategy and limitations. *Science* 193:1030–1032.

Whitcomb, R. F., C. S. Robbins, J. F. Lynch, B. L. Whitcomb, M. K. Klimkiewicz, and D. Bystrak. 1981. In *Forest Island Dynamics in Man-Dominated Landscapes*, ed. R. L. Burgess and D. M. Sharpe, 126–205. New York: Springer-Verlag.

Weisbrod, A. R. 1976. Insularity and mammal species number in two national parks. In *Proceedings of the First Conference on Scientific Research in National Parks*, vol. 5.2, ed. M. Linn, 83–87. Washington, D.C.: U.S. Department of Interior.

Wilcove, D. S. 1985. Nest predation in forest tracts and the decline of migratory song birds. *Ecology* 66:1211–1214.

Wilcox, B. A. 1980. Insular ecology and conservation. In *Conservation Biology: An Evolutionary-Ecological Perspective*, ed. M. E. Soulé and B. A. Wilcox, 95–117. Sunderland, Mass.: Sinauer.

———. 1984a. *In situ* conservation of genetic resources: Determinants of minimum area. In *National Parks, Conservation, and Development: The Role of Protected Areas in Sustaining Society*, ed. J. A. McNeely and K. R. Miller, 639–647. Washington, D.C.: Smithsonian Institution Press.

———. 1984b. Concepts in conservation biology: Applications to the management of biological diversity. In *Natural Diversity in Forest Ecosystems: Proceedings of the Workshop*, ed. J. L. Cooley and J. H. Cooley, 155–172. Athens, Ga.: Institute of Ecology.

Wilcox, B. A., and D. D. Murphy. 1985. Conservation strategy: The effects of fragmentation on extinction. *Amer. Nat.* 125:879–887.

Willis, E. O. 1974. Populations and local extinctions of birds on Barro Colorado Island, Panama. *Ecol. Monogr.* 44:153–169.

———. 1979. The composition of avian communities in remnant woodlots in southern Brazil. *Papeis Avalsos Zool.* 33:1–25.

Willis, E. O., and E. Eisenmann. 1979. A revised list of birds on Barro Colorado Island. Panama. *Smithsonian Contr. Zool.* 219:1–31.

Wilson, E. O., and E. O. Willis. 1975. Applied biogeography. In *Ecology and Evolution of Communities*, ed. M. L. Cody and J. M. Diamond, 522–534. Cambridge, Mass.: Harvard University Press.

Wright, S. J., and S. P. Hubbell. 1983. Stochastic extinction and reserve size: A focal species approach. *Oikos* 41:466–476.

Ziswiler, V. 1967. *Extinct and Vanishing Animals.* 2d ed. New York: Springer-Verlag.

Commentary

Edward F. Connor

At present, the conservation of genetic resources is largely a task of conserving the individuals and populations of plants and animals that carry certain genes and gene combinations. This is because we are largely ignorant of the genetics of most organisms, because knowledge about phenotypes is much easier to gather than knowledge about genotypes, and because no single set of quantitative population genetic rules governing the loss and maintenance of genetic variation is applicable to all populations. This is not to say that the principles of population genetics and detailed genetic studies are not useful in the conservation of genetic resources (cf. Vrijenhoek et al., 1985). Rather, because of the expense of gathering sufficiently detailed genetic data to allow the application of population genetic principles to particular populations, a population genetical approach will

77

be secondary to population ecological approaches, and will be most useful when attempting to conserve small populations where the consequences of inbreeding and loss of genetic variation for population persistence will be most severe.

If the conservation of genetic resources is largely the ecological task of ensuring population persistence, how is it to be accomplished? In his paper Wilcox suggests that this task will be largely empirical, combining ideas from population ecology with knowledge of the habitat requirements and natural history of the species to be conserved, rather than being based on a set of general principles derived from particular ecological theories. I concur, but I am even less confident than Wilcox that ecological theory will provide a significant contribution to this process.

Wilcox outlines three major steps in this ecological approach to conserving genetic resources: (1) to define the resource—the populations or species to be conserved, (2) to develop a pluralistic conservation strategy employing traditional "nature reserves," as well as more wide-ranging habitat preservation techniques such as easements, and (3) to actively manage the protected populations and their habitats to ensure continued persistence. The first step is self-evident and needs no further amplification. However, the latter two constitute the substance of conservation of both the diversity of life and genetic variation, and require further comment.

PLANNING FOR POPULATION PERSISTENCE

Wilcox correctly stresses the need for a pluralistic approach to ensuring population persistence. Reliance on traditional, *intense* conservation techniques such as nature reserves and national parks not only is myopic but is likely to be totally inadequate for many species. Nature reserves, large enough to maintain viable populations of species with large home ranges and low population densities, cannot be established in all instances. Therefore, the persistence of these species will require a combination of intense conservation techniques, such as nature reserves, and more *diffuse* conservation measures, such as managing and securing easements for areas outside of reserves.

The formulation of a conservation strategy, once the conservation goal has been established, involves (1) determining what mix of intense and diffuse conservation measures will ensure population persistence, (2) designing the specific conservation entities (nature reserves, management areas, and easements), and (3) determining where to undertake these measures. Scientific input to the development of conservation strategies has focused on siting and designing nature reserves, with only limited attention to conservation measures applied outside of nature reserves.

Diamond (1975) proposed a series of design and siting guidelines for nature reserves that were formulated with the goal of ensuring the continued persistence of the largest number of species possible (Wilcox, Figure 2). He based these guidelines on two empirical generalizations concerning geographical patterns of species diversity on islands: that large geographical areas contain more species than small areas, and that areas near to sources of colonizing organisms also contain more species than remote areas. Often these generalizations are confused with one particular explanation for these patterns, the equilibrium theory of island biogeography (MacArthur and Wilson, 1967). However, subsequent studies have shown that only Diamond's principle A, which is identical to the first generalization, is actually true (Simberloff and Abele, 1976; Abele and Connor, 1979; Higgs, 1981; Simberloff and Gotelli, 1984; Blouin and Connor, 1985). Neither island biogeography theory nor the empirical observation of the dependence of species diversity on area and isolation yields unambiguous guidelines about how to design a nature reserve to conserve the greatest diversity of species.

As for siting nature reserves, several other authors have recommended siting reserves in areas that are hypothesized to have been "forest refugia" during times of Pleistocene glaciations (Prance, 1977; Gentry, 1979; Oren, 1982; Myers, 1982; Lovejoy, 1982). The rationale is that these areas are presumably less susceptible to disruption by climatic change and therefore will increase the probability of species persisting for long periods. As a result, these areas should also contain a high diversity of species. While this would indeed be a good strategy if we could reliably recognize such "refugia," those "refugia" that have been proposed have little basis in evidence (Beven et al., 1984). Nevertheless, this concept acknowledges that habitats and landscapes, as well as populations, are dynamic, and that we must plan nature reserves cognizant of the size and frequency of environmental disturbances that may affect them in the future (Shugart and West, 1981).

Wilcox acknowledges these attempts to develop guidelines for siting and designing nature reserves, but echoes the view that a more empirical approach tailored to the specific goal of a conservation project should be used. For example, Terborgh and Winter (1983) illustrate that an empirical study of the biogeography of the species to be conserved is the best route to intelligent decisions about siting nature reserves. They map the geographical distributions of birds in the Andes of Colombia and Ecuador. With these maps, one can locate those areas with the greatest number of species or that contain particular species or groups of species, depending on the goal of the conservation project. Within these areas, more detailed surveys are necessary to aid in selecting specific sites for nature reserves. No current ecological theory could predict the observed pattern of species distribution or species diversity. However, the simple compilation of biogeographic data provides a sound basis to begin the process of selecting potential sites for nature reserves.

Probably the most useful concept for conservation that has emerged from population ecology is the notion of the minimum viable population (MVP) size (Shaffer, 1981; Shaffer and Samson, 1985). This is the size of a population necessary to ensure persistence for a specified period with a *known probability*. The application of this concept is largely empirical, requiring the development of a stochastic demographic model for the species to be conserved that incorporates realistic mortality and natality schedules. However, simulation studies using a range of natality and mortality schedules may provide results that are at least applicable to some circumscribed set of species. Further development of these and other more complicated models may prove quite useful.

MANAGING FOR POPULATION PERSISTENCE

Most of the scientific input to conservation has focused on *planning* for population persistence, rather than managing populations once they are protected. The assumption has been that management is unnecessary. However, the realization that the persistence of many species is vulnerable to fire and other environmental disturbances has changed this attitude. The persistence of species requires not only a minimally viable population but also that their habitat and other resource requirements be met. In some instances this may be achieved, or be more rapidly or confidently achieved, only by active management. For example, the red cockaded woodpecker requires old-growth pine with heart rot in order to establish nest cavities. Managing pine forests to ensure a continuing supply of suitable nest trees may be the only way to ensure the persistence of this species. However, usually we do not know if some sort of management would be useful, or how to manage landscapes to achieve a specific conservation goal. If our goal were to ensure the persistence of all bird, mammal, and herb species in deciduous woodlands in eastern North America, would managing these largely second-growth forests be useful? Thus far, habitat management has been largely idiosyncratic, oriented to conserving single species. However, Dueser et al. (1986) outline a synthetic approach to managing habitats and landscapes to ensure the persistence of groups of species that may be useful for different habitats. More research on the habitat relations of plants and animals, and how to manage landscapes to create the habitats necessary to ensure the persistence of their constituent species, is certainly needed.

BEYOND POPULATION PERSISTENCE TO GENE PERSISTENCE

As described above, the ecological approach to ensuring population persistence requires that the protected populations have the demographic capacity to persist, with some known probability, and that adequate habitat be available continuously. This may be sufficient to ensure the persistence of large populations; but when attempting to conserve small populations, it may also be necessary to manage the genetic structure of the population. Just how to manage the genetic structure of such populations is likely to be somewhat idiosyncratic, but we may be able to develop some guidelines that depend on the breeding system and level of genetic heterogeneity in the population to be conserved. However, the observation that some species suffer from "outbreeding depression" serves to warn us that promoting gene flow and attempting to manage for high levels of genetic heterogeneity within populations may not be advisable in all instances.

Besides the issue of ensuring that particular populations persist, the conservation of genetic resources also requires that we establish a specific goal concerning which *genes* we wish to persist. If our goal is to maintain genetic diversity rather than to ensure the persistence of certain genes or gene combinations, our methods are likely to be quite different. Many of the issues that have arisen concerning how to design nature reserves to conserve species diversity will surface again in discussions of how to conserve genetic diversity. Is it wise to conserve several smaller populations, each individually meeting an MVP criterion, or to conserve only a single large population with a more stringent MVP criterion? Will either of these approaches be better at ensuring species persistence and preserving genetic variation? If multiple populations are preserved, should gene flow be prohibited or promoted between populations? The answers to these questions will depend on, among other things, how genetic variation is arrayed between and within populations. In any event, more detailed knowledge of the genetic structure of the populations to be conserved will be necessary. Furthermore, these data will probably only provide specific answers for particular situations, not simple universal rules for conserving genetic resources.

REFERENCES

Abele, L. G., and E. F. Connor. 1979. Application of island biogeography theory to refuge design: Making the right decision for the wrong reason. *Proceedings of the First Conference on Scientific Research in National Parks*, vol. 1, ed. R. M. Linn, 89–94. Washington, D.C.: U.S. Department of Interior.

Beven, S., E. F. Connor, and K. Beven. 1984. Avian biogeography in the Amazon basin and the biological model of diversification. *J. Biogeogr.* 11:383–399.

Blouin, M. S., and E. F. Connor. 1985. Is there a best shape for nature reserves? *Biol. Conserv.* 32:277–399.

Diamond, J. M. 1975. The island dilemma: Lessons of modern biogeographic studies for the design of natural reserves. *Biol. Conserv.* 7:129–146.

Dueser, R. D., H. H. Shugart, and E. F. Connor. 1986. The dynamic landscape approach to habitat management. In *Wilderness and Natural Areas in the Eastern United States: A Management Challenge*, ed. D. L. Kulhavy and R. N. Conner, 4–16. Nacogdoches, Tex.: Stephen F. Austin State University.

Gentry, A. 1979. Extinction and conservation of plant species in tropical America: A phytogeographical perspective. In *Systematic Botany, Plant Utilization and Biosphere Conservation*, ed. I. Hedberg, 100–126. Stockholm: Almqvist and Wikell.

Higgs, A. J. 1981. Island biogeography theory and nature reserve design. *J. Biogeogr.* 8: 117–124.

Lovejoy, T. E. 1982. Designing refugia for tomorrow. In *Biological Diversification in the Tropics*, ed. G. Prance, 673–680. New York: Columbia University Press.

MacArthur, R. H., and E. O. Wilson. 1967. *The Theory of Island Biogeography*. Princeton, N.J.: Princeton University Press.

Myers, N. 1982. Forest refuges and conservation in Africa—with some appraisal of survival prospects for tropical moist forest throughout the biome. In *Biological Diversification in the Tropics*, ed. G. Prance, 650–672. New York: Columbia University Press.

Oren, D. C. 1982. Testing the refuge model for South America: A hypothesis to evaluate discrepancies in refuge number across taxa. In *Biological Diversification in the Tropics*, ed. G. Prance, 601–607. New York: Columbia University Press.

Prance, G. T. 1977. The phytogeographic subdivisions of Amazonia and their influence on the selection of biological reserves. In *Extinction Is Forever*, ed. G. T. Prance and T. S. Elias, 195–213. New York: N.Y. Botanic Garden.

Shaffer, M. L. 1981. Minimum population sizes for species conservation. *BioScience* 31: 131–134.

Shaffer, M. L., and F. B. Sampson. 1985. Population size and extinction: A note on determining critical population sizes. *Amer. Nat.* 125:144–152.

Shugart, H. H., and D. C. West. 1981. Long-term dynamics of forest ecosystems. *Amer. Sci.* 69:657.

Simberloff, D. S., and L. G. Abele. 1976. Island biogeography theory and conservation practice. *Science* 191:285–286.

———. 1982. Refuge design and island biogeographic theory: Effects of fragmentation. *Amer. Nat.* 120:41–50.

Simberloff, D. S., and N. Gotelli. 1984. Effects of insularisation on plant species richness in the prairie-forest ecotone. *Biol. Conserv.* 29:27–46.

Terborgh, J., and B. Winter. 1983. A method for siting parks and reserves with special reference to Colombia and Ecuador. *Biol. Conserv.* 27:45–58.

Vrijenhoek, R. C., M. E. Douglas, and G. K. Meffe. 1985. Conservation genetics of endangered fish populations in Arizona. *Science* 229:400–402.

Commentary

Peter Ashton

Wilcox's comprehensive and thoughtful review demands few additional comments. I would merely add the following.

A clear definition of the time scale envisioned is mandatory in preservation management. Wilcox rightly implies that any alteration of the area of species ranges will influence, at a statistical level, the likelihood of extinction. Such changes are, of course, unavoidable. They therefore force more careful consideration of what the result of a particular "preservation action" is. In endangered taxa not bound in precise mutualistic relationships (such as keystone mutualists and mobile links), it may well be necessary to establish careful criteria for choice not only of which species merit preservation but even which varieties or character states.

Plants differ in important respects from the majority of animals, in ways that affect their proneness to extinction. They are sessile, and therefore dependent on local climatic conditions for movement and establishment or, in species-rich systems in particular, on mutualistic associations with animals for their seed and pollen dispersal. Where such mutualisms are species specific they impose an added impediment to dispersal. This effect is greatest when the mutualistic associations are obligatory and highly specific, and when the dispersal of diaspores is very restricted, as seems to be the case in some tropical trees and their obligatory symbiont mycorrhiza. Artificial reduction of population densities, such as occurs to mature biota in tropical forests as a consequence of selective exploitation, may be all that is needed to cause extinction of such species.

The great diversity of systems of reproduction in plants is also pertinent to preservation strategies. Polyploidy allows the possibility of sympatric speciation by preventing backcrossing to parental species. Apomixis, by overcoming the need for a mate (at the cost of reduced genetic variability), subverts the reduction in fecundity normally caused by low population density. Thus, where intensity of competition is density dependent, some plant species may persist indefinitely even under very low population densities if the environment does not change significantly. These considerations emphasize the intractability of definition of entities such as species and demes among plants. They also underline the need for a vastly increased rate of inventory of the world's biota before in situ preservation can be managed on a scientific basis.

The MVP approach may have little more than theoretical interest, except for those few species that we have designated as imperative to preserve. The large-scale biogeographic approach, using existing systematic and ecological information, must be the basis for establishing priorities in the foreseeable future. In due course, a method of successive approximation can be applied as policy definition becomes more precise. Antonovics (this volume) presents some additional considerations that might be useful when this stage is reached.

Ecosystems with a history of being subject to natural perturbations are those that are most resilient (Orians and Kunin, this volume). But such ecosystems are generally low in species richness, except where perturbations are moderate or highly predictable, such as in grazed grasslands. Thus simulation of the natural perturbations may have to be incorporated into the management procedures for such environments. Some of the most perturbation-free equable climates for life are the oceanic tropical lowland climates, the cloud zones of tropical and sub-tropical mountains, and Mediterranean regions. These areas are where species richness of plants, insects, and many other groups is greatest. The reefs of the western Pacific provide a marine analogue. The richness is associated with a high proportion of "mobile links" and "keystone mutualists" among the species components. Island biogeographic effects are therefore likely to be particularly important in these ecosystems and will push up minimum area requirements.

There are several points that Wilcox's paper does not address. For example, he does not consider land races and archaic agricultural practices. Taxa and geno-types conserved in these have a particularly important role as genetic bridges between crop taxa and more distantly related wild species. In plant conserva-tion, more experience has been gained in the conservation of these than any other group of similar entities (Weissinger, this volume). They provide a valuable paradigm, since active management is patently required.

The media have convinced some people that biotechnology can replace the need for conservation of genetic diversity. This argument needs to be evaluated through a comprehensive analysis of the likely interdependence of biotechno-logical achievement and conservation of biological diversity. Will biotechnology enable us to reconstruct gene complexes? Antonovics seems pessimistic. Con-versely, Weissinger raises the important point that DNA sequences can be pre-served unaltered in dead desiccated specimens at ambient temperatures. What, in the future, will museum collections be able to contribute to the conservation of valued character states, through insertion of relevant genetical material from an extinct taxon into the genome of a surviving relative?

Summary of the Discussion

Margery L. Oldfield

During the workshop on *in situ* conservation of genetic resources, the pan-elists focused their comments and criticisms on four major concerns related to the issues raised in Wilcox's paper: (1) the validity and robustness (generality) of applied island biogeography and other ecological principles for the design of reserves; (2) the extent to which different types of human-induced habitat alterations or environmental perturbations may enhance, rather than diminish, diversity; (3) the value of *in situ* conservation of traditional agroecosystems—

that is, mechanisms for conserving the dynamic evolutionary processes within these human-modified ecosystems that affect the evolution and maintenance of genetic resources of economically useful crop (and livestock) species; and (4) the proximate and ultimate goals for management programs and projects and the basic and applied research issues relevant to *in situ* conservation methods that merit greater academic attention. Each major concern is discussed in detail in the following sections.

BIOGEOGRAPHICAL AND ECOLOGICAL HYPOTHESES AND IN SITU CONSERVATION

This part of the discussion centered on "species-area relationships" and geometric principles for design of *in situ* reserves. Extrapolating from species-area curves, some participants concluded that increasing fragmentation and insularization or compartmentalization of reserves (see Wilcox and Murphy, 1985) might actually be beneficial rather than detrimental, since smaller reserves should maximize the number of species conserved per unit area of habitat. Whether or not this is true depends on both species composition and the degree of species overlap between reserves. Human-induced habitat fragmentation has typically decreased global or between-habitat species diversity in the past. Increased insularization and decreased habitat size within reserves has affected localized mainland or island endemics and "area-sensitive" species, such as those confined to climax vegetation, mobile or highly dispersed taxa, and large-bodied, long-lived vertebrates (Wilcox and Murphy, 1985; see also Ehrenfeld, 1970; Ziswiler, 1967). In contrast, comparatively slow climatic and geological changes occurring on an evolutionary time scale have been postulated as important natural selection processes that, in concert with natural habitat fragmentation, contributed to a gradual accumulation of species over time (Simpson and Haffer, 1978). Clearly, the best management strategies depend on the time frame of reference.

Habitat fragmentation will increase species diversity if (and only if) speciation rates exceed extinction rates over the long run (but see Nilsson and Nilsson, 1983, regarding potential sampling errors). However, human-induced extinction rates appear to have outstripped speciation rates over the last few centuries (Ziswiler, 1967; Oldfield, 1984).

It was pointed out that the species-area relationships are very crude measures of expectations of extinctions following habitat fragmentation processes. "Boundary effects" may significantly influence the extent to which reserves are exposed to and affected by external land (or water) use alternatives. Yet it would be a mistake to disregard species-area relationships until more predictive models have been developed (however, see Boecklen and Gotelli, 1984). The equilibrium model of island biogeography, with its application to the study of species extinctions, has the distinct advantage that baseline data already exist for a broad range of habitat sizes, from extremely small reserves to very large land masses (such as subcontinents).

The issue of small versus large reserves and maximization of species diversity per unit area of habitat is part of a larger set of interrelated issues concerning reserve design and site selection. For example, maintenance of different species assemblages might be facilitated by the establishment of a series of small, iso-lated reserves selected within many biogeographical provinces or specifically for maintenance of different or unique habitat types. Greater species diversity would be maintained than if there were a few large reserves representing fewer biogeo-graphical provinces. On the other hand, if the locations of large reserves were also carefully chosen on the basis of identical site selection principles, the large reserves would still conserve more species than the small reserves, although more land area would be required.

Ecologists and biogeographers are not able to predict or assess the minimum level of biological "provincialization" that may be beneficial, below which in-creasing fragmentation of reserves will exacerbate losses in species richness. Only very rarely do we have the opportunity to observe experimentally the impacts of periodic long-term ecological stresses. For example, the widespread drought in the early 1980s throughout the tropical forests of the Philippines, and portions of Borneo and Mindanao, resulted in the death of a large fraction of dominant canopy trees. Even when ecologists are so fortunate as to be present to witness the acute effects of rare ecological or geological events, the requisite baseline data needed to investigate them are typically lacking. In the absence of such data, ecologists cannot accurately determine whether a few large reserves will fare better than many small reserves in the face of such large-scale, infrequent ecological stresses.

Another factor to be considered in evaluating the relative worth of large and small reserves is the extent to which the area to be preserved has already been subjected to human-induced disturbance and habitat fragmentation. For example, British species diversity is maximized when habitats are increasingly fragmented, but habitats within the British Isles became fragmented centuries ago, whereas many areas of the tropics have only recently been affected by in-tensive development. As a consequence, nearly all of the "area-sensitive" or other extinction-prone species have already been eliminated from the British Isles. Thus it is important to consider the history of human perturbations or habitat alterations within a particular region or biogeographical province (e.g., see Olson and James, 1982, 1984) when designing *in situ* preservation strategies.

DIVERSITY—REDUCING VERSUS
DIVERSITY—ENHANCING HUMAN ACTIVITIES

A significant amount of time during the discussion was devoted to the extent to which various types of human disturbance or modes of intervention could be employed as management tools for enhancing species or genetic diversity within reserves. The same disturbance or intervention activity can be viewed as "beneficial" or "detrimental" depending both on philosophical grounds and on the time scales of concern (e.g., short-term versus long-term considerations or contemporary ecological versus evolutionary or geological time scales). Much of this literature and its relevance to *in situ* conservation has been analyzed and summarized (e.g., see Greenway, 1967; Ziswiler, 1967; Ehrenfeld, 1970; Myers, 1979; Ehrlich and Ehrlich, 1981; Soulé, 1983; Oldfield, 1984). Three types of human intervention were discussed in the context of diversity-enhancing versus diversity-reducing human activities: (1) intentional introductions of exotic species; (2) potential application of the intermediate disturbance hypothesis for habitat and population management purposes; and (3) the role of human intervention in creating and maintaining genetic diversity within crop and livestock populations.

Artificial introductions of exotic species were discussed first. Such introductions often produce short-term economic benefits, but they may have adverse ecological effects in the long run. Exotic organisms typically benefit from ecological or evolutionary advantages in their interactions with native species. Indigenous species often lack the requisite chemical, physiological, behavioral, or other defense mechanisms necessary to buffer the impacts of exotic predators, competitors, pathogens, or pests. For example, the European Brown Trout (*Salmo trutta*), an exotic sport fish that is generally considered to be a beneficial introduction into the United States, is believed to be an important cause of decline of many native trout species (Taylor et al., 1984).

Competitive exclusion of native species by introduced exotics is only one consequence of such perturbations. Extinctions or severe reductions in the population sizes of native species also occur when exotic predators, diseases, or disease-carrying vectors are introduced. Exotic species often have a competitive, predatory, or pathogenic advantage in interactions with native flora and fauna whenever they are introduced without their own coevolved diseases, predators, or competitors.

Endemic species on islands appear to be especially vulnerable to introductions. Thousands of species confined to island habitats are already extinct or are currently threatened with extinction because of: (1) competition from domestic livestock and other nonnative species; (2) predation by introduced carnivores, such as fish, rats, snakes, mongoose, and domesticated animals including cats, dogs, pigs, and goats; (3) exotic diseases, such as VVND (Newcastle disease) of birds and various diseases of fish associated with the live animal trade, avian malaria which has particularly harmed the Hawaiian honeycreepers (family Drepanididae), and rinderpest which has adversely affected many species of African ungulates; and (4) hybridization with closely related exotic species, for example, the genetic swamping of *Salmo clarki lewis*, the westslope cutthroat trout, via hybridization with rainbow trout (*S. gairdneri*) introduced from the East (Leary et al., 1984).

Introductions of exotic species do not always produce detrimental ecological (or economic) effects, as is illustrated by the introduction of nonnative game birds to North America. However, prior to these exotic introductions, most of the native game bird species that might have been competitively affected by them had already suffered severe population reductions as a result of hunting and of habitat alterations.

In sum, even though some introductions of species do not appear to have resulted in detrimental ecological consequences, it is difficult and costly to ascertain on an a priori basis which introductions will be economically *and* ecologically successful. More research is needed to improve our ability to predict the impact introductions will have on native species, and whether they can be controlled or eradicated by ecologically consonant methods if they attain pest status in their new environment(s).

The second major point of discussion centered on the "intermediate disturbance hypothesis" and its potential applications in management. A body of literature already exists on the topic of human intervention to maintain habitat diversity, and therefore species diversity. The Forest Service and some other government agencies, such as the Bureau of Land Management and the Fish and Wildlife Service, have instituted habitat management projects to increase overall species diversity within wilderness areas or reserves or to maintain populations of rare or threatened species. The Forest Service, on its own initiative, has attempted to protect a diversity of successional stages of habitat, whereas the National Park Service has tended to emphasize protection of climax successional stages. Thus the former agency tends to have an implicit goal of maximizing overall species diversity sometimes at the expense of the area-sensitive or climax habitat species (e.g., the ivory-billed woodpecker, the spotted owl), while the latter agency's implicit goal is to maintain climax habitats in order to sustain the most vulnerable, area-sensitive species (e.g., the grizzly bear), even though this may decrease overall species diversity within a particular park or reserve.

This contrast led to a discussion about the objectives or goals of conservation. It was noted that the only way to conserve area-sensitive species that tend to be extremely vulnerable to human-induced extinction is to maintain sizable tracts of habitat. Therefore, if one's goal is to maximize global species diversity, some sizable tracts of natural habitat within each biogeographical region will have to be maintained. On the other hand, if one's goal is to maximize species diversity in each park or reserve, smaller tracts of habitat managed for a diversity of successional stages may be preferable.

Nevertheless, management practices that encourage or sustain early successional species may be needless efforts, because those species are likely to survive anyway. Mention was made of preliminary results of the World Wildlife Fund study of isolated patches of Amazon rain forest habitat (see Lovejoy et al., 1984, cited in Wilcox and Murphy, 1985).

Another issue discussed in conjunction with the intermediate disturbance hypothesis focused on communities of organisms that were once maintained by fires or other periodic, natural perturbations that have recently been eliminated or controlled. For example, intentional exclusion of natural fires in the United States prior to 1960 had already resulted in the invasion of roughly 70 million acres of former grasslands by mesquite (*Prosopis* spp.). Moreover, some commercially or recreationally important timber trees, such as Douglas-fir (*Pseudotsuga menziesii*), longleaf pine (*Pinus palustris*), and redwood (*Sequoia dendron sempervirens*), are fire-adapted species (i.e., they cannot persist in the absence of fires that periodically remove competing climax species). Prescribed fires have long been used as a management tool in longleaf pine forests in the southeastern United States. In such cases, the prohibition or removal of natural disturbances that tend to maintain habitat heterogeneity is actually a form of human intervention that can lead to losses of particular disturbance-adapted species or even to elimination of an entire community of organisms (e.g., grasslands).

Finally, human intervention may be employed to enhance the genetic diversity of crops or livestock and domesticated animals. Throughout the history of plant and animal domestication, humans have frequently brought geographically or ecologically distinct species, subspecies, races, varieties, or populations into reproductive contact with one another. This can increase genetic diversity via heterosis within the recipient population, but it may also lead to outbreeding depression and "hybrid breakdown." On the other hand, the intentional movement of organisms to suitable new habitats has sometimes resulted in losses of species or distinct subspecies as a consequence of hybridization with other geographically distinct species or races. These activities have, over the short run, frequently led to reductions in genetic or species diversity.

An example concerning U.S. corn varieties was discussed at length. Lancaster sharecropper, an open-pollinated variety of corn, was commonly used in the United States until around the turn of the century. It was later removed from cultivation for genetic improvement. Eventually it was lost after it was removed from its cultural environment. Farmers growing this variety had apparently frequently exchanged small lots of seed over a very broad geographical area, thus maintaining heterosis. Similar modes of human intervention to maintain crop genetic diversity are believed to have contributed to the evolution of corn as a crop plant in the principal center of corn diversity, Mexico and Guatemala. Maintenance and creation of valuable crop (and livestock) genetic resources is often dependent on maintenance of *in situ* genetic reservoirs in regions characterized by indigenous, traditional agricultural practices. Thus some of the participants felt that *in situ* conservation of traditional agricultural systems may contribute to preservation of genetic diversity more than is generally realized.

IN SITU CONSERVATION OF TRADITIONAL AGROECOSYSTEMS

Traditional agriculture and livestock husbandry systems have been more thoroughly investigated by anthropologists, or agricultural and other applied scientists, than by ecologists and evolutionary biologists. Thus ecological principles that may guide the selection, design, or management of *in situ* crop genetic reservoirs simply do not exist, even in the few countries where reserves have been established, such as the USSR, Sri Lanka, and India (Prescott-Allen and Prescott-Allen, 1983). Site selection was based solely on distributions of wild relatives of edible or medicinal crop plants. The problem is that the continuing survival and maintenance of economically useful crop (and livestock) genetic resources is dependent on human cultivation and husbandry. If *Homo sapiens* is omitted from these ecosystems, most domesticated species would be unable to survive. The great bulk of our rapidly disappearing crop and livestock genetic resources that are useful to plant and animal breeders is located within man-modified rather than natural ecosystems (Oldfield, 1984; Prescott-Allen and Prescott-Allen, 1983). Therefore, when the human cultures that utilize these resource species disappear, or are substantially altered, the primitive cultivars and obscure livestock breeds that they sustain usually disappear as well. Genetic losses within these areas have reached such proportions in recent decades that the problem is generally referred to as "genetic erosion" (e.g., see Oldfield, 1976, 1984; Prescott-Allen and Prescott-Allen, 1983).

An important international vehicle for conservation of traditional agroecosystems is the UNESCO Man and the Biosphere Program's Project 8. The criteria and guidelines for selection of the world biosphere reserves specifically include habitats resulting from *traditional* patterns of land use (Batisse, 1982).

Some time was devoted to strategies for the *in situ* conservation of traditional agricultural systems, in areas of high crop diversity. Subsidies or other means of financially supporting traditional farmers within crop gene centers should be considered by the developed nations that require a continuous supply of these valuable genetic resources. This option would both protect crop genetic diversity *in situ* and lend support to the life-styles that many traditional farmers in the developing nations might actually prefer. People in the developing nations face very complex problems and environmental hazards that influence their choices

90

of agricultural production options. In the tropics, diseases and pests pose year-round problems; petroleum by-products such as fertilizers and pesticides are costly and difficult to obtain; and as people are forced to move into marginal or harsh environments, they have increased needs for novel (especially minor) crops or genetically improved major crops.

The choice of which traditional agricultural systems to preserve will be complicated. It must be remembered that even though the developed nations focus their agricultural research attention on a few major crops, roughly a century ago nearly all of these edible plants were minor crops that had recently been wrested from the world's centers of diversity. It is difficult to predict which minor crop of today will be a major crop in the future.

Nonetheless, some people believe that there are insurmountable political, socioeconomic, and technical problems associated with *in situ* conservation of traditional agricultural systems. It will be difficult to attempt to preserve these man-modified ecosystems with their human cultural traditions and their primitive cultivars intact, because these systems involve people with a way of life that is bound to change if modern agricultural innovations or technologies continue to expand worldwide. However, *in situ* conservation methods do not necessarily entail prohibition of the use of modern high-yield crop varieties within traditional farming systems. For example, the Andean peasants grow their native potato varieties for home cooking and consumption, while they still devote other fields to the production of the white European or American potato varieties that they export to nearby city markets.

GOALS AND FUTURE RESEARCH FOR IN SITU CONSERVATION

It was repeatedly highlighted in the discussions that there are often conflicting goals or aims for *in situ* conservation. For example, as noted previously, the selection, design, and management of *in situ* reserves depend, in part, on conservation objectives. Goals may be maximizing global or overall species diversity versus maximizing species diversity per reserve or habitat area. Qualitative aspects of species diversity, such as species composition within reserves or species overlap among reserves, are sometimes considered more important than more quantitative or numerical counts, especially since the goal of maintaining current global diversity may be unattainable. Since genetic diversity, and the resources we obtain by exploiting this diversity, can be utilized both intraspecifically and interspecifically, perhaps focusing on genetic diversity at both the specific and subspecific levels is more appropriate than relying on species diversity alone.

Difficulties may arise with respect to conservation objectives in part because the concept "species diversity" may pose problems when applied to conservation issues. Conservation is an anthropocentric, resource-oriented concept; but species diversity is instead a factual, measurable biological concept. Human-induced perturbation or habitat alterations may profoundly affect the periphery of reserves (i.e., boundary effects may have an overriding influence on reserve design and management). Thus perhaps our objectives should reflect or be oriented more toward processes (ecological, evolutionary, economic, etc.) rather than quantitative measures such as species-area relationships.

With respect to guidelines and suggestions for future research, the issues emphasized by workshop participants included: (1) basic bioinventories for reserve management purposes; (2) further research on principles for reserve selection and design; (3) population biology and mathematical ecology research aimed at facilitating our understanding of human-induced versus natural extinction processes; (4) studies on demographic stochasticity and the concept of minimum viable population (MVP) to facilitate conservation of threatened taxa; (5) investigations aimed at delineating ecological and genetic principles that will enable managers to avoid both inbreeding and outbreeding depression when utilizing heterosis (via intra- or interspecific hybridization) for conservation or reserve management purposes; and (6) mechanisms for maintaining the genetic integrity of species threatened with extinction due to genetic swamping as a consequence of hybridization with introduced, exotic taxa.

REFERENCES

Batisse, M. 1982. The biosphere reserve: A tool for environmental conservation and management. *Environ. Conserv.* 9:101–112.

Boecklen, W. J., and N. J. Gotelli. 1984. Island biogeographic theory and conservation practice: Species-area or specious-area relationships? *Biol. Conserv.* 29:63–80.

Ehrenfeld, D. W. 1970. *Biological Conservation.* New York: Holt, Rinehart and Winston.

Ehrlich, P. R., and A. H. Ehrlich. 1981. *Extinction: The Causes and Consequences of the Disappearance of Species.* New York: Random House.

Greenway, J. C., Jr. 1967. *Extinct and Vanishing Birds of the World.* New York: Dover.

Leary, R. F., F. W. Allendorf, S. R. Phelps, and K. L. Knudsen. 1984. Introgression between westslope cutthroat and rainbow trout in the Clark Fork River drainage, Montana. *Proc. Mont. Acad. Sci.* 43:1–18.

Lovejoy, T. E., J. M. Rankin, R. O. Bierregaard, Jr., K. S. Brown, Jr., L. H. Emmons, and M. E. Van der Voort. 1984. Ecosystem decay of Amazon forest fragments. In *Extinctions,* ed. M. H. Nitecki, 296–325. Chicago: University of Chicago Press.

Myers, N. 1979. *The Sinking Ark: A New Look at the Problem of Disappearing Species.* Oxford and New York: Pergamon Press.

Nilsson, S. G., and I. N. Nilsson. 1983. Are estimated species turnover rates on islands largely sampling errors? *Amer. Nat.* 121:595–597.

Oldfield, M. L. 1976. The utilization and conservation of genetic resources: An economic analysis. M.S. thesis, The Pennsylvania State University, University Park.

———. 1981. Tropical deforestation and genetic resources conservation. *Stud. Third World Soc.* 14:277–345.

————. 1984. *The Value of Conserving Genetic Resources*. Washington, D.C.: U.S. Department of the Interior, National Park Service.

Olson, S. L., and H. F. James. 1982. Fossil birds from the Hawaiian Islands: Evidence for wholesale extinction by man before Western contact. *Science* 217:633–635.

————. 1984. The role of Polynesians in the extinction of the avifauna of the Hawaiian Islands. In *Quaternary Extinctions: A Prehistoric Revolution*, ed. P. S. Martin and R. G. Klein, 892. Tucson: University of Arizona Press.

Prescott-Allen, R., and C. Prescott-Allen. 1983. *Genes from the Wild: Using Wild Genetic Resources for Food and Raw Materials*. London: International Institute for Environment and Development.

Simpson, B. B., and J. Haffer. 1978. Speciation patterns in Amazonian forest biota. *Ann. Rev. Ecol. Syst.* 9:497–518.

Soulé, M. E. 1980. Thresholds for survival: Maintaining fitness and evolutionary potential. In *Conservation Biology: An Evolutionary-Ecological Perspective*, ed. M. E. Soulé and B. A. Wilcox, 151–169. Sunderland, Mass.: Sinauer.

————. 1983. What do we really know about extinction? In *Genetics and Conservation: A Reference for Managing Wild Animal and Plant Populations*, ed. C. M. Schonewald-Cox, S. M. Chambers, B. MacBryde, and W. L. Thomas, 111–124. Menlo Park, Calif.: Benjamin/Cummings.

Taylor, J. N., W. R. Courtenay, Jr., and J. A. McCann. 1984. Known impacts of exotic fishes in the continental United States. In *Distribution Biology and Management of Exotic Fishes*, ed. W. R. Courtenay, Jr., and J. R. Stauffer, 322–373. Baltimore: Johns Hopkins University Press.

Templeton, A. R., H. Hemmer, G. Mace, U. S. Seal, W. M. Shields, and D. S. Woodruff. 1986. Local adaptation, coadaptation, and population-boundaries. *Zoo. Biol.* 5(2): 115–126.

Wilcox, B. A., and D. D. Murphy. 1985. Conservation strategy: The effects of fragmentation on extinction. *Amer. Nat.* 125:879–887.

Ziswiler, V. 1967. *Extinct and Vanishing Animals*. New York: Springer-Verlag.

3

Genetically Based Measures of Uniqueness

Janis Antonovics

The understanding of organic diversity has been a major continuing goal of biology. The earliest component of such endeavors is systematics, which already in pre-Darwinian times had recognized the major discontinuities whereby organic diversity could be classified and quantified. Much of the pre-Darwinian legacy (for example, Ray distinguishing monocots from dicots, and Lamarck separating the vertebrates from the invertebrates) is still with us. Taxonomic divisions, then and now, were seen as reflective of a natural order, with the proviso that, in these post-Darwinian times, the epithet "natural" is reflective not of inherent design but of evolutionary divergence and history. However, the sense of early accomplishment that came with pigeonholing organic diversity into phyla, classes, orders, families, and species was not long-lived. Even Linnaeus became concerned with variation and with hybridization among his different species. Later, evolutionists were quick to point out that the difficulties of applying a rigorous species concept were not necessarily to be found in the deficiencies of taxonomy, but in the continuing dynamic nature of the evolutionary process. By the middle of this century, the exploration and clarification of intraspecific variation had become a major goal of experimental taxonomists and evolutionary biologists. It was evident that the variation in every species had a complex structure: subspecies, races, and ecotypes were the rule rather than the exception. And even if discontinuities were hard to recognize and name, clines and genetic variants were commonplace and thus became a major focus for understanding the genetics of the evolutionary process.

At the beginning of this century, however, experimental genetics developed relatively independently of evolutionary biology. Indeed, experimental geneticists regenerated, albeit subconsciously, a disturbingly archaic, typological view of the species. They spoke of a "wild type" and of "mutants," as if there was a species ideal from which deviations were aberrant forms of great experimental value, but of only passing natural interest.

The advent of electrophoresis, and the discovery of large amounts of genetic variation at the protein level, overthrew the idea of a wild type; and the problem of organic diversity at the intraspecific level fell squarely into the laps of population geneticists. We now know that about 30 to 40 percent of the genes in a population may exist in different allelic states. Given that an individual might contain 10,000 or so different genes, the potential number of gene combinations is enormous. If each of the 23 chomosome pairs in the human genome carried a single polymorphic locus, containing two alleles, the number of possible genotypes would be 3^{23} or 9.4×10^{10}—a number well in excess of the current world population! The discovery of electrophoretic variation introduced much complexity, but also some simplification. Unified measures of difference and diversity became possible, with their units being the units of population genetics, namely allelic frequency. Such measures could be applied to every population, and comparisons could be made across species, genera, and higher taxa.

Most recently the problem of genetic diversity has also become a central concern of molecular biologists. Large-scale DNA sequencing is now a plausible reality. The genome was formerly visualized as a "string of beads," but now it is clear that the enormity of the information content present among the beads pales by comparison with the information content within each bead. Each "gene" may contain several thousand DNA base pairs, and this sequence may vary in complex ways. Base pairs may be substituted, added, and deleted. Whole regions may duplicate, transpose, and invert. Furthermore, those translated coding regions that we have imagined as the repositories of genetic information and which produce the structural genes analyzable by electrophoresis, turn out to be minor regions, interrupted by untranslated sequences, and flanked by large regions concerned with control and other more important or nonexistent functions (we don't yet know). The string has become as important as the beads.

Thus within each of the several million species on the earth, there are several thousand genes, each of which in turn may have several thousand base pairs in it and around it. It is the purpose of this paper to outline how such diversity can be measured and how uniqueness can be identified. How this is done at a higher taxonomic level is well understood, and is the core of modern systematics. I will not attempt to review this. Instead I will concentrate on variation at and below the species level. Already from a brief history of the subject, we have come to several important conclusions: (1) No species is genetically uniform. (2) The concept of a "wild type" is erroneous: allelic variants are abundant and commonplace. (3) The number of possible genetic combinations usually far exceeds the number of individuals in a population (or in a species): every individual is genetically unique, and its particular genotype may never reappear.

These conclusions may seem self-evident. But they are still resisted when taxonomists search for a "type" and when they deem it worthwhile to pursue endless discussions on "What is a species?" These conclusions are also regularly ignored by ecologists who use the species as their unit of description.

Given this variability, several important issues are posed:

1. How does one categorize and measure this enormous diversity and uniqueness? Are the measures we propose useful as a basis for conservation strategies?

2. What should our conservation strategies be at the intraspecific level? Why should we conserve genetic variants?

3. Finally, are genetic resources renewable? Mutations occur continuously,

and mutation rates can be enhanced by mutagenic agents. Can genetic variants, once lost, be regenerated?

Resolution of these issues is complicated by the diversity of genetic systems. Species may be outbred or inbred, monogamous or polygamous, and composed of mobile or sedentary individuals. These different modes affect *both* the amount and the structuring of the genetic variation. I deal primarily with the conventional genetics of more or less outbred species, but at times I digress to more special cases.

MEASUREMENT OF GENETIC DIVERSITY

Measurement of diversity includes not only its overall quantification but also an analysis of its structuring. Simple measures of diversity do not tell us how variation is "structured." For example, genetic variability could be caused either by one group being quite different from all the others (i.e., unique) or by each group being equally different from every other. In these two cases, our strategies for conserving the extant variation would be quite different. In the former case, preservation would entail sampling only the unique isolate, and the relatively homogeneous remainder of the population. In the latter case, each unique subpopulation would have to be sampled.

The analysis and conceptualization of differences within and among populations is in principle identical regardless of whether we are considering a "population" to be a local deme, geographical race, subspecies, species, or higher taxonomic group. Genetic differences can be measured in terms of phenotypic traits, allelic frequencies, or DNA sequences. We will consider each of these measures in turn.

Phenetic Diversity

Phenetic diversity is based on measures of phenotypes, without close regard to the underlying allelic structure. Under this category we usually include readily measured morphological and physiological characteristics. Since different phenotypes may also arise from purely environmental effects, they may not necessarily reflect genetic differences. Experimental approaches are required to distinguish rigorously genetic from environmental effects on phenotypic variance. A key component of such experiments is sampling contrasting populations and rearing them under "standard conditions."

Where phenotypic differences are substantial, it is often assumed that they represent, at least in part, genetic differences. This has been common practice in studies of animal populations (e.g., Johnston and Selander, 1971); whereas in plant populations, cultivation under standard conditions has been considered a prerequisite for inferring genetic effect (Briggs and Walters, 1984). This probably reflects our perception that plants are much more plastic phenotypically than animals. However, even in plants, it is remarkable that in nearly every case where well-established populations living in contrasting habitats have been studied, their phenotypic differences have been shown to have at least a partial genetic basis (Heslop-Harrison, 1964; Langlet, 1971).

Increasing resolution of genetic effects can be achieved using the methods of quantitative genetics (Falconer, 1981; Mather and Jinks, 1982). Such methods, while giving substantial insight into the amount of genetic variation in a trait, cannot identify individual loci or their alleles. These techniques are also difficult to apply on a large scale, and therefore may not be feasible for a practicing conservation biologist. However, differences among populations can be studied using "transplant" or "common garden" experiments, and such experiments should be encouraged. Where this cannot be done, distinct phenotypic differences among populations can cautiously be considered to reflect, at least in part, genetic differentiation.

Phenetic diversity is usually measured by the variance of a particular trait. Where several traits are measured simultaneously, multivariate measures of variance (such as the determinant of the variance-covariance matrix) are used. Partitioning variance into between- and within-group effects is normally done by analysis of variance and variance component analysis.

Assessment of the structure of such phenetic variance is the province of numerical taxonomy (Sneath and Sokal, 1973). It involves the establishment of a "similarity matrix" that expresses the degree of similarity among groups. This similarity matrix usually takes the form of a correlation matrix, for closely related groups, indicating the correlations between each pair of populations with regard to a set of measured traits. This similarity matrix may then be subjected to a cluster analysis (which identifies groups at various levels of similarity) or an ordination analysis (which places the groups in a reduced hyperspace such that their distance from each other can be visualized readily). With closely related groups, principal component analysis on the correlation matrix is often used. Distances between groups can then be measured in terms of their differences in principal component scores. These distances (perhaps with the aid of a further cluster analysis) can then be used to identify sets of similar groups, and to describe how the overall variation is partitioned.

The use of these approaches is illustrated in the work of Adams (1977), who used chemical and morphological characteristics to assess variation among bean (*Phaseolus vulgaris*) varieties. Principal component analysis identified clusters of similar varieties, and the resulting groupings were consistent with genetic relationships based on pedigree analysis (Figures 5 and 6). Such multivariate methods are being used with increasing frequency in the analysis of crop resource data (e.g., Martinez et al., 1983; Witcombe and Rao, 1976; Murphy and

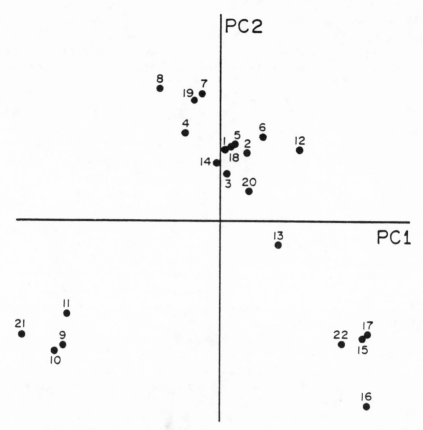

FIGURE 5. Relationship between phenetic distance derived by principal component analysis, varietal membership, and degree of genetic relationship among bean cultivars (after Adams, 1977). The display shows twenty-two cultivars on principal axes 1 and 2 derived from principal component analysis of eighteen chemical and agronomic characteristics.

Key to cultivars: (1) Red Mexican U.I. 34; (2) Red Mexican U.I. 36; (3) Gr. Northern U.I. 31; (4) Gr. Northern U.I. 59; (5) Gr. Northern U.S. 1140; (6) Gr. Northern Neb. 1; (7) Pinto U.I. 111; (8) Pinto U.I. 114; (9) Cal. Dk. Red Kid.; (10) Royal Red Kid.; (11) Idaho Lt. Red Kid.; (12) Big Bend Red Mex.; (13) Cal. Sm. Wh. 59; (14) Black Turtle Soup; (15) Sanilac Navy; (16) Gratiot Navy; (17) Seafarer Navy; (18) Gr. Northern U.I. 61; (19) Gr. Northern Tara; (20) Idaho Fl. Sm. Wh.; (21) Redkote Kidney; (22) Michelite 62 Navy.

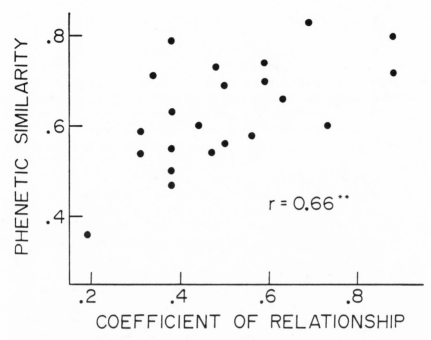

FIGURE 6. Comparison of phenetic similarity ($= 1-$ distance) based on six principal component scores derived from principal component analysis of eighteen chemical constituent characteristics (see Figure 5) with relationship coefficients derived from ancestry patterns among seven cultivars.

Witcombe, 1981) and in the analysis of morphological variation in natural populations (Hamrick, 1975; Clay and Antonovics, 1985; Johnston and Selander, 1971).

Phenetic traits have the advantage that they are easily measured, and their ecological or practical utility is either obvious or can be readily inferred. They have the disadvantage that their genetic basis is difficult to assess precisely, and standardized comparisons are difficult when populations or taxa are measured for qualitatively different traits.

Allelic Diversity

Measures of allelic diversity require knowledge of the allelic composition at individual loci. This information is generally obtained using electrophoretic analysis of enzyme variants. Allelic variants may also be recognizable as overt polymorphisms, but currently electrophoretic techniques are the most efficient for analyzing allelic variation at a large number of loci (usually 10 to 30). Diversity may be underestimated if there are "hidden alleles" (i.e., amino acid changes that do not affect electrophoretic mobility). For particular loci, this hidden variability may be substantial (Singh et al., 1976), although when averaged over all loci, the increase in heterozygosity or number of alleles is relatively small, on the order of 20 percent (Ayala, 1982). It is commonly assumed, however, that if populations are studied using identical techniques, the relative differences are reasonably accurate, since there is little reason to believe a priori that markedly more hidden variation is present in some populations than others.

The degree to which electrophoretic methods provide measures of genetic variation that are "representative" of the entire genome remains a matter of debate, since only genetic variation at "structural" loci is measurable. Variation is known to be present in "regulatory" loci (Ayala and McDonald, 1980; Laurie-Ahlberg et al., 1980), yet generalized methods of assessing allelic variation at such loci are not available. Moreover, variation at the DNA level that does not affect amino acid composition is not detectable.

Allelic diversity is measured at the individual level as the percentage of loci heterozygous in that individual. At the population level, allelic diversity is measured in number and frequency of alleles at a given locus, and fraction of loci that are polymorphic. The more alleles, the more equitable their frequencies, and the more loci that are polymorphic, the greater the genetic diversity. Average expected heterozygosity (the probability that two alleles sampled at random will be different) is a commonly used overall measure. This measure is equal to the average number of heterozygotes per locus assuming Hardy-Weinberg proportions.

The partitioning of allelic variation within and among populations is generally carried out using the Shannon-Weaver diversity index, and by means of Wright's F-statistics. The Shannon-Weaver index measures the "information content" provided by the allelic variation, at both within- and between-group levels (Lewontin, 1972). F-statistics use the difference between observed and expected heterozygosities as a measure of within-individual, within-population, and between-population estimates of genetic substructuring (see Weir and Cockerham [1984] for a recent treatment).

Assessment of the structure of genetic variance among populations is carried out using measures of "genetic distance." The most commonly used measure is that of Nei (1975, 1978). First, one calculates genetic identity, I, a measure of the probability that two alleles chosen at random from two different populations will be identical, relative to the probability that these same alleles chosen from within each population will be identical. Genetic identity is then converted to genetic distance, D, by averaging over all loci, and using the relationship $D = -\ln I$. Nei (1975, 1978) has shown that D reflects the average number of allelic substitutions that have occurred between two populations since they diverged, based on the assumption that divergence has been by random spread of neutral mutations. Assuming that mutation rates within the populations are similar, and that the chosen loci are representative of the genome as a whole, D is proportional to time since the populations diverged.

Other measures of genetic distance use the kinship (or coancestry) coefficient, f. This is defined as the probability that alleles sampled from two individuals (or populations) are identical by descent (i.e., have been derived by replication from the same ancestral gene). Cavalli-Sforza and Bodmer (1971) show that if two populations are assumed to have diverged only by genetic drift, with no mutation, the time of divergence is proportional to $-\ln(1 - f)$. This measure is therefore more appropriate than that of Nei when dealing with small populations that have diverged over relatively short periods. Precise methods of estimating the coancestry coefficient from gene frequency data have recently been described by Reynolds et al. (1983). The measure of Nei will probably continue to be the most commonly used because of its intuitive appeal, its well-established status, and its rough and ready translation into average number of allelic substitutions since population divergence.

Detection of allelic variation by electrophoresis has the advantage that it can be precisely quantified to provide comparative measures of genetic variation. However, it has the disadvantage in that it may not be representative of variation in the genome as a whole. It has the further disadvantage that its functional significance or selective importance is generally not known. Particular alleles may have specific phenotypic effects and may be subject to selection (e.g., Koehn et al., 1980), but in most cases electrophoretic variants can be used successfully as neutral markers of past historical events, or neutral markers to assess breeding system events. One of the most elegant demonstrations of such use is provided by the studies of Ward and Neel (1970), who showed a strong correlation between genetic relationship and village history in South American Indian tribes (Figures 7 and 8).

Sequence Diversity

Two techniques are currently used for assessing diversity in DNA sequences. The first involves the use of restriction enzymes that recognize a particular small base sequence (usually four to six bases), and cleave DNA at this site. Any changes in base composition may either prevent cleavage by the restriction enzyme at this site or generate new restriction sites, thereby producing a different fragment length. Differences in length of DNA fragments produced by the action of a particular restriction enzyme therefore reflect changes in base sequence, and can be detected electrophoretically. The use of several such enzymes permits "sampling" of the DNA for changed bases. Since the restriction enzymes are usually of bacterial origin, there is no a priori reason to suppose any one recognition site will be more abundant or evolve in a different manner from any other. This technique is most applicable to small DNA molecules, because in any large DNA molecule the restriction sites are likely to be so numerous as to

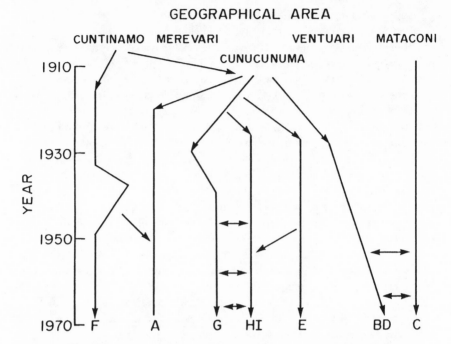

FIGURE 7. Relationship between village history and genetic distance among villagers of the Makiritare tribe (after Ward and Neel, 1970). The diagram shows the historical relationships and migration patterns during sixty years leading to the establishment of seven Makiritare villages. The time scale is shown on the left and the five main geographical areas at the top.

produce a confusing continuum of fragment lengths. Therefore, most analyses at the population-intraspecific level have involved mitochondrial genomes in animals and chloroplast genomes in plants (e.g., Clegg et al., 1984). Differences in DNA sequence are usually estimated in terms of the average number of pairwise differences in base pair composition, assuming that each altered restriction site represents a single base pair change (see Table 1 for an example in *Peromyscus*).

Numerous genes have been cloned and sequenced, but such techniques only now are being developed to a sufficient degree that they can be applied to studies of the same gene in a substantial number of individuals within a species. In one such study (Kreitman, 1983) of the Adh locus in the fruitfly *Drosophila melanogaster*, where previously only two electrophoretic variants were known, a large number of other polymorphisms (forty-two) were revealed at the DNA level which did not result in amino acid changes (Figure 9 and Table 2). Given this complexity and functional diversity at the DNA level, the analysis of DNA sequence data is rapidly developing into a sophisticated branch of genetic statis-

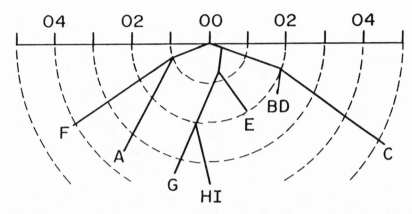

FIGURE 8. Genetic networks derived from the pairwise genetic distances (see Figure 7) based on eleven loci. The networks are plotted to scale on polar coordinates with units of genetic distance read along the radius of the diagram.

TABLE 1

Estimated genetic distance in base substitutions per nucleotide between *Peromyscus* populations at various stages of evolutionary divergence.

Comparison	Mean Distance Between Collections	Four Base Enzymes	Six Base Enzymes
Within geographic locality			
P. leucopus	500 feet	—	0.000
P. maniculatus	¼ mile	—	<0.010
P. polionotus	1 mile	0.004	0.002
Between geographic localities			
P. maniculatus	1,000 miles	0.042	0.031
P. polionotus	80–500 miles	0.006	0.015
Sibling species			
P. maniculatus vs.			
P. polionotus		0.166	0.132
Nonsibling species			
P. leucopus vs.			
P. maniculatus and P. polionotus		—	0.214

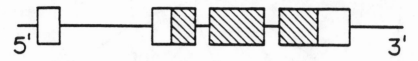

⊢ 100 B.P.

FIGURE 9. DNA sequence polymorphisms among eleven cloned *Drosophila melanogaster* alcohol dehydrogenase (Adh) genes from five natural populations, and representing two electrophoretic variants, fast (F) and slow (S) (after Kreitman, 1983). The diagram shows the structure of the Adh gene. Boxes are Exons 1–4, among which are Introns 1–3. The protein coding region is shaded. Scale = 100 base pairs.

tics (Weir, 1983). However it will probably be several years before standardized methods for dealing with DNA sequence data are developed.

Comparison of Phenetic, Allelic, and Sequence Data

At the intraspecific level, there is in general a broad concordance between phenetic and allelic measures of diversity. Populations are less different from each other than are races, and these in turn are less different than subspecies and species. This concordance appears not to hold particularly well with regard to higher taxonomic levels. Thus species differences within genera may reflect different degrees of electrophoretic divergence depending on the group being considered (Avise and Aquadro, 1982); and large morphological differences may reflect small genetic distances (Cherry et al., 1978). Even at the population level, discordant estimates of divergence have been obtained using various methods. Usually the most common discordance is the observation of substantial morphological or chromosomal differentiation, but relatively little isozyme differentiation (Jain et al., 1980; Turner, 1974; Avise et al., 1975; Heywood and Levin, 1984). Clearly, large morphological and chromosomal changes may occur without substantial evolution at the electrophoretic level. Often discordance is claimed, but the data are too limited either with regard to electrophoretic or morphological data to provide a rigorous comparison (e.g., Snyder and Linton, 1984; Giles, 1984). At other times, discordance is claimed where agreement between electrophoretic and taxonomic data is certainly not perfect, but is substantial (Heywood and Levin, 1984).

It is often difficult to reach firm conclusions about concordance, because the types of data collected in morphological and electrophoretic studies differ in units and scale. Comparison is therefore tenuous, and null hypotheses of what would constitute no difference are difficult to formulate. Too often authors seem eager to excite the reader by emphasizing the existence of discordance, rather than clearly arguing how such discordance might be manifest and how it might be quantified.

104

TABLE 2

DNA sequences (variable sites only) of fast and slow electrophoretic alleles of Adh from isochromosomal lines sampled for several localities. The reference nucleotide sequence on the top line is the most common Adh-S nucleotide at each of the polymorphic sites.

Strain Origin	Enzyme Mobility	5' Flanking	Exon 1	Intron 1	Exon 2	Intron 2	Exon 3	Intron 3	Exon 4	A∇C	3' Flanking
		CCG		CAATATGG∇C∇G	C T	AC	CCCC	GGAAT	CTCC·ACTAG	A∇C	AGC∇C∇T∆
Seattle	S	. . .		. . . A T	. .	. .	T T . A	C A . T A	A C	. . .	. . . . . . O
Florida	S	. . C		. . . .	. .	. .	T T . A	C A . T A	A C	. . .	. . . . . . O
Africa	S	. . .		. . . .	. .	. .	. . . .	. . . . .	. . . . A .	. . .	. . T O . 1 A .
France	S	. . .		. . . .	. .	G T	. . . .	. . . . .	. . . . A .	−1	T A . . . . . .
Florida	S	. . C		A G . . . A . T C . . .	A G	G T	. . . .	. . . . .	. . . .	C 3 .	. . . .
Japan	S	. . C		. . . .	. G	. .	. . . .	. . . . .	. . . .	C 4 .	. . . .
Florida	F	. . C		A G . . . A . T C O G O	. G	. .	. . . .	. . . . .	. . T . T . C A .	C 4 .	. . . . T . . .
France	F	T G C		A G . . . A . T C O G O	. G	. .	. . . .	. . . . .	. G T C T C C . .	C 4 G	. . . .
Seattle	F	T G C		A G . . . A . T C O G O	. G	. .	. . . .	. . . . .	. G T C T C C . .	C 4 G	. . . .
Africa	F	T G C		A G . . . A . T C O G O	. G	. .	. . . .	. . C . .	. G T C T C C . .	C 5 G	. . . .
Japan	F	T G C		A G G G G A . . O . . T	. G	. .	. A . .	. . . . .	. G T C T C C . .	C 4 .	. . . . −1 . .
No. of polymorphic sites		3	0	11	1 1	2	4	5	7	2	5
Average no. of nucleotides		63	87	620	70 99	65	405	70	204	178	767

∇/△ = insertion/deletion polymorphisms, where numbers are the differences in homopolynucleotide run lengths compared with the consensus sequence.

• Thr-Lys amino acid replacement polymorphism. All other polymorphisms are either silent or noncoding.

With both phenetic and allelic measures, the overall picture is one of continuity of difference from population to species. Highly divergent populations that have been given the rank of races, ecotypes, or subspecies are more different from each other than are populations within these groups, and their differences approach those found among species. In particular instances, certain closely related species may be less differentiated than races within another species.

GENETIC UNIQUENESS

The Uniqueness of Individual Alleles and Traits

Studies of allelic variation among populations within a geographical region show that common alleles are generally "universal" (i.e., found in all populations). Usually only rare alleles are confined to one or few populations. Such "private" alleles may be restricted in distribution or they may be ubiquitous but rarely sampled in the average population survey because of their low frequency. The value and importance of such rare alleles are difficult to assess. Where they are relatively universal, they are powerful indicators of gene flow (Slatkin, 1985); otherwise they may be rare because they are disadvantageous, because they are novel mutations, or because they occur as a result of chance effects. Where unique alleles are found, only rarely can a strong case be made for their preservation. Usually their effect on the phenotype is not known, nor does their rarity provide any clue to their origin. Their collection may be of value in increasing the range of allelic variants as genetic markers, or for functional studies of enzyme expression.

A similar picture emerges from studies of individual phenetic traits: frequency distributions of these traits usually overlap extensively. Unique phenotypes commonly represent developmental or genetic abnormalities (Spencer, 1947). The preservation of such "odd" variants is hard to justify.

Within species, very few alleles or phenotypic traits are unique. Uniqueness characteristically arises only when there are distinct habitat differences, where populations are isolated, when there is inbreeding, or where populations have divergent histories (Table 3). While the richness of allelic variation may lead us to think that all evolutionary avenues are open to species, this is not true, because limits to selection are imposed by the availability of appropriate variation. In studies of the adaptation of plant species to mine tailings, for example, certain species fail to adapt, and their failure can be directly attributed to a lack of genetic variance for tolerance (Bradshaw, 1983). Different populations within species may likewise have an uneven distribution of genetic variance.

The Uniqueness of Allelic and Trait Combinations

It is harder to assess the importance of unique *combinations* of alleles and traits. At a phenetic level, combinations of character states appear to be very

106

TABLE 3
Examples of the different sources of ecological adaptation
in plant breeding programs (after Bradshaw, 1983).

Sources	Adaptations
From original gene pools	
Potato	Blight resistance within *Solanum tuberosum*
Alfalfa	Spotted aphid resistance
Sugar beet	Sugar content
Rye	Reduced height
From other gene pools—other cultivars	
Barley	Yellow dwarfness from Abyssinian cultivars
Wheat	Dwarfing genes from Japanese cultivars
Grapes	Root aphid resistance from American material
Cotton	Blackarm resistance from African cultivars
From other gene pools—other species	
Oats	Mildew resistance from *Avena ludoviciana*
Bread wheat	Stem rust resistance from *Triticum dicoccum*
Bread wheat	Eye spot resistance from *Aegilops ventricosa*
Rice	Grassy-stunt resistance from *Oryza nivara*
Delphinium	Red flower color from *Delphinium cardinale*
Potato	Blight resistance from *Solanum demissum*

important. Thus closely related species are usually distinguished on the basis of character combinations rather than on single "key characters"; and such combinations are often frustratingly ambiguous, as anyone who has ever tried to use a key knows. Within species, populations are distinguished by character complexes (Clausen and Hiesey, 1958), which may occur at a very low level, in outbreeders (Antonovics and Bradshaw, 1970) as well as in inbreeders (Hamrick, 1975). Even within one population, such complexes may have considerable integrity (Clay and Antonovics, 1985; Morishima et al., 1984). Moreover, many characters of value in plant breeding do not appear as single genes, in isolation from other traits, but as character complexes. For example, Qualset (1975) found that resistance to yellow dwarf mosaic virus in barley was found only in Ethiopia, and there it was associated with many other distinct morphological traits.

It is pertinent to ask whether such complexes are "unique," in the sense of being difficult to reassemble from component traits, or component alleles. If these complexes are the result of coordinate changes at many loci, and if these many loci are themselves interspersed and closely linked, then reassembling the complex may be very difficult or nearly impossible. The difficulty of disentangling such complexes is illustrated in the work of Davies (1971), who showed that correlated responses to selection for abdominal and thoracic bristle number in *Drosophila* were due to linkage of many interspersed loci, determining bristle number in either of the two body regions. Pleiotropic loci (affecting both characters simultaneously) were relatively rare.

Unique and important combinations may occur in all populations. In outbreeding species, however, such combinations are likely to be ephemeral and readily broken down. Combinations are therefore likely to have clear identity only where populations also show differentiation. This view is supported by the difficulty of finding linkage disequilibrium (nonrandom allelic associations) among all but extremely closely linked loci ("supergenes") in outbreeders. In inbreeders the situation is quite different, since heterozygotes and hence opportunities for recombinational events are far fewer. It is here that most character complexes have been identified (Hamrick, 1975). Unfortunately there have been very few studies on whether the component characters or alleles of such complexes do indeed enhance fitness (singly or in combination), or whether the complex is the result of chance associations resulting from "hitchhiking" of neutral characters on a few particular traits that are under selection.

The evolutionary literature is replete with references to "coadapted gene complexes," and such complexes have been postulated to play a major role in the evolutionary process. However, such complexes have only rarely been studied in depth. They have been inferred either from observed allele and character associations or from experimental studies showing loss of fitness following interpopulation crosses (Dobzhansky, 1950; Price and Waser, 1979). The development of conceptual frameworks and methods for the study of allelic associations, and for the study of evolution in correlated traits, is currently a very active area of evolutionary biology. But our current knowledge is inadequate to assess fully the importance and uniqueness of complex traits at the intraspecific level. Uniqueness in trait and allele combinations remains the least understood of the categories we consider.

GENETIC CONSERVATION AND UNIQUENESS

Four major goals in genetic conservation, listed in increasing order of complexity and conceptual difficulty, can be recognized. The first is preservation of relatives of useful species, as potential sources of useful variation for breeding programs. The second is preservation of genetic variation to maintain viability in rare species, particularly in captive populations. The third is maintenance of genetic variation for continued evolution of wild species. The fourth goal is preservation of genetic variants within wild species for aesthetic reasons and for as yet unforeseen potential utility.

TABLE 4

Strategies for sampling extant genetic variation from natural
populations (after Marshall and Brown, 1975).

Model	Allelic Profile				
	1	2	3	4	5
Allele					
A_1	0.25	0.76	0.63	0.49	0.80
A_2	0.25	0.20	0.23	0.22	0.005
A_3	0.25	—	0.09	0.12	0.05
A_4	0.25	—	—	0.07	0.05
Remainder	—	0.04	0.05	0.10	0.05
Sample size (n)	19	15	37	43	80
Probability (P)	1.00	1.00	1.00	0.99	0.98

Sample sizes (n) required to be 95 percent certain of obtaining at least one copy of each common
allele (frequency > 0.05) and probability of achieving this objective if $n = 100$ for five contrasting
allelic profiles.

Preservation of Relatives of Useful Species

It is widely recognized that there has been serious loss of genetic variability in
relatives of our cultivated plants and animals. The process of domestication—
with its associated founder effects, severe inbreeding, strong selection, and rapid
varietal turnover—has resulted in a rapid erosion of genetic variability. The
preservation of wild relatives and crop plants has been a matter of international
concern over the past twenty-five years. A number of symposia have addressed
this issue, and the principles of genetic conservation are now well established
(Frankel and Bennett, 1970; Frankel and Hawkes, 1975; Holden and Williams,
1984).

Much of the genetic uniqueness and variability of crop species is concentrated
in "centers of diversity," and much effort continues to be spent on collecting
from these areas and characterizing them genetically so as to further aid sampling
strategies (e.g., Witcombe and Rao, 1976).

Marshall and Brown (1975) have calculated the numbers of individuals that
need to be collected within a population to have a reasonable chance of sampling all the common alleles. They find that sample sizes of fifty should suffice in
most cases (Table 4) and that vast collections per population are not necessary.
Samples of larger size may be useful, however, in providing baseline information
from which future sampling strategies may be further improved. Larger samples
give a clearer view of gene frequencies and also permit analysis of character
complexes and associations (Qualset, 1975).

109

TABLE 5

Theoretical values of optimal number of plants to sample per site and optimal number of sites to sample per day for a range of genetic models.

Population		Modern Cultivars (1)	Primitive Cultivars (2)	Wild Relatives (3)	Outbreeding Species (4)	(5)
P		0.01	0.05	0.10	0.50	0.75
p		0.95	0.20	0.05	0.05	0.05
a/b ratio			Number of plants per site (n)			
25		1	10	15	30	36
100		2	15	39	50	55
a	b		Number of sites per day (N)			
25	1	18	14	12	9	8
50	0.5	10	8	7	6	6

P = proportion of total variation represented by population.

p = proportion of genetic variation per population represented by an individual.

a = amount of effort expended for each site visited.

b = amount of effort expended to sample each plant.

Marshall and Brown (1975) use data on the distribution of genetic variation within and among populations to calculate how sampling resources should be divided between, among, and within population samples. It is quite feasible to collect most of the extant variation in a region in a relatively short period (Tables 5 and 6), since, among outbreeders at least, most of a region's variation can be found within populations. Sampling extant genetic variation is therefore not a daunting task.

In the relatives of a domesticated species, it is clear that uniquely different genes and gene combinations are more likely to be found among relatives that are distant either historically, geographically, or taxonomically. However, genes from more widely divergent populations are correspondingly more difficult to transfer back to the domesticated variety. For any particular domesticated species we can conceptualize an optimum allelic or phenetic distance, which is determined by the trade-off between increased probability of finding uniqueness and decreased probability of gene transfer. This optimum distance will not be constant, but will increase as developing technology permits more efficient gene transfer from distant relatives. Indeed, transspecific gene transfer is now one of the major goals of biotechnology.

TABLE 6

Number of days required to collect 95 percent of variation using a maximally efficient sampling procedure under a range of genetic models.

Population		Modern Cultivars (1)	Primitive Cultivars (2)	Wild Relatives (3)	Outbreeding Species	
					(4)	(5)
P		0.01	0.05	0.10	0.50	0.75
p		0.95	0.20	0.05	0.05	0.05
a	b	Days to collect 95% of variation				
25	1	16	5	5	1	1
50	0.5	30	8	5	1	1

Preservation in Rare and Captive Populations

Rare or captive species generally live in small populations. As such they are subject to (1) loss of alleles by chance effects (Franklin, 1980), (2) inbreeding due to mating among close relatives, and (3) selection for performance in the captive environment (i.e., "domestication"). Loss of alleles will lead to decreased genetic variance, increased disease susceptibility (Antonovics and Ellstrand, 1984), and decreased probability of future evolutionary response should conditions change. Inbreeding may lead to reduced vigor and the appearance of abnormal types (Ralls and Ballou, 1983; Senner, 1980), while selection under domestication will lead to preservation of unnatural traits and decreased probability of survival in the wild. In such situations the goals should be twofold: maintenance of variation and elimination of the deleterious effects of inbreeding or domestication (Templeton and Read, 1983).

Preservation of Evolutionary Potential

The dangers of arrested evolution are inherent in small or permanently subdivided populations, because such populations may be genetically depauperate. Large, interconnected natural populations usually contain abundant genetic variation that permits long-term continued evolutionary change. It has been argued that long-term evolutionary considerations must also be an aspect of any genetic conservation strategy (Frankel and Soulé, 1981). However, it is unclear how such long-term considerations can be implemented practically. Structuring preserved populations so that genetic variation is maintained may be best for the maintenance of long-term evolutionary potential. Seldom are preserves designed in such a way as to make this possible.

Preservation of Genetic Variants in Wild Species

From a genetic perspective it is hard to escape the conclusion that the use of the species as a unit of conservation is coarse, artificial, and arbitrary. It is nonetheless difficult to come up with alternative, ready-made solutions. The enormity of genetic diversity precludes a glib simplistic approach to its conservation. The Noah's ark approach of species conservation cannot be used at the genotype or allelic level. Limits, and the choices that these limits entail, require that criteria be developed whereby the value of a particular variant can be judged. The criteria for preserving genetic variants "for their own sake" will be complex. Genetic data are likely to be only a small component of the decision process. It is also for this reason that electrophoretic data are likely to be much less useful than phenetic data. Thus it would be totally arbitrary to designate a particular genetic distance as a criterion beyond which different populations should be preserved. Such a criterion does not take into account discordances of genetic and morphological distances and does not consider that we usually do not know what electrophoretic variants represent biologically; it is also simplistic, since other criteria of aesthetics, potential utility, and so forth, are ignored. For example, different mimicry morphs in butterflies are determined by different alleles at only one locus (albeit perhaps a complex switch gene), yet on aesthetic grounds it would be extremely hard to advocate forgoing their preservation.

An example which illustrates several of these issues is provided by two "species" of sandwort, *Arenaria alabamensis* and *A. uniflora,* found in the Carolinas. The former is extremely rare and restricted to a few granite outcrops. Recent studies by Wyatt (1984) have revealed that *A. alabamensis* represents a polyphyletic group of independently derived inbred populations, resembling each other in floral morphology (related to their inbreeding syndrome) but each resembling the local *A. uniflora* populations in other morphological and electrophoretic traits. Therefore, Wyatt (1984) has advocated that on taxonomic grounds, *A. alabamensis* should not be considered a distinct monophyletic species. Should this preclude efforts for its conservation? The arguments for its preservation are strong: it is morphologically distinct; its populations have enormous scientific interest because of the large breeding system differences among them and because they are a paradigm for one type of speciation process. Moreover, preserving these populations would essentially involve preserving its unique habitat and other associated rare species. While arguments for preservation of *A. uniflora* var. *alabamensis* are compelling, a species-centric view would doom it to extinction.

Very often a subspecies or genetic variant may be as interesting and valuable as a species. Some genetic variants are remarkable in themselves (such as albino tigers in the zoo). Others may be of potential commercial value (such as chemical races of lichens, variants of thyme with unique flavors, or metal tolerance races for recolonizing waste spoils from mining operations). The potential uses of wild species are legion (Myers, 1983). Yet such utility is often not fully realized unless specific traits are amplified by breeding programs that use the extant genetic variance or that of close relatives.

How can such a diversity of goals and issues be accommodated? First, what might be the generalized "political" procedures that we could use for genetic preservation? Second, what criteria would be used or how would these criteria be applied?

Genetic Impact Statements

By analogy with the process whereby some species can be declared "endangered," we can also envisage species to be "genetically endangered." Assessment of such a status would require a "genetic impact statement." Indeed, it is now an accepted practice to provide such a statement for any species that is endangered, captive, or existing in small isolated populations (Schonewald-Cox et al., 1983), although such statements as yet have no legislative imperative. Such a "genetic impact statement" could identify justifiable reasons for the preservation of genetic variants within a species, whether these variants be subspecies, races, particular populations, or particular genotypes. Justification for such special attention might be: (1) genetic uniqueness (Is the population morphologically and genetically distinct?); (2) current and potential commercial utility (Is the population a member of a species of commercial importance, or is it a relative of a commercially important species?); (3) research importance (Is the population of research interest for evolutionary biologists, ecologists, agronomists?); (4) aesthetic importance (Is the population of interest to collectors, of interest to naturalists, a source of wonder and curiosity to the public?).

Strong cases for genetic conservation could be made on one or any combination of these factors. Clearly uniqueness per se is insufficient; every individual is genetically unique and genetically ephemeral. Every population, separated from others spatially or distinct ecologically, will be different. Criteria other than purely genetic measures have to be invoked. I would anticipate that in the process of establishing priorities for saving the world's genetic resources, certain species might be sacrificed for genetically unique populations.

RENEWABILITY OF UNIQUENESS

Genetic variation can be regenerated by mutation; however, mutation rates are extremely low. Therefore, large populations, long time periods, and specific methods of selecting rare phenotypes are necessary for recovery of mutant types. Even with the use of mutagenic agents, large numbers of individuals and great expenditure of effort are necessary because such mutagens cause high mortality. Many recessive mutations are only detectable as segregants in subsequent generations, and even then particular desired mutations may be rare. Where general phenetic traits are concerned, spontaneous mutation appears to be a useful steady source of variation. This is presumably because many such traits are affected by numerous loci, thus giving substantial rates of mutation per generation per character (Sprague et al., 1960; Mukai, 1964). There is the further problem that the frequency of induced mutations may not reflect the frequency of spontaneous mutations; evidence for this comes from the classic fine-structure mapping of the R II locus in phage by Benzer (1955), and from simple considerations of molecular mechanisms of mutation, as well as from studies of mutation breeding in crop plants. Therefore, although a large number of crop varieties produced by mutation breeding have been released (for review, see Gottschalk and Wolff, 1983), such "mutation breeding" has rarely been proposed as a complete substitute for genetic preservation. This is not surprising, because much extant variation in natural populations is probably maintained and influenced by selection processes that have acted on the spontaneous spectrum of mutations over a considerable period. Moreover, selection has also built up gene and character combinations that would be almost impossible to reconstruct de novo, given even a relatively low level of complexity.

CONCLUSIONS

1. Populations contain large amounts of genetic variation at a genotypic level; every individual in an outbreeding population is genetically unique.

2. Differences among individuals, populations, and species can be measured in terms of phenetic, allelic, or sequence diversity. Such measures are concordant in reflecting a continuum of increased divergence from the population to species level, but may be quite discordant in particular cases, and are not easily generalizable across highly divergent taxa.

3. Unique alleles and traits tend to be found in populations that are divergent geographically, ecologically, and historically.

4. Unique character combinations ("character complexes") are perhaps common at the intraspecific level. In populations of outbreeders, unique combinations are quickly broken up, but they may persist in inbreeding species. The latter species are especially likely to show substantial geographic variation in which sets of characters are fixed. The genetic basis of such complex traits is not well understood: it is therefore difficult to assess whether such complexes can be easily regenerated.

5. Unique alleles and gene combinations have the greatest importance in relation to useful "domesticated" species. It is here that "genetic erosion" has been severe because of bottleneck effects, inbreeding, and selection. Wild relatives and primitive races are therefore a useful source of variability in breeding programs.

6. If species are endangered or captive, special efforts may have to be made to preserve allelic variation and to avoid inbreeding.

7. A strong case can be made for preserving variants at the intraspecific level. The justification can be made in terms of aesthetics, actual and potential utility, and research interest.

8. Genetic resources are partly renewable by mutation, spontaneous or induced. Mutation spectra for spontaneous and induced mutations (using different mutagens) are not identical. The use and identification of induced mutations require large populations and extensive resources: the process is therefore unlikely to be useful as a mode of genetic recovery in small populations.

9. It is clear that the use of the species as a unit of genetic conservation is inadequate. Genetic uniqueness can occur within species, useful variants are often restricted to particular populations or regions, and genetic variants per se may have a potential utility.

10. A "genetic impact statement" should become an integral part of protection plans for endangered species, and could serve as a mechanism whereby a species or population could be declared "genetically endangered."

11. Mechanisms should be developed that permit preservation of potentially valuable variants and populations at the intraspecific level.

ACKNOWLEDGMENTS

I wish to thank Deborah Roach and Doug Gill for helpful suggestions on earlier drafts of this paper.

REFERENCES

Adams, M. W. 1977. An estimation of homogeneity in crop plants with special reference to genetic vulnerability in the dry bean, *Phaseolus vulgaris* L. *Euphytica* 26:655–679.

Antonovics, J., and A. D. Bradshaw. 1970. Evolution in closely adjacent plant populations. VIII. Clinal patterns at a mine boundary. *Heredity* 25:349–362.

Antonovics, J., and N. C. Ellstrand. 1984. Experimental studies on the evolutionary significance of sexual reproduction. I. A test of the frequency dependent selection hypothesis. *Evolution* 38:103–115.

Avise, J. C., and C. F. Aquadro. 1982. A comparative summary of genetic distances in the vertebrates: Patterns and correlations. *Evolution. Biol.* 15:157–185.

Avise, J. C., R. A. Lansman, and R. O. Shade. 1979. The use of restriction endonucleases to measure mitochondrial DNA sequence relatedness in natural populations. I. Population structure and evolution in the genus *Peromyscus*. *Genetics* 92:279–295.

Avise, J. C., J. J. Smith, and F. J. Ayala. 1975. Adaptive differentiation with little genetic change between two native California minnows. *Evolution* 29:411–426.

Ayala, F. J. 1982. The genetic structure of species. In *Perspectives on Evolution*, ed. R. Milkman, 60–82. Sunderland, Mass.: Sinauer.

Ayala, F. J., and J. F. McDonald. 1980. Continuous variation: Possible role of regulatory genes. *Genetica* 52/53:1–15.

Ayala, F. J., M. L. Tracey, D. Hedgecock, and R. Richard. 1974. Genetic differentiation during the speciation process in *Drosophila*. *Evolution* 28:576–592.

Benzer, S. 1955. Fine structure of a genetic region in bacteriophage. *Proc. Nat. Acad. Sci. (USA)* 41:344–354.

Bradshaw, A. D. 1983. The importance of evolutionary ideas in ecology and vice versa. In *Evolutionary Ecology*, ed. B. Shorrock, 1–25. Oxford: Blackwell.

———. 1984. Ecological significance of genetic variation between populations. In *Perspectives on Plant Population Ecology*, ed. R. Dirzo and J. Sarukhan, 213–228. Sunderland, Mass.: Sinauer.

Briggs, D., and S. M. Walters. 1984. *Plant Variation and Evolution*. 2d ed. Cambridge: Cambridge University Press.

Bryant, E. H., H. van Dijk, and W. van Delden. 1981. Genetic variability of the face fly, *Musca autumnalis* De Geer, in relation to a population bottleneck. *Evolution* 35:872–881.

Cavalli-Sforza, L. L., and W. F. Bodmer. 1971. *The Genetics of Human Populations*. San Francisco: Freeman.

Cherry, L. M., S. M. Case, and A. C. Wilson. 1978. Frog perspective on the morphological difference between humans and chimpanzees. *Science* 200:209–211.

Clausen, J., and W. M. Hiesey. 1958. *Experimental Studies on the Nature of Species. IV. Genetic structure of ecological races*. Carnegie Institution of Washington Publication 615. Washington, D.C.

Clay, K., and J. Antonovics. 1985. Quantitative variation of progeny from chasmogamous and cleistogamous flowers in the grass *Danthonia spicata*. *Evolution* 39:335–348.

Clegg, M. Y., A. M. D. Brown, and P. R. Whitfield. 1984. Chloroplast DNA diversity in wild and cultivated barley: Implications of genetic conservation. *Genet. Res.* 43:339–343.

Coyne, J. A. 1976. Lack of genetic similiarity between two sibling species of *Drosophila* as revealed by varied techniques. *Genetics* 84:593–607.

Davies, R. W. 1971. The genetic relationship of two quantitative characters in *Drosophila melanogaster*. II. Location of the effects. *Genetics* 69:363–375.

Dobzhansky, T. 1950. Genetics of natural populations. XIX. Origin of heterosis through natural selection in populations of *Drosophila pseudoobscura*. *Genetics* 35:288–302.

Falconer, D. S. 1981. *Introduction to Quantitative Genetics*. 2d ed. London: Longman.

Frankel, O. H., and E. Bennett, eds. 1970. *Genetic Resources in Plants: Their Exploration and Conservation*. Oxford: Blackwell Scientific Publications.

Frankel, O. H., and J. G. Hawkes, eds. 1975. *Crop Genetic Resources for Today and Tomorrow*. IBP 2. Cambridge: Cambridge University Press.

Frankel, O. H., and M. E. Soulé, eds. 1981. *Conservation and Evolution*. Cambridge: Cambridge University Press.

Franklin, I. R. 1980. Evolutionary change in small populations. In *Conservation Biology: An Evolutionary-Ecological Perspective*, ed. M. E. Soulé and B. A. Wilcox, 135–149. Sunderland, Mass.: Sinauer.

Giles, B. E. 1984. A comparison between quantitative and biochemical variation in the wild barley *Hordeum murinum*. *Evolution* 38:34–41.

Gottschalk, W., and G. Wolff. 1983. *Induced Mutations in Plant Breeding*. New York: Springer-Verlag.

Hamrick, J. L. 1975. Correlations between quantitative characters and enzyme genotypes in *Avena barbata*. *Evolution* 29:438–442.

Heslop-Harrison, J. 1964. Forty years of genecology. *Adv. Ecol. Res.* 2:159–247.

Heywood, J. S., and D. A. Levin. 1984. Allozyme variation in *Gaillardia pulchella* and *G. amblyodon* (Compositae): Relation to morphological and chromosomal variation and to geographical isolation. *System. Bot.* 9:448–457.

Highton, R., and T. P. Webster. 1976. Geographic protein variation and divergence in populations of the salamander, *Plethodon cinereus*. *Evolution* 30:33–45.

Holden, J. H. W., and J. T. Williams. 1984. *Crop Genetic Resources: Conservation and Evaluation*. London: Allen and Unwin.

Jain, S. K., L. Wu, and K. R. Vaidya. 1980. Levels of morphological and allozyme variation in Indian amaranths: A striking contrast. *J. Hered.* 71:283–285.

Johnston, R. F., and R. K. Selander. 1971. Evolution in the house sparrow. II. Adaptive differentiation in North American populations. *Evolution* 25:1–28.

Koehn, R. K., R. I. E. Newell, and F. Immerman. 1980. Maintenance of an amino-peptidase allele frequency cline by natural selection. *Proc. Nat. Acad. Sci. (USA)* 77: 5385–5389.

Kreitman, M. 1983. Nucleotide polymorphism at the alcohol dehydrogenase locus of *Drosophila melanogaster*. *Nature* 304:412–417.

Langlet, O. 1971. Two hundred years of genecology. *Taxon* 20:653–722.

Laurie-Ahlberg C. C., G. Maroni, G. C. Bewley, J. C. Lucchesi, and B. S. Weir. 1980. Quantitative genetic variation of enzyme activities in natural populations of *Drosophila melanogaster*. *Proc. Nat. Acad. Sci. (USA)* 77:1073–1077.

Lewontin, R. C. 1972. The apportionment of human diversity. *Evolution. Biol.* 6:381–397.

Marshall, D. R., and A. H. D. Brown. 1975. Optimum sampling strategies in genetic conservation. In *Crop Genetic Resources for Today and Tomorrow*, ed. O. H. Frankel and J. G. Hawkes, 53–80. IBP 2. Cambridge: Cambridge University Press.

Martinez, O. J., M. M. Goodman, and D. H. Timothy. 1983. Measuring racial differentiation in maize using multivariate measures standardized by variation in F2 populations. *Crop Sci.* 23:775–781.

Mather, K., and J. L. Jinks. 1982. *Biometrical Genetics*. 3d ed. London: Chapman and Hall.

Morishima, H., Y. Sano, and H. J. Oka. 1984. Differentiation of perennial and annual types due to habitat conditions in the wild rice *Oryza perennis*. *Plant System. Evol.* 144: 119–135.

Mukai, T. 1964. Genetic structure of natural populations of *Drosophila melanogaster*. I. Spontaneous mutation rate of polygenes controlling viability. *Genetics* 50:1–19.

Murphy, P. J., and J. R. Witcombe. 1981. The analysis of plant genetic resources data by multivariate techniques. FAO/UNEP IBPGR Technical Conference on Crop Genetic Resources, Rome.

Myers, N. 1983. *A Wealth of Wild Species*. Boulder, Colo.: Westview Press.

Nei, M. 1975. *Molecular Population Genetics and Evolution*. New York: Elsevier.

117

————. 1978. The theory of genetic distance and the evolution of human races. *Japanese J. Human Genetics* 23:341–369.

Price, M. V., and N. M. Waser. 1979. Pollen dispersal and optimal outcrossing in *Delphinium nelsoni*. *Nature* 277:294–297.

Qualset, C. O. 1975. Sampling germplasm in a center of diversity: An example of disease resistance in Ethiopian barley. In *Crop Genetic Resources for Today and Tomorrow*, ed. O. H. Frankel and J. G. Hawkes, 81–96. IBP 2. Cambridge: Cambridge University Press.

Ralls, K., and J. Ballou. 1983. Extinction: Lessons from zoos. In *Genetics and Conservation: A Reference for Managing Wild Animals and Plant Populations*, ed. C. M. Schonewald-Cox, S. M. Chambers, B. MacBryde, and W. L. Thomas, 164–184. Menlo Park, Calif.: Benjamin/Cummings.

Reynolds, J., B. S. Weir, and C. C. Cockerham. 1983. Estimation of the coancestry coefficient: Basis for a short-term genetic distance. *Genetics* 105:767–779.

Schonewald-Cox, C. M., S. M. Chambers, B. MacBryde, and W. L. Thomas, eds. 1983. *Genetics and Conservation: A Reference for Managing Wild Animal and Plant Populations*. Menlo Park, Calif.: Benjamin/Cummings.

Senner, J. W. 1980. Inbreeding depression and the survival of zoo populations. In *Conservation Biology: An Evolutionary-Ecological Perspective*, ed. M. E. Soulé and B. A. Wilcox, 209–224. Sunderland, Mass.: Sinauer.

Singh, R. S., R. C. Lewontin, and A. A. Felton. 1976. Genetic heterogeneity within electrophoretic "alleles" of xanthine dehydrogenase in *Drosophila pseudoobscura*. *Genetics* 84:609–629.

Slatkin, M. 1985. Rare alleles as indicators of gene flow. *Evolution* 39:53–65.

Sneath, P. H. A., and R. R. Sokal. 1973. *Numerical Taxonomy: The Principles and Practice of Numerical Classification*. San Francisco: Freeman.

Snyder, T. P., and M. C. Linton. 1984. Population structure in black flies: Allozymic and morphological estimates for *Prosimulium mixtum* and *P. fuscum* (Diptera: Similiidae). *Evolution* 38:942–956.

Soulé, M. E., and B. A. Wilcox, eds. 1980. *Conservation Biology: An Evolutionary-Ecological Perspective*. Sunderland, Mass.: Sinauer.

Spencer, W. P. 1947. Mutations in wild populations of *Drosophila*. *Advances in Genetics* 1:359–402.

Sprague, G. F., W. A. Russell, and L. H. Penny. 1960. Mutations affecting quantitative traits in the selfed progeny of doubled monoploid maize stocks. *Genetics* 45:855–866.

Templeton, A. R., and B. Read. 1983. The elimination of inbreeding depression in a captive herd of Speke's gazelle. In *Genetics and Conservation: A Reference for Managing Wild Animal and Plant Populations*, ed. C. M. Schonewald-Cox, S. M. Chambers, B. MacBryde, and W. L. Thomas, 241–261. Menlo Park, Calif.: Benjamin/Cummings.

Turner, B. J. 1974. Genetic divergence of Death Valley pupfish species: Biochemical vs. morphological evidence. *Evolution* 28:281–294.

Ward, R. H., and J. V. Neel. 1970. Gene frequencies and micro-differentiation among the Makiritare Indians. IV. A comparison of a genetic network with ethno-history and migration matrices; a new index of genetic isolation. *Amer. J. Human Genetics* 22:538–561.

Weir, B. S. 1983. *Statistical Analysis of DNA Sequence Data.* New York and Basel: Marcel Dekker.

Weir, B. S., and C. C. Cockerham. 1984. Estimating *F*-statistics for the analysis of population structure. *Evolution* 38:1358–1370.

Witcombe, J. R., and A. R. Rao. 1976. The genecology of wheat in a Nepalese centre of diversity. *J. Applied Ecol.* 13:915–924.

Wyatt, R. 1984. The evolution of self-pollination in granite outcrop species of *Arenaria* (Caryophyllaceae). I. Morphological correlates. *Evolution* 38:804–816.

Commentary

David S. Woodruff

In his paper Janis Antonovics provides a comprehensive survey of methods used to identify and conserve genetic variants. He views the goals of genetic conservation as the preservation of genetic variants of economically, scientifically, or aesthetically useful species by conserving their populations and those of closely related species. He argues that genetic considerations are likely to play only a minor role in decisions affecting which desirable variants or unique populations are to be saved. Antonovics concludes that the classic Noah's ark approach to species conservation is inappropriate for the preservation of intraspecific variants and that viewing the species as the unit for conservation is inadequate.

Antonovics's emphasis on intraspecific variant conservation reflects his expertise as a botanist interested in population-level adaptations. My own interest in species-level phenomena in animals leads me to slightly different views. As a zoologist I am more comfortable with the biological evolutionary species concept than are many botanists (e.g., Levin, 1979). Species are more easily recognized in many groups of animals than they are in plants, where phenotypic plasticity, polyploidy, and asexual reproduction often obscure phylogenetic relationships. Even in the genus *Cerion*, regarded by various authorities as the most difficult land snails to classify, species can be recognized using a combination of genetic, morphological, and biogeographic criteria (Woodruff and Gould, 1981). Also, my work on the characterization of species in taxonomically difficult groups of animals has led me to downplay the significance of intraspecific variation (Wilson and Brown, 1953). I believe that geneticists will play a significant role in the conservation of biological diversity, and that they can contribute to the management of variants, populations, and species. Species, far from being inadequate conservation units, are, in my opinion, key natural units upon which many conservation decisions will focus. Moreover, most of the present legal framework for the conservation of diversity focuses on species-level taxa; for example, the Endangered Species Act of 1973 (1982), CITES (Convention on International Trade in Endangered Species of Wild Fauna and Flora), Strategy Conference on Biological Diversity (U.S. Department of State, 1982), and World Conservation Strategy (IUCN, 1980; Thibodeau and Field, 1984).

Nonetheless, I agree with Antonovics that conservation genetics is currently a population-level science and that management will have to focus at the population level. Regrettably our understanding of species-level genetics is in its infancy. We need to know much more about the significance of coadapted gene complexes (Templeton, 1981; Lande, 1983; Templeton et al., 1986), genetic structuring of populations (Brussard, 1984), and the processes of genetic differentiation in small populations (Carson and Templeton, 1984) before we can hope to develop a species-level approach to genetic conservation.

THE MEASUREMENT OF DIVERSITY AND UNIQUENESS

The maintenance of genetic diversity is probably the most fundamental issue in population genetics (Clarke, 1979; Cook, 1984; Hamilton, 1984), and it has a direct bearing on conservation biology, since variability determines a population's ability to respond to environmental change and coevolutionary challenges. Several points, however, deserve additional mention. First, Antonovics had inadvertently omitted karyotype analyses from his list of techniques relevant to the measurement of genetic uniqueness. Studies of chromosomal polymorphism within and between species have been very important in evolutionary biology (White, 1978) and are proving useful to conservationists (see Briscoe et al., 1982, for cases involving rock wallabies; Benirschke, 1983, for primates and deer; and Ryder, 1984, for zebras).

The stress placed on multivariate approaches to phenetic variation in Antonovics's paper is also appropriate. Single variable analyses have led to incorrect decisions regarding conservation priorities. The strength of multivariate covariation analyses is well illustrated by studies of the variable land snail *Cerion*, where this approach has enabled us to synonymize hundreds of taxonomically confused morphospecies (Woodruff and Gould, 1981; Gould and Woodruff, 1978, 1986).

Single gel electrophoresis on starch or acrylamide slabs is one of the most useful genetic techniques available to the conservation biologist (see Milkman, 1982; Nei and Koehn, 1983; Oxford and Rollinson, 1983, for recent reviews). For data on levels of genetic variability in over 1,000 species of organisms see Nevo et al. (1984) and Hamrick (1983). The problem of hidden (undetectable) variation, once thought to conceal the presence of two of three alleles, is not extensive enough to grossly alter estimates of variation based on approximately twenty loci and thirty to fifty individual organisms from a population (Ayala, 1982; Selander and Whittam, 1983). Allozymic variation is now characterized routinely in established laboratories, but it is important to remember that the genes surveyed constitute a very small fraction of the whole genome. Regulatory genes may well be more important than structural genes in determining a population's evolutionary uniqueness, but they remain difficult to study. Conservation biologists should note progress in determining regulatory gene divergence based on the study of developmental schedules of gene expression in F_1 hybrids (Philipp et al., 1983b; Parker et al., 1985). Such considerations are relevant to the manager's problem of outbreeding depression and the handling of organisms from natural hybrid zones (Woodruff, 1979, 1981; Barton and Hewitt, 1983; Philipp et al., 1983a).

The relationship of enzyme variation, genetic distance, and taxonomic separation is very complex (Avise, 1983; Thorpe, 1983). It is clear that Nei's distance (D) or identity values can rarely be interpreted simply—as absolute proof of conspecificity or of significant divergence. Nevertheless, within a higher taxonomic category (genus or family) they provide the conservationist with the best available measure of genetic differentiation. Nonetheless, they do not always work. Examples of difficulties include (1) $D = 0$ in green and white lacewing species that differ by perhaps only three loci (Tauber and Tauber, 1977), and (2) $D = 1.5$ for two sibling species of salamanders (Highton and Larson, 1979). Similarly, the relationship between D and time of divergence (t) is more complex than originally conceived by those who believed in the regularity of molecular clocks. Avise and Aquadro (1982) found a twentyfold difference in divergence rates based on D in five classes of vertebrates. Fortunately, this problem should be of little concern to conservation biologists who are typically involved with closely related populations.

Antonovics notes that sequence diversity—as measured by restriction enzyme mapping of mtDNA, chloroplast DNA, and rDNA—is in its infancy. Results obtained to date generally parallel those obtained with cheaper allozyme techniques (e.g., Sytsma and Schaal, 1985; Ashley and Wills, 1987). Yet, because mtDNA is inherited maternally, it affords the manager a powerful tool for detecting interpopulation hybridization—a natural phenomenon that may occur at much higher frequency in "managed" populations inadvertently made up of several geographic races or sibling species (Harlan, 1983; Carr and Dodd, 1983; Cade, 1983).

Another approach to the measurement of genetic uniqueness mentioned by Antonovics involves pedigree analysis and estimation of kinship. Such techniques are useful only for those few populations for which pedigree data are available. Nevertheless, improved algorithms for microcomputer users are being developed and this approach should become more popular (Ballou, 1983).

I agree with Antonovics's conclusion that unique and important combinations of loci occur in every individual and in all populations of a species. Furthermore, in outbreeding species, individual uniqueness is ephemeral, since the genome passes through meiotic recombination each generation. There is, then, no simple way of defining or of preserving genetic uniqueness, and Antonovics concludes that other factors will be more important in setting conservation goals. Although the latter may have been true in the past, I see no reason why genetics cannot make more significant contributions to conservation in the future. Present tendencies of conservationists to insist on preserving every local "variety," and zoo curators to get credit for the number of "types" they exhibit, indicate the extent to which genetics has been ignored. In the remainder of this commentary I will try to indicate areas where geneticists can make major contributions.

HOW CAN GENETICISTS CONTRIBUTE TO BIOLOGICAL CONSERVATION?

Proper Identification of Biologically Meaningful Conservation Units

Antonovics defines the goals of genetic conservation in terms of the preservation of variants and populations of useful species and their relatives. I take a broader view: the primary goal should be the maintenance of evolutionarily significant units as communities of species. Accordingly, I would add to Antonovics's list of species deserving attention the "keystone" species (sensu Orians and Kunin) and the coevolved pollinators and mutualists of useful species. Oldfield (1981, 1984) discusses the importance of this broader ecological view. Unfortunately, neither approach helps us decide which species or variants are useful. This is a serious problem and we run the risk of overloading the ark with naturally rare species, endemics, and specialists before we have developed appropriate selection criteria (Western, 1985; Norton, 1986). The charismatic megavertebrates will undoubtedly receive more than their fair share of our resources for some time to come.

Once a taxon has been identified in general terms as needing conservation, geneticists can define biologically meaningful units for in situ and ex situ management. The literature contains numerous examples of cases where past conservation efforts failed to recognize such biologically meaningful units as the following ones.

1. Until very recently, Bornean and Sumatran orangutans (Pongo pygmaeus) were managed as a single species. It is now clear, however, that the two island populations are chromosomally distinct (Seuanez et al., 1979) and reputable zoos are excluding hybrids from their breeding programs.

2. Northern and southern isolates of the African white rhinoceros (Ceratotherium simum) are widely regarded as being only regional varieties of a formerly widespread species. Recently, however, George et al. (1983) discovered that the ctDNA of representatives of the northern and southern races show more differences than one would expect for conspecific populations. Unfortunately, this

finding may have come too late, since less than fifteen northern white rhinos survive in Garamba, Zaire.

3. The Sonoran topminnow (*Poeciliopsis occidentalis*) is an endangered species in Arizona. Studies of allozyme variation throughout the range of this species revealed that the wrong stocks are being employed in restocking efforts (Vrijenhoek et al., 1985). Largemouth bass have been mismanaged on a larger scale (Philipp et al., 1983a). Future management will hopefully pay more attention to the genetic variability of hatchery fish and the natural variation of the species they represent.

Other cases where genetic considerations are playing a role in the management of captive populations are discussed by Benirschke (1983). Surprisingly, very few of the species managed by zoos are well defined genetically. This is evident from the number of ongoing debates concerning endangered mammals where genetic data are badly needed for management decisions. Most of these debates involve the question whether individuals from geographically different populations should be mixed or maintained separately. Obviously, the greater the geographic distance between source populations, the more likely hybridization will result in outbreeding depression and a decline in fitness. Unfortunately, relevant genetic data are simply unavailable to the managers of sea otters in California and Alaska, of Asian elephants from Sri Lanka and India, and of Sumatran rhinos from Sabah and peninsular Malaysia. Although past practice would have involved mixing such geographic races indiscriminately, it is now recognized that the integrity of evolutionarily significant units should be maintained whenever possible. This leads us back to the central problem of deciding which populations to conserve. It is clear, for example, that zoos cannot afford to maintain self-sustaining populations of all eight subspecies of tiger. Similarly, how does one decide which populations of black rhino to concentrate on? The 9,000 surviving animals representing seven subspecies are scattered across seventeen African countries in over fifty isolated populations (Western and Vigne, 1985). Efforts to conserve every odd variant of a species are unjustifiable economically and may be unnecessary biologically. Hopefully, genetic data will be obtained before too many more decisions to preserve either "generic" or "pure" populations are made.

To reiterate, population genetic characterization is a prerequisite for effective conservation, be it at the level of the population, geographic race, or species. Morphology, the criterion used traditionally to define management units, is often unreliable. However, as Antonovics points out, morphology, coupled with genetic characterization, provides the conservationist with a means of defining evolutionarily significant units. The strength of the combined use of phenetic and genetic techniques is well illustrated in the following three cases.

1. The cephalopod *Nautilus pompilius* is widely distributed in the Indo-Pacific region and is quite variable in size and coloration. The larger-shelled animals

from Palau in Micronesia were recently recognized as a morphologically distinct species, N. *belauensis* (Saunders, 1987). The possibility remained, however, that they were just a locally large race of N. *pompilius*. This possibility was ruled out when studies of allozyme variation revealed a much greater difference between N. *pompilius* and N. *belauensis* ($D > 0.30$) than between several populations of the former (Woodruff et al., 1983; Woodruff and Carpenter, 1986).

2. In salamanders of the genus *Plethodon*, morphological divergence is unrelated to the amount of electrophoretically detectable genetic differentiation occurring among populations. Recent studies have shown that P. *dorsalis* is actually two allopatrically distributed species that are virtually identical in appearance but differ dramatically ($D = 1.5$) in 80 percent of their structural genes (Highton and Larson, 1979). The same authors report that three geographically isolated morphologically defined "subspecies" of P. *nettingi* have actually diverged as much as several well-characterized sympatric species.

3. Traditional taxonomy, based entirely on shell characteristics, would badly mislead a conservationist working with the West Indian land snails of the genus *Cerion*. On small New Providence Island in the Bahamas conchologists had described seventy-one living "species," many of which have since gone extinct as a result of the recent growth of the city of Nassau. Gould and Woodruff (1986), in contrast, found that there were really only two, imperfectly isolated species on the island. These species are characterized on the basis of the pattern of multivariate morphometric character *covariation*, allozyme variation, and biogeography. These taxa are not threatened, since they are common elsewhere in the Bahamas.

These examples show how easy it is to focus on biologically inappropriate units. In the future, geneticists should be able to make significant contributions to conservation biology by defining and characterizing evolutionarily meaningful units for management purposes.

Genetically Sound Breeding Plans for Closely Managed Populations

I will mention this important area only briefly, since numerous books are available on the application of genetics to breeding domesticated species in addition to the chapters and papers in the recent conservation biology literature (Soulé and Wilcox, 1980; Frankel and Soulé, 1981; Schonewald-Cox, et al., 1983; and especially Ralls and Ballou, 1986). Among the general principles appropriate for most closely managed populations are the following ones.

1. Increase population size as fast as possible to avoid loss of genetic variance and allelic diversity through genetic drift. The founder event itself may or may not have a profound effect on the genetics of the survivors. One species of mammal that exists only in captivity—the Arabian oryx (*Oryx leucoryx*)—has apparently lost most of its original genetic variability (Woodruff and Ryder,

1986). The challenge for managers is to retain as much of the remaining variation as possible.

2. Avoid inbreeding, because its associated costs may be significant (Ralls and Ballou, 1983). When this is impractical, as with Speke's gazelle, where the founding population numbered only four animals, it is possible to reduce the deleterious effects of inbreeding depression by careful selection of mating pairs (Templeton and Read, 1983). This has apparently happened naturally in species with a long history of inbreeding, such as Père David's deer (*Elaphurus davidianus*).

3. Maximize the effective population size (N_e) by subdividing the population and equalizing the contribution of all the founders. This latter principle, probably the most powerful weapon in the hands of captive breeders (Frankel and Soulé, 1981:40), is seldom employed.

4. Exchange enough genes each generation to maintain qualitative or quantitative genetic similarity between various subpopulations. Application of this principle would involve far more planning and cooperation among conservation organizations than now occurs.

5. Increase the variability of previously mismanaged populations by introducing more alleles from the wild. The endangered Indian lion is represented in U.S. zoos by the descendents of only three full sib pairs. The problem is even more serious in the case of major crop plants; for example, the North American soybean industry was launched in 1930 with six plants from one site in China (IUCN, 1980; King, 1984).

6. Avoid outbreeding depression. Crossing two differently coadapted populations may result in an increase in gametic incompatibility, zygotic and embryonic inviability, and hybrid mortality in the F_1, F_2, or backcross generations (Shields, 1982; Templeton et al., 1986). Such outbreeding depression in heterogeneous populations can severely hamper breeding programs (Benirschke, 1983). In at least one case, where Turkish and Nubian ibex were mixed with the Tatra Mountain ibex in Czechoslovakia, the hybrids were so poorly adapted that the entire population went extinct (Greig, 1979; Templeton et al., 1986). Methods should be discussed for the detection of outbreeding depression, for distinguishing inbreeding from outbreeding depression, and for the management of outbreeding depression. Ideally, the genetic characterization of evolutionarily meaningful conservation units through the study of natural populations will result in less disruption of coadapted gene complexes in closely managed situations in the future.

7. Avoid artificial selection for phenotypic conformity, since this will undoubtedly affect other traits and ultimately reduce the population's fitness. Highly selected, "domesticated" populations are notoriously maladapted for life under natural conditions. Norway brown trout, for example, do very well in hatcheries

but are useless as stock for reintroduction into the wild. Cutthroat trout raised in one hatchery for fourteen years had a 57 percent reduction in the proportion of polymorphic loci (Allendorf and Phelps, 1980). We need to develop techniques of genetic reinvigoration of remnant populations with genetic disease and highly selected inbred populations.

These genetic principles enable a conservationist to preserve the genetic diversity present in a closely managed population. Their application should increase the chances of the managed population's sustained long-term propagation and suitability for reintroduction into the wild.

Several of these principles are aimed at reducing the rate at which genetic diversity is lost in small populations. In this connection it should be stressed that there is no single "right" amount of genetic variation for a population or a species. Optimal levels of genetic polymorphism and heterozygosity vary with mating systems and demographic history (Soulé, 1980; Beardmore, 1983; Selander, 1983). Some species have paradoxically low levels of variation, such as the cheetah (O'Brien et al., 1985), the northern elephant seal (Bonnel and Selander, 1974), various gastropods (Selander and Ochman, 1983), and Torrey pines (Ledig and Conkle, 1983). Others show considerable geographic variation in levels of genetic variation. Isolated peripheral populations are often less variable than continuously distributed central populations (Briscoe et al., 1982; Brussard, 1984). The challenge for the geneticist is to characterize the variability of the population or species to be conserved and assist the manager in maintaining the variability.

Unfortunately, the application of these genetic management principles is by no means routine. Although ecological principles have long guided conservation practice, it is only recently that genetic principles were considered relevant (Frankel, 1982). This arose because of a shift from short-term to long-term conservation issues. The Species Survival Plan program of the American Association of Zoological Parks and Aquariums represents a pioneering attempt to incorporate genetic principles into the management of selected species in the collections of participating institutions. Although there are SSPs for only about forty species of endangered animals, the lessons learned with these will undoubtedly be transferred to other less closely managed species.

Integration of Genetic Principles in the Design and Management of Nature Reserves

As the costs of maintaining self-sustaining populations of free-ranging organisms are often two to four orders of magnitude less than the costs of captive propagation (Western, 1985), the need to further improve field management techniques should be obvious. Geneticists can contribute to ongoing discussions concerning the determination of minimal viable population size (MVP) of selected species. To quote Shaffer (1981:132), "A minimum viable population for any given species in any given habitat is the smallest isolated population having

126

a 99% chance of remaining extant for 1000 years despite the foreseeable effects of demographic, environmental, and genetic stochasticity, and natural catastrophes." The genetic determinants of MVP sizes are still unclear: "their resolution hinges primarily on a better knowledge of the breeding structure and genetic variability . . . and, most importantly, the role of genetic variability in population growth and regulation" (p. 134). Factors relevant to designing or modifying nature reserves stem from considerations of this type. Reserve acquisition can be based on genetic criteria (Hopper et al., 1982), and genetic arguments for the need for multiple reserves have yet to be put into practice. Another problem requiring attention concerns the empirical determination of optimal levels of gene flow and interpopulation hybridization (e.g., James, 1982; Harlan, 1983). Incorrect decisions can be costly (Hall-Martin, 1984). Finally, the conservation of taxa threatened by hybridization with neighboring taxa presents still other challenges for geneticists (Briscoe et al., 1982; Novak, 1982).

The increased application of genetics to the management of wild plant and animal populations presupposes greater cooperation between laboratory-bound geneticists and traditionally field-oriented conservation biologists. The various specialist groups of the International Union for the Conservation of Nature/Species Survival Commission (IUCN/SSC) facilitate such interdisciplinary planning. Oldfield (1984) discusses several other national and international organizations (including the National Council on Gene Resources, and the National Academy of Sciences 1978 report on germ plasm conservation) and suggests ways of strengthening conservation legislation relevant to our genetic heritage. Antonovics's call for genetic impact statements marks another positive step in the increased involvement of geneticists in conservation.

EPILOGUE

At the beginning of this commentary I suggested that Antonovics and I differ in our view of the goals of conservation genetics: he emphasizes the preservation of representative samples of genetic diversity and the preservation of existing variants of domesticated and useful wild species; I am more concerned with the future survival of wild species and the need to conserve genetically coadapted populations representative of such taxa. This dichotomy of opinion is representative of the variety of opinion about conservation genetics in the extensive literature on that subject. I use the term "conservation genetics" for the emerging applied science that seeks to define the genetic conditions for the continued evolution of wild biota. Antonovics's view is clearly closer to that of Bennett (1965), who introduced the near synonym "genetic conservation" for the preservation of genetic material used by plant breeders. This dichotomy, however, is only one of relative emphasis (Frankel, 1970, 1982, 1983; Frankel and Soulé, 1981). Both conservation genetics and genetic conservation are based on a common set of scientific principles and have important roles to play in conservation biology.

REFERENCES

Allendorf, F. W., and S. R. Phelps. 1980. Loss of genetic variation in a hatchery stock of cutthroat trout. *Trans. Amer. Fish. Soc.* 109:537–543.

Ashley, M., and C. Wills. 1987. Analysis of mitochondrial DNA polymorphisms among Channel Island deer mice. *Evolution* 41:854–863.

Avise, J. C. 1983. Protein variation and phylogenetic reconstruction. In *Protein Polymorphism: Adaptive and Taxonomic Significance*, ed. G. S. Oxford and D. Rollinson, 103–130. London: Academic Press.

Avise, J. C., and C. F. Aquadro. 1982. A comparative summary of genetic distances in the vertebrates: Patterns and correlations. *Evolution. Biol.* 15:151–185.

Ayala, F. J. 1982. The genetic structure of species. In *Perspectives on Evolution*, ed. R. Milkman, 60–82. Sunderland, Mass.: Sinauer.

Ballou, J. 1983. Calculating inbreeding coefficients from pedigrees. In *Genetics and Conservation: A Reference for Managing Wild Animal and Plant Populations*, ed. C. M. Schonewald-Cox, S. M. Chambers, B. MacBryde, and W. L. Thomas, 509–520. Menlo Park, Calif.: Benjamin/Cummings.

Barton, N. H., and G. M. Hewitt. 1983. Hybrid zones as barriers to gene flow. In *Protein Polymorphism: Adaptive and Taxonomic Significance*, ed. G. S. Oxford and D. Rollinson, 341–360. London: Academic Press.

Beardmore, J. A. 1983. Extinction, survival, and genetic variation. In *Genetics and Conservation: A Reference for Managing Wild Animal and Plant Populations*, ed. C. M. Schonewald-Cox, S. M. Chambers, B. MacBryde, and W. L. Thomas, 125–151. Menlo Park, Calif.: Benjamin/Cummings.

Benirschke, K. 1983. The impact of research on the propagation of endangered species in zoos. In *Genetics and Conservation: A Reference for Managing Wild Animal and Plant Populations*, ed. C. M. Schonewald-Cox, S. M. Chambers, B. MacBryde, and W. L. Thomas, 402–413. Menlo Park, Calif.: Benjamin/Cummings.

Bennett, E. 1965. Genecological aspects of plant introduction and genetic conservation. *Scottish Plant Breeding Station Record* 1965:27–113.

Bonnel, M. L., and R. K. Selander. 1974. Elephant seals: Genetic variation and near extinction. *Science* 184:908–909.

Briscoe, D. A., J. H. Calaby, R. L. Close, G. M. Maynes, C. E. Murtagh, and G. B. Sharman. 1982. Isolation, introgression and genetic variation in rock-wallabies. In *Species at Risk: Research in Australia*, ed. R. H. Groves and W. D. L. Ride, 73–87. Berlin: Springer-Verlag.

Brussard, P. F. 1984. Geographic patterns and environmental gradients: The central-marginal model in *Drosophila* revisited. *Ann. Rev. Ecol. Syst.* 15:25–64.

Cade, T. J. 1983. Hybridization and gene exchange among birds in relation to conservation. In *Genetics and Conservation: A Reference for Managing Wild Animal and Plant Populations*, ed. C. M. Schonewald-Cox, S. M. Chambers, B. MacBryde, and W. L. Thomas, 288–309. Menlo Park, Calif.: Benjamin/Cummings.

Carr, A. F., and C. K. Dodd. 1983. Sea turtles and the problem of hybridization. In *Genetics and Conservation: A Reference for Managing Wild Animal and Plant Populations*, ed. C. M. Schonewald-Cox, S. M. Chambers, B. MacBryde, and W. L. Thomas, 277–287. Menlo Park, Calif.: Benjamin/Cummings.

Carson, H. L., and A. R. Templeton. 1984. Genetic revolutions in relation to speciation phenomena: The founding of new populations. *Ann. Rev. Ecol. Syst.* 15:97–131.

Clarke, B. 1979. The evolution of genetic diversity. *Proc. Roy. Soc. London B* 205:453–474.

Cook, L. M. 1984. The problem. In *Evolutionary Dynamics of Genetic Diversity*, ed. G. S. Mani, 1–12. Berlin: Springer-Verlag.

Frankel, O. H. 1970. Variation, the essence of life. Sir William Macleay Memorial Lecture. *Proc. Linn. Soc. N.S.W.* 95:158–169.

―――. 1982. The role of conservation genetics in the conservation of rare species. In *Species at Risk: Research in Australia*, ed. R. H. Groves and W. D. L. Ride, 159–162. Berlin: Springer-Verlag.

―――. 1983. Foreword. In *Genetics and Conservation: A Reference for Managing Wild Animal and Plant Populations*, ed. C. M. Schonewald-Cox, S. M. Chambers, B. MacBryde, and W. L. Thomas. Menlo Park, Calif.: Benjamin/Cummings.

Frankel, O. H., and M. E. Soulé, eds. 1981. *Conservation and Evolution*. Cambridge: Cambridge University Press.

George, M., L. A. Puentes, and O. Ryder. 1983. Genetische Unterscheide zwischen den Unterarten des Breitmaulnashorns. In *Internat. Studbook African Rhinoceros*, ed. H. G. Klos and R. Frese, 60–67. Berlin: Zoologischen Garten.

Gould, S. J., and D. S. Woodruff. 1978. Natural history of *Cerion* VIII: Little Bahama Bank—A revision based on genetics, morphometrics and geographical distribution. *Bull. Mus. Comp. Zool.* 148:371–415.

―――. 1986. Systematics of *Cerion* on New Providence Island: A radical revision. *Bull. Amer. Mus. Nat. Hist.* 182:391–490.

Greig, J. C. 1979. Principles of genetic conservation in relation to wildlife management in southern Africa. *S. Af. J. Wildlife Res.* 9:57–78.

Hall-Martin, A. 1984. Kenya's black rhinos in Addo, S. Africa. *Af. Elephant and Rhino Group Newsletter* 3:11.

Hamilton, W. D. 1984. Pathogens as causes of genetic diversity in their host populations. In *Population Biology of Infectious Diseases*, ed. R. M. Anderson and R. M. May, 269–296. New York: Springer-Verlag.

Hamrick, J. L. 1983. The distribution of genetic variation within and among plant populations. In *Genetics and Conservation: A Reference for Managing Wild Animal and Plant Populations*, ed. C. M. Schonewald-Cox, S. M. Chambers, B. MacBryde, and W. L. Thomas, 500–508. Menlo Park, Calif.: Benjamin/Cummings.

Harlan, J. R. 1983. Some merging of plant populations. In *Genetics and Conservation: A Reference for Managing Wild Animal and Plant Populations*, ed. C. M. Schonewald-Cox, S. M. Chambers, B. MacBryde, and W. L. Thomas, 267–276. Menlo Park, Calif.: Benjamin/Cummings.

Highton, R., and A. Larson. 1979. The genetic relationships of the salamanders of the genus *Plethodon*. *System. Zool.* 28:579–599.

129

Hopper, S. D., N. A. Campbell, and G. F. Moran. 1982. *Eucalyptus caesia,* a rare mallee of granite rocks from south-western Australia. In *Species at Risk: Research in Australia,* ed. R. H. Groves and W. D. L. Ride, 46–61. Berlin: Springer-Verlag.

International Union for the Conservation of Nature and Natural Resources (IUCN). 1980. *World Conservation Strategy.* International Union for the Conservation of Nature and Natural Resources/United Nations Environment Programme/World Wildlife Fund. Gland, Switzerland.

James, S. H. 1982. The relevance of genetic systems in *Isotoma petraea* to conservation practice. In *Species at Risk: Research in Australia,* ed. R. H. Groves and W. D. L. Ride, 63–71. Berlin: Springer-Verlag.

King, F. W. 1984. Preservation of genetic diversity. In *Sustaining Tomorrow: A Strategy for World Conservation and Development,* ed. F. R. Thibodeau and H. H. Field, 41–55. Hanover, N.H.: Tufts-University Press of New England.

Lande, R. 1983. The response to selection on major and minor mutations affecting a metrical trait. *Heredity* 50:47–65.

Ledig, F. T., and M. T. Conkle. 1983. Gene diversity and genetic structure in a narrow endemic, Torrey pine (*Pinus torreyana* Parry ex Carr.). *Evolution* 37:79–85.

Levin, D. A. 1979. The nature of plant species. *Science* 204:381–384.

Loveless, M. D., and J. L. Hamrick. 1984. Ecological determinants of genetic structure in plant populations. *Ann. Rev. Ecol. Syst.* 15:65–95.

Milkman, R., ed. 1982. *Perspectives on Evolution.* Sunderland, Mass.: Sinauer.

Nei, M., and R. K. Koehn, eds. 1983. *Evolution of Genes and Proteins.* Sunderland, Mass.: Sinauer.

Nevo, E., A. Beiles, and R. Ben-Schlomo. 1984. The evolutionary significance of genetic diversity: Ecological, demographic and life history correlates. In *Evolutionary Dynamics of Genetic Diversity,* ed. G. S. Mani, 13–213. Berlin: Springer-Verlag.

Norton, B. G., ed. 1986. *The Preservation of Species: The Value of Biological Diversity.* Princeton, N.J.: Princeton University Press.

Novak, R. 1982. The red wolf recovery program. In MAB Symposium: Application of genetics to the management of wild plant and animal populations. Speakers' abstracts. August 9–13, 1982, Washington, D.C.

O'Brien, S. J., et al. 1985. Genetic basis for species vulnerability in the cheetah. *Science* 227:1428–1434.

Oldfield, M. L. 1981. Tropical deforestation and genetic resources conservation. *Stud. Third World Soc.* 14:277–345.

———. 1984. *The Value of Conserving Genetic Resources.* Washington, D.C.: U.S. Department of the Interior, National Park Service.

Oxford, G. S., and D. Rollinson, eds. 1983. *Protein Polymorphism: Adaptive and Taxonomic Significance.* London: Academic Press.

Parker, H. R., D. P. Philipp, and G. S. Whitt. 1985. Relative developmental success in interspecific *Lepomis* hybrids as an estimate of gene regulatory divergence between species. *J. Exp. Zool.* 233:451–466.

Philipp, D. P., W. F. Childers, and G. S. Whitt. 1983a. A biochemical genetic evaluation of the northern and Florida subspecies of largemouth bass. *Trans. Amer. Fish. Soc.* 112:1–20.

Philipp, D. P., H. R. Parker, and G. S. Whitt. 1983b. Evolution of gene regulation: Isozymic analysis of patterns of gene expression during hybrid fish development. *Isozymes* 10:193–238.

Ralls, K., and J. Ballou. 1983. Extinction: Lessons from zoos. In *Genetics and Conservation: A Reference for Managing Wild Animal and Plant Populations*, ed. C. M. Schonewald-Cox, S. M. Chambers, B. MacBryde, and W. L. Thomas, 164–184. Menlo Park, Calif.: Benjamin/Cummings.

Ralls, K., and J. Ballou, eds. 1986. Genetic management of captive populations. *Zoo Biology* 5:81–238.

Ryder, O. A. 1984. Molecular cytogenetics of the Equidae. III. Cytological localization of heterochromatin and satellite DNA in Hartmann's mountain zebra, *Equus zebra hartmannae* (Matschie, 1898). In *One Medicine*, ed. O. A. Ryder and M. L. Byrd, 119–127. Berlin: Springer-Verlag.

Saunders, W. B. 1987. The species of living *Nautilus* and their distribution. *Veliger* 24: 8–18.

Schonewald-Cox, C. M., S. M. Chambers, B. MacBryde, and W. L. Thomas, eds. 1983. *Genetics and Conservation: A Reference for Managing Wild Animal and Plant Populations*. Menlo Park, Calif.: Benjamin/Cummings.

Selander, R. K. 1983. Evolutionary consequences of inbreeding. In *Genetics and Conservation: A Reference for Managing Wild Animal and Plant Populations*, ed. C. M. Schonewald-Cox, S. M. Chambers, B. MacBryde, and W. L. Thomas, 201–215. Menlo Park, Calif.: Benjamin/Cummings.

Selander, R. K., and H. Ochman. 1983. The genetic structure of populations as illustrated by molluscs. *Isozymes* 10:93–123.

Selander, R. K., and T. S. Whittam. 1983. Protein polymorphism and the genetic structure of populations. In *Evolution of Genes and Proteins*, ed. M. Nei and R. K. Koehn, 89–114. Sunderland, Mass.: Sinauer.

Seuanez, H. N., H. J. Evans, D. E. Martin, and J. Fletcher. 1979. An inversion of chromosome 2 that distinguishes between Bornean and Sumatran orangutans. *Cytogenetics and Cell Genetics* 23:137–140.

Shaffer, M. L. 1981. Minimum population sizes for species conservation. *BioScience* 31: 131–134.

Shields, W. M. 1982. *Philopatry, Inbreeding and the Evolution of Sex*. Albany: State University of New York Press.

Soulé, M. E. 1980. Thresholds for survival: Maintaining fitness and evolutionary potential. In *Conservation Biology: An Evolutionary-Ecological Perspective*, ed. M. E. Soulé and B. A. Wilcox, 151–169. Sunderland, Mass.: Sinauer.

Soulé, M. E., and B. A. Wilcox, eds. 1980. *Conservation Biology: An Evolutionary-Ecological Perspective*. Sunderland, Mass.: Sinauer.

Sytsma, K. J., and B. A. Schaal. 1985. Phylogenetics of the *Lisianthius skinneri* (Gentianaceae) species complex in Panama utilizing DNA restriction fragment analyses. *Evolution* 39:594–608.

Tauber, C. A., and M. J. Tauber. 1977. Sympatric speciation based on allelic changes at three loci: Evidence from natural populations in two habitats. *Science* 197:1298–1299.

Templeton, A. R. 1981. Mechanisms of speciation: A population genetic approach. *Ann. Rev. Ecol. Syst.* 12:23–48.

Templeton, A. R., H. Hemmer, G. Mace, U. S. Seal, W. M. Shields, and D. S. Woodruff. 1986. Local adaptation, coadaptation, and population-boundaries. *Zoo. Biol.* 5(2): 115–126.

Templeton, A. R., and B. Read. 1983. The elimination of inbreeding depression in a captive herd of Speke's gazelle. In *Genetics and Conservation: A Reference for Managing Wild Animal and Plant Populations*, ed. C. M. Schonewald-Cox, S. M. Chambers, B. MacBryde, and W. L. Thomas, 241–261. Menlo Park, Calif.: Benjamin/Cummings.

Thibodeau, F. R., and H. H. Field. 1984. *Sustaining Tomorrow: A Strategy for World Conservation and Development.* Hanover, N.H.: Tufts-University Press of New England.

Thorpe, J. P. 1983. Enzyme variation, genetic distance and evolutionary divergence in relation to levels of taxonomic separation. In *Protein Polymorphism: Adaptive and Taxonomic Significance*, ed. G. S. Oxford and D. Rollinson, 131–152. London: Academic Press.

U.S. Department of State. 1982. *Proceedings of the U.S. Strategy Conference on Biological Diversity.* November 16–18, 1981. Department of State Publication 9262. Washington, D.C.

Vrijenhoek, R. C., M. E. Douglas, and G. K. Meffe. 1985. Conservation genetics of endangered fish populations in Arizona. *Science* 229:400–402.

Western, D. 1985. The role of captive populations in global conservation. In Primates: The Road to Self-sustaining Populations. Zoological Society of San Diego Conference, June 1985. Manuscript.

Western, D., and L. Vigne. 1985. The deteriorating status of African rhinos. *IUCN/SSC African Elephant and Rhino Specialist Group Newsletter.*

White, M. J. D. 1978. *Modes of Speciation.* San Francisco: W. H. Freeman.

Wilson, E. O., and W. L. Brown. 1953. The subspecies concept and its taxonomic application. *System. Zool.* 2:97–111.

Woodruff, D. S. 1979. Postmating reproductive isolation in *Pseudophryne* and the evolutionary significance of hybrid zones. *Science* 203:561–563.

———. 1981. Towards a genodynamics of hybrid zones: Studies of Australian frogs and West Indian land snails. In *Essays on Evolution and Speciation in Honor of M. J. D. White*, ed. W. R. Atchley and D. S. Woodruff, 171–197. Cambridge: Cambridge University Press.

———. 1987. Fifty years of interspecific hybridization: Genetics and morphometrics of a controlled experiment on the land snail *Cerion* in the Florida Keys. *Evolution* 41: 1022–1045.

Woodruff, D. S., and M. P. Carpenter. 1986. Genetic variation and phylogeny in *Nautilus*. In *Nautilus: The Biology and Paleobiology of a Living Fossil*, ed. W. B. Saunders and N. H. Landman, 65–83. New York: Plenum Press.

Woodruff, D. S., and S. J. Gould. 1981. Geographic differentiation and speciation in *Cerion*: A preliminary discussion of patterns and processes. *Biol. J. Linn. Soc. London* 14:389–416.

Woodruff, D. S., M. Mulvey, W. B. Saunders, and M. P. Carpenter. 1983. Genetic variation in the cephalopod *Nautilus belauensis*. *Proc. Nat. Acad. Sci.* (USA) 136: 147–153.

Woodruff, D. S., and O. A. Ryder. 1986. Genetic characterization and conservation of endangered species: Arabian oryx and père David's deer. *Isozyme Bull.* 19:35.

Summary of the Discussion

Lawrence Riggs

Gordon Orians opened the session by focusing attention on how genetic knowledge and research techniques might be applied in determining what genetic materials are valuable, in identifying criteria for selecting materials for preservation, and in selecting technologies appropriate for preservation of valuable materials. The group discussion continued for over two hours, ranging across many topics and referring to examples drawn from the experience of many of the participants.

SELECTION CRITERIA

Deborah Rabinowitz raised two provocative issues for discussion based on Janis Antonovics's observations that the genetic criteria for uniqueness are frequently not adequate justification for preservation, and that information about phenetic traits (morphological, physiological, or behavioral) may be more useful than genotypic data or composite indices.

Antonovics pointed out that the measures providing the most accurate specification of genetic information (e.g., DNA sequence data) are the most difficult to relate to their phenotypic effects, while methods affording a less accurate description of genetic substrates (e.g., phenetics) are better understood biologically. Isozyme data may fall somewhere in between.

The most difficult task for the geneticist, according to Antonovics, is to decide the level at which uniqueness should be measured. There are many unknowns in the mapping of genetic differences onto phenotypic differences. Single allelic variants are unique, and many phenotypic variants are distinctly different. Nearly all individuals in an outbreeding species are unique, but this level of resolution is not very useful in setting priorities for conservation. Systematists are still struggling to develop general methods that will distinguish species and subspecies in unequivocal ways on the basis of multiple trait differences. Techniques for recognizing important discontinuities in organic diversity below the species level are still being developed. Distributional information and historical data must be integrated with genetic and phenetic data on a case-by-case basis.

In concert with Antonovics, David Woodruff observed that genetics is not a sufficiently mature science to meet the challenges posed by management situations. Nevertheless he challenged assertions in Antonovics's paper that (1) genetic data are likely to be only a small component of the decision process and that (2) species may be inadequate units for conservation.

Other topics potentially relevant to the selection problem were broached but not pursued in detail; for example, the possibility of using electrophoretic markers to aid the selection of genetic materials for preservation was mentioned. Such applications would require research on the association, if any, between the genetic markers and phenetic data. Arthur Weissinger outlined a series of questions that must be addressed: How much electrophoretic variation exists in isozymes? Is it important? Is it useful? If electrophoretic variation is not associated with phenotypic differences within any environment in which the organism is found, it may not represent anything important for conservation. Difficulties in using aggregate indices such as Nei's D in setting conservation priorities were also discussed briefly. Gardner Brown pointed out that the Nei index assumes that all allelic variants of a gene are equally important. An economic analysis carried out using those indicies would thus have to make the untenable assumption that physical changes in phenotype are equivalent for all allelic substitutions. While Nei's index may be the most useful one for evolutionary biologists seeking to calibrate divergence times at the lower end of the time scale, it may not be appropriate for setting criteria for genetic resource preservation. Either weighting criteria must be developed to permit disaggregation of allelic frequency data, rendering them useful to the selection problem, or different indices will have to be developed.

COADAPTED GENE COMPLEXES

The lack of strong evidence supporting the existence of coadapted gene complexes was noted by Antonovics, and this stimulated considerable discussion. Many of the participants saw adaptive gene combinations or coadapted complexes as a potentially important feature of genetic uniqueness, and one that might carry resource value whether or not detailed genetic information was available. Antonovics suggested that the probability of successfully assembling even ten to twenty-five genes in efforts to reconstruct gene complexes is very low, given current understanding and technology.

Arthur Weissinger outlined a study being implemented by Pioneer Hi-Bred International which will test the possibility of coadaptation between nuclear and cytoplasmic DNA in corn. The use of cytoplasmic sterility factors introduced to lower costs of hybrid seed production in corn in the late 1960s led to homogeneity of a portion of the genome apparently associated with disease resistance. In response to the severe corn blight epidemic of 1970, corn breeders ceased using single-cell derived cytoplasms and began research to determine how much cytoplasmic DNA variation might exist and how it might be deployed in commercial plantings to reduce vulnerability to disease outbreaks. Weissinger outlined studies documenting fairly limited genetic variability in breeding stock cytoplasms and experiments for which sixteen combinations of recognizable nuclear and cytoplasmic types were created by a complex series of backcrosses. He described the scope of testing efforts necessary to distinguish coadaptation from environmental effects on yield parameters, and asked whether participants considered this kind of experiment possible with other organisms and effective in testing for the existence of coadapted gene combinations.

The population geneticists who were present recognized the technical feasibility of the approach described and its potential for addressing some questions of interest in evolutionary biology (e.g., for the study of male sterility mechanisms in natural plant populations, plant-pathogen interactions, etc.). However, the complexities of sorting out the possible combinations in populations considerably more variable and less well studied than breeding stocks of corn limit its application. It was concluded that, without specific knowledge of the genes involved in the expression of observable traits and of their location on chromosomes, and without appropriate measures of fitness, testing of coadaptation hypotheses would not be productive. *Drosophila* geneticists may be close to being able to conduct such experiments, given present knowledge of the *Drosophila* genome. In the meantime, the controlled recombination of cytoplasmic and nuclear variants may be the only experimental means to approach the issue of coadapted gene complexes. Basic population geneticists may find it difficult to conduct comparable studies, given the scope of the problem, and the fact that they rarely have access to the resources large firms such as Pioneer Hi-Bred have.

As a response to several questions, there was a review of the evidence supporting the validity of the coadapted gene complex concept. It was pointed out that the notion was developed at a time when DNA structure was represented by the "string of beads" analogy and gene expression was thought to be the simple result of translation and transcription. Since the picture has become more complicated with the recognition of extranuclear genes, the possibility of interspecific gene transfer through retroviral infection, and the observation of posttranscriptional changes in gene products, it has seemed less intuitively reasonable that genomic organization is necessarily adaptive, or that its structure is stable. Much of the evidence suggesting coadapted organization is weak. For example, from studies of parthenogenetic races of *Drosophila mercatorum*, Alan Templeton inferred the existence of alternative adaptive combinations of genes from the fact that members of each parthenogenetic line mate successfully with others of their own line but not with members of the parental population or other lines. While similar findings between species were advanced by Dobzhansky many years ago to explain hybrid dysgenesis and to support the plausibility of the coadapted gene complexes, such a criterion is not sufficient to test the hypothesis. As we know from a variety of studies, many other mechanisms can account for the failure of hybrid or between-line crosses, and not all represent adaptive responses.

Woodruff summarized the principal approach used by population geneticists to infer the existence of nonrandom associations of genes, the measurement of linkage disequilibrium; this makes no reference to adaptive explanations. A statistical test compares the observed associations of alleles at different marker

135

loci with that expected under the random segregation of alleles in an infinitely large population at demographic equilibrium. Strong associations of genes are indicated by significant departures from the null hypothesis, or "disequilibrium." Strong disequilibrium can be interpreted as evidence that the alleles under examination confer a higher (or lower if the alleles are negatively correlated) average fitness when they are associated in the genome. The argument for existence of coadapted gene complexes stems, in part, from the finding that allelic variants are often nonrandomly segregating. Other inferential data and examples were discussed. It was recalled that inbreeding has been credited with maintaining coadapted gene complexes in many genetically homogeneous species, and that outbreeding depression is often explained as the breaking up of gene complexes.

Without direct evidence for the existence and function of a coadapted gene complex, this concept can be applied to conservation efforts only indirectly. Woodruff offered an example based on studies of isozymic variability in populations of the slipper limpet, a bottom-dwelling mollusc found in San Diego harbor and on the coast of China. Much lower levels of variability in the Chinese population suggest its evolutionary derivation from North American populations. If asked to preserve the species for posterity, Woodruff would give priority to the North American population on the grounds that all of the variability (the isozymic variability at least) would be represented. However, if asked to address preservation of both forms, he would consider that sufficient evolutionary time had elapsed between divergence of the two populations that genetic coadaption could have developed in each, and they should be preserved separately rather than pooled in a single heterogeneous and interbreeding group.

Concern was voiced that the action of transposable elements ("jumping genes") might render the concept of coadapted gene complexes completely useless in conservation planning. Douglas Gill provided a cursory overview, mentioning the existence of substantial evidence for the movement of large segments of genetic material and for limitations on the locations of extraction and insertion sites within the genome. Movement of genetic material within an individual genome has been documented, and it is already known that the expression of genetic material may change depending on its location in the genome (the positional effect on eye color in *Drosophila*, for example). The determination of what genetic material is worth saving could be made impossibly complex: because genotypes are ephemeral, determination of the genotype might depend on what part of the individual organism one is examining. While no one in the group could provide a specialist's insight, it was felt that both the species concept and the notions of genotypic stability and germ line continuity are likely to be maintained as research progresses in this area. In addition to recommending that geneticists stay abreast of developments from studies of "selfish DNA," viral elements, and so forth, Antonovics suggested that retroviruses and other transposable elements should, as a class, be deserving of a high priority for genetic conservation.

The take-home messages regarding the role of coadapted gene complexes in genetic conservation were the following: (1) the concept is relevant, though not necessarily critical to the definition of fundamental population units in nature (stocks, strains, varieties, etc.), and (2) to manage or sustain coadapted gene complexes in the absence of detailed information, as many individuals as possible should be sampled from natural populations.

SAMPLING VERSUS RECONSTRUCTION OF GENE COMBINATIONS

With seeming consensus that gene combinations, adaptive or otherwise, constitute one of the most "useful" aspects of genetic material, the question of how such combinations might best be preserved or otherwise obtained received considerable attention. Several comments by Margery Oldfield and Christine Schonewald-Cox suggested that collection from natural populations might be the best and perhaps the only practical way to obtain samples of adaptive genetic combinations. Maximum access to such material might be maintained through *in situ* management of populations.

There was no support for the notion that techniques of mutation breeding could be effective in reconstructing important gene combinations. However, Oliver Ryder argued that developing biotechnologies eventually will enable us to "cut and paste" genotypes to order, provided that a template is available. He suggested that samples of DNA could be preserved at very low temperatures until capabilities to isolate, identify, and study unique sequences eventually allow reintroduction of gene combinations into viable strains. Subsequently, Michael Rosenzweig suggested that tissue culture techniques might already be able to speed the process of recombining genes.

Antonovics highlighted several observations pertinent to the conservation of unique genetic combinations. Despite the incredibly large number of unique combinations generated by reproduction in outbreeding organisms, not all possible gene combinations occur in nature. Discontinuities ease the job of the taxonomist, but complicate life for the plant breeder, who often must search for traits to incorporate into commercial varieties from relatives of crop species. In most cases one can only guess at the existence of adaptive or "right" gene combinations from measures indicating nonrandom association of genes. Only in the most intensively studied species can we identify the genetic basis of particular traits and systematically assemble the desired gene combinations, using available technologies.

137

IMPACTS OF INBREEDING AND OUTBREEDING

The discussion of inbreeding and outbreeding effects revealed a considerable maturation of views on the application of genetics in this area. Whereas only a few years ago scientists and managers unequivocally condemned practices that might result in inbreeding, both groups have become more sophisticated. Descriptions of the mating systems in natural populations have shown certain levels of inbreeding to be acceptable, perhaps even desirable, and management of captive populations has provided examples in which promotion of outbreeding has been detrimental.

Citing Selander's work on Old World and New World species of slugs, Schonewald-Cox indicated the growing evidence that inbred and apomictic species of organisms have been highly successful colonists of natural habitats. She suggested that the often narrow and well-defined ecological requirements of such species could make them especially useful to managers attempting to install or intentionally modify biological communities in nonpristine habitats. She observed that the diversity of ways in which natural populations are genetically structured might suggest a broader range of possible management prescriptions than has been considered heretofore.

Ryder cited Alan Templeton's work on captive populations of zoo animals as evidence that inbreeding and outbreeding depression are temporal phenomena that can be alleviated by innovative management of subpopulations and some demographic good luck. Selective breeding of closely related individuals may be able to maintain numbers while homozygous recessive alleles are purged from the population. Recognition of genetically incompatible cytotypes of some cryptic species (e.g., spider monkeys) can help the manager avoid ineffectual crossbreeding.

Bruce Wilcox summarized Templeton's ongoing study of the collared lizard, *Crotaphytus collaris*, in the Ozarks as an example of how management might use knowledge of inbreeding and outbreeding effects. Habitat glades in the Ozarks are being managed as unique ecological areas, and efforts are being made to restore habitats and species. The problem in regenerating collared lizard populations was one of availability of source stocks. Lizard populations were extinct in many areas and very small in others. Templeton used allozyme markers and mitochondrial DNA analysis to examine population differentiation, and discovered fixed monomorphic differences between populations. This suggested that the risk of outbreeding depression might be high and that local adaptation might be more important than heterozygosity in establishing new populations. The data permitted "appropriate" breeding stocks to be selected for reestablishment of populations on particular glades. The assumptions behind the genetic models and the overall management strategy are being tested by implementation of his recommendations.

Templeton's work was seen by several in the group as exemplary of the kind required to bring state-of-the-art genetics to bear on management problems. All too frequently basic research with application to management problems is not communicated to those who may adapt even its tentative findings to management and decision making.

PROFESSIONAL PRIORITIES AND ECONOMIC CHOICES

For one economist listening to the first two hours of the group's deliberations, the various issues and arguments boiled down to a question of how to allocate limited monetary resources to activities that might be undertaken by professional scientists. Roger Noll summarized his understanding of the key issues in a decision theoretic framework. He recognized two sources of value in genetic materials. Direct values, manifested by physical or behavioral traits and products derived from organisms, seemed relatively easy to measure. Even more ephemeral qualities, such as recreational or aesthetic values, could be quantified by a direct valuation approach. Indirect or instrumental values, on the other hand, seemed to occupy the core of most of the morning's discussion and to represent a much more difficult, and interesting, problem.

Given that social and environmental conditions are not deterministic over relatively long periods, that there are many possible future states of the world, and that it is not possible to predict precisely what species or genetic variants will be needed, how should decisions be made? In this context, the instrumental value of genetic diversity would be its effect on adaptability of species to possible future environmental changes. Given a budgetary constraint on genetic preservation activities, the decision process might follow three steps: (1) decide which species to save on the basis of priorities established by direct valuation analysis (products or traits needed or desired at present and in the foreseeable future), (2) determine how much genetic "variety" (nonobservable diversity measured by biologists in laboratories) will be required to ensure survivability of the species selected, and (3) determine how much to invest in improving abilities to measure variety in a way that will meaningfully link variety to expression. The issue that became the focus of subsequent discussion was whether budgetary allocations should shift toward the funding of more research (figuring out how to better map genetic variability onto survivability and physical characteristics) or application of existing knowledge to management, planning, and policy development.

Noll's reading of the situation was that biologists and geneticists understand relatively little with confidence and that given a limited pool of funds and qualified workers, relatively more effort should be devoted to research. On this basis he suggested that preparation of genetic impact statements might deflect resources and talent from more important tasks.

139

Challenged by Noll's remarks, Antonovics and others recounted examples of theoretical relationships and general observations in which geneticists do have confidence, and of contextual data that might influence professional judgments in particular management situations. For example, if a species is normally outbred in nature, imposition of inbreeding is likely to produce symptoms of inbreeding depression, including lower fecundity and lower survival. Although the exact consequences cannot be predicted for particular species, there is ample evidence to predict trends. Suggested examples of other "safe" generalizations included: (1) genetically uniform populations of organisms are susceptible to disease epidemics; (2) biological control is more effective if the organism being controlled is not highly variable genetically; and (3) population size is a critical parameter of survival for endangered species.

Several more detailed illustrations of how genetic and biological information might be used to recommend management action were given in the course of the discussion. Woodruff summarized work on topminnows found in springs in south central Arizona, where the species is endangered by habitat disturbance and an introduced predator, and in northern Sonora, where it is common. Restocking of unoccupied springs from a hatchery population is being attempted by the Arizona Department of Fish and Game, but the plan might be improved using recent genetic data. A survey of electrophoretic variation in populations from both areas, by Robert Vrijenhoek (1985), indicates that a large portion of overall genetic variability in the species is due to differences between geographic clusters of populations (53 percent) and between the populations themselves (26 percent). On the basis of these data, ecological information, and general considerations on adaptation and inbreeding, Vrijenhoek recommended specific modifications of the restocking plan.

Riggs summarized an effort to integrate data from a variety of studies to recommend priorities for sampling Douglas-fir, an economically important forest tree species (NCGR, 1982). Evidence that local adaptation could be important to timber yield and viability of forest ecosystems provided the justification for establishing a high priority for sampling and conserving the genetic diversity of this common and widespread species. In the absence of a comprehensive study of genetic variation in Douglas-fir, data from a number of independent studies were used to develop a decision-making framework. Information on variation in isozymes, monoterpenes, and a large set of phenotypic parameters was available. Sampling priorities could have been established on any one of these traits alone, but the results would not have been in agreement. Variation patterns shown by the three different groups of genetic parameters were compared in an effort to develop a rationale for sampling some portions of the spectrum of genetic variability more closely than others.

There were various responses to the obvious need to establish priorities for research and conservation activities subject to budgetary constraint. It was sug-

140

gested that the effort to generate genetic impact statements on common species might be wasted unless there was already evidence that variants of such species possessed direct value or were endangered in some way. On the other hand, the present lack of any legal or administrative mechanism for recognizing and preserving genetic variants bothered several in the group. Someone observed that a movement among plant and animal systematists to split taxa so that formal species designation would confer protection on subspecific variants could be viewed as a necessary response to this situation. The possibility that impact statements could contribute at least a preliminary step in the process of direct valuation was also suggested. For example, the utility of metal tolerant races of plants for reclamation of mining spoils might be communicated in a genetic impact statement on certain species possessing such tolerance.

Again Noll imposed the perspective of the decision maker, pointing out that the core problem would not be in regard to endangered species, or common species of obvious economic importance, but in making trade-offs in expenditures on species in the middle—those neither common nor endangered and not yet evaluated for potential direct values. Under this proviso the biologists focused again on questions requiring further research. Illustrative examples of such questions were given, and more complete lists of research topics were solicited from participants later in the workshop. Reservations were expressed throughout the discussion about both the likelihood that information needed for development of conservation strategies could be supplied by research as rapidly as it is needed and the ability of policy makers to proceed without considerably more research. Stated in the starkest terms, the alternatives seemed to be to conduct research at the risk of deferring action until too late or to take action based on inadequate knowledge that has a high probability of being incorrect.

Antonovics enunciated three take-home recommendations from his perspective as a research biologist with conservation interests: (1) let other applied scientists worry about the common species, (2) require genetic impact statements for rare and endangered species, and (3) place priority on research for species in the intermediate category.

GENETIC IMPACT STATEMENTS

Antonovics suggested to the group that an institutionalized mechanism, a genetic impact statement, for considering these issues in a manner similar to that used in developing environmental impact statements would be beneficial in improving the state of knowledge and making better use of existing knowledge.

Hal Salwasser speculated that if genetic impact statements were required, a demand might be created for new kinds of information and ways of explaining cause-and-effect relationships. He noted that there might actually be a synergistic effect between impact statement preparation and research funding. Eric

Charnov suggested advantages for "intellectual opportunists," adducing, as an example, past discrepancies between the funding levels for coyote control and the study of factors important to coyote population dynamics.

Observations on the effects of EIS (environmental impact statement) and EIR (environmental impact report) requirements over the past fifteen years supported arguments both pro and con. William Kunin pointed out that voluminous EIS and EIR studies had been used for little else than fulfilling NEPA requirements. Without provisions for monitoring or follow-up studies, their conclusions could not be tested, and modification of development projects or management activities could not be justified. Riggs observed that constraints of time, budget, and personnel might often prevent the basic research biologist from pursuing those questions of greatest personal interest, and that past behavior of agencies made significant funding of associated basic research unlikely.

Comments on the complex and voluminous nature of EIS/EIR studies drew remarks from economists Brown and Noll. They noted that these features were endogenous—that is, explainable by the internal responses of organizations to procedural changes in public policy. Noll pointed out that this is itself an area of scientific inquiry, and recommended a recent publication by Serge Taylor (1984), describing how EIS requirements changed decision-making procedures within the Bureau of Land Management. He outlined the sequence of predicted responses: the procedural complexity that results from the legislated requirements increases the probability that disputes will involve litigation, and this in turn increases the need to anticipate information and documentation requirements, as well as the likelihood that appropriate expertise will be brought in earlier in the decision-making process. With this background, some of the costs and benefits of instituting genetic impact statements could be anticipated. A benefit would be that genetic principles and data would be considered at an earlier stage in the decision-making process. One negative effect might be a deflection of resources from other activities (on-the-ground management for example) and an increase in the overall expense of carrying out particular projects. Whether, on balance, the effect on preserving genetic resources would be good or bad is not easily predicted. Noll stated that such procedural changes are not to be recommended lightly; they call for the same kind of expertise and attention to detail that the impact statement itself would require in order to be worthwhile.

INSTITUTIONAL ROLES

Discussions of genetic impact statements, allocation of professional and monetary resources, and coordination of conservation efforts each focused attention on institutional roles in genetic conservation. Woodruff asserted that institutional impediments hamper progress on conservation, and he provided several examples. He observed that the species survival planning efforts of the Ameri-

142

can Association of Zoological Parks and Aquariums (AAZPA) and the survival service commissions of the International Union for the Conservation of Nature (IUCN) are limited by organizational priorities and the energy of volunteer professionals. He pointed out that conservation-related genetics falls through the cracks within and between funding agencies. NSF does not fund applied biology as a rule, while the National Institutes of Health (NIH) supports studies of the biology of animal models used in the study of certain diseases, leaving the investigation of exotic species to the National Science Foundation (NSF). Competitive grants that can be applied to conservation questions are available, but, Woodruff suggested, there is far from enough support in the areas requiring answers, and too much of the present funding goes to "fire fighting" (e.g., determining what to do with the last remaining California condors).

In the area of research funding, Ruth Shaw suggested that interest and action by scientists participating in the workshop might pave the way to designation of special funding categories and review panels in NIH and NSF programs. Schonewald-Cox noted that interest in similar goals had been expressed at a meeting of the Association of Systematic Collections and at the Second Conference on Conservation Biology, both held in 1985, and suggested that these and other groups might join forces to establish and lobby for priorities in pertinent areas. The establishment of a Society for Conservation Biology, and the creation of a new journal in which to publish conservation-oriented biological research, appeared to be consistent with many of the objectives voiced at the Lake Wilderness workshop. However, designation of separate review categories in research funding programs could backfire by reducing funds available to other important basic research topics and by increasing administrative rigidity.

An alternative to seeking formal recognition of conservation biology from traditional basic research funding sources, as Antonovics suggested, might be to push for sponsorship of conservation-related research by management agencies such as the Forest Service and the Bureau of Land Management. In a rather critical discussion of the present state of affairs, varied opinions were expressed about the roles and mandated responsibilities of the various agencies. Two points of view emerged as issues requiring further attention. On the one hand, some felt that the agencies should recognize the importance of basic research to their own ability to improve and adapt administrative guidelines and management procedures, and that they should fund competitive grants programs on that basis. Others argued that conservation biology should be recognized as a valid and important discipline by funding agencies traditionally supporting research.

A need for institutional support of activities other than research was indicated in comments by Schonewald-Cox, Riggs, Kunin, and others. In contrast to the present situation in the health professions, most biological scientists are asked to wear many hats—to serve on review panels, to seek their own funding, to conduct research, and to transmit their findings to audiences of varying levels of sophistication. Several examples of work that might benefit from more concerted professional attention were cited: (1) survey studies evaluating species and biological phenomena in ways that would permit interim management action on the basis of tentative "fuzzy" classification or generalizations; (2) integration of findings from diverse disciplines for the purpose of impact evaluation or other management and policy objectives; and (3) translation of research findings into a form easily applied by managers or planners. At least one commenter (Schonewald-Cox) called for institutional support of such activities.

An example of a relatively new institutional arrangement for bringing scientific and technical expertise to bear on planning efforts was outlined by Willa Nehlsen. She indicated how wildlife and fisheries management activities administered by the Bonneville Power Administration (BPA) in the Columbia River basin could be improved through adaptive management concepts introduced to the Northwest Power Planning Council by Kai Lee. The council operates under legislative mandate to integrate economic, social, scientific, and other considerations into resource plans recommended for implementation by BPA. One goal is to continuously test assumptions and interpretations of data used to justify management actions by careful design and execution of monitoring programs. It is hoped that recommendations to BPA's Fish and Wildlife Program will influence the way that BPA conducts its research programs and allocates money to research projects.

RESERVE DESIGN

Various comments throughout the discussion touched on the application of genetics to reserve design and management. Wilcox noted that data on species distributions and genetic variability could be used in identifying important locations for reserves and parks. Others did not hold out much promise for such efforts, pointing out that little is known about geographic patterns of overall genetic diversity and that, in any case, biologists have little to say in the selection of reserve locations.

Observing that the system of parks and reserves in the United States was established with little or no reference to distributions of species or species variants, Woodruff suggested that biologists must focus their efforts on recommendations to modify park boundaries and acquire the little remaining available land that might be most important to preservation of high priority species or populations. Ensuring that sufficient habitat is preserved and properly managed to maintain viable populations will also be a high priority, and, as was noted by both Woodruff and Schonewald-Cox, may be particularly critical for those species that are normally wide-ranging and outbreeding in nature.

REFERENCES

National Council on Genetic Resources (NCGR). 1982. Douglas-fir genetic resources: An assessment and plan for California. Department of Food and Agriculture, State of California.

Taylor, S. 1984. *Making Bureaucracies Think: The Environmental Impact Statement Strategy of Administrative Reform.* Stanford: Stanford University Press.

Vrijenhoek, R. 1985. Conservation genetics of endangered fish populations in Arizona. *Science* 229:400–402.

4

Ecological Uniqueness and Loss of Species

Gordon H. Orians and William E. Kunin

"When *I* use a word," Humpty Dumpty said, in rather a scornful tone, "it means just what I choose it to mean—neither more nor less."

"The question is," said Alice, "whether you *can* make words mean so many different things."

"The question is," said Humpty Dumpty, "which is to be master—that's all."

Lewis Carroll: *Through the Looking Glass*

When a species goes extinct, the alleles it alone possesses are irrevocably lost, and so are the values to human society of those alleles, actual and potential. Many ecological properties of species, however, while ultimately grounded in genetics, are not readily predictable from that perspective. The objective of this paper is to explore ways of determining ecological role uniqueness of species. Ecologically, we may call a species or population unique if it cannot be replaced in its role(s) by some other species or population. Since replacements can be more or less complete, uniqueness in this sense is a matter of degree.

Among sexually reproducing organisms, each individual is unique. This individual variability may be of ecological significance, especially in predator-prey interactions where chemical differences among prey individuals influence their palatability to predators and pathogens. Despite the increasing recognition of the importance of uniqueness at this level, we concentrate on the species, or well-marked geographical races of species, as our unit because it is the loss of such units that we wish to evaluate in ecological terms.

UNIQUENESS AND MULTIPLE CURRENCIES

Given the wide array of different "roles" performed by any species, and the disparate units in which they are tallied, a discussion of overall ecological uniqueness requires a common denominator into which all roles can be translated. We have chosen to use as that common currency the effect that the species in question has on the populations of other species. The removal of a highly "ecologically unique" species would therefore result in major population shifts, perhaps even extinctions, in many other members of its community. This operational definition of uniqueness in terms of effects on other populations differs from other uses of the concept. For example, by this definition rare organisms are generally less unique than common ones (Allen and Forman, 1976), while in normal usage rarity often connotes a sense of uniqueness. In addition, our notion of ecological or functional uniqueness ignores aesthetic or economic considerations, which must be part of a complete measure of value.

The focus of the problem posed in this paper is the deletion sensitivity of natural communities. If one or more species are lost from a system, what will be the consequences for the remaining ones? After providing an introduction to the kinds of "roles" that organisms play in ecosystems, we summarize the state of development of formal deletion sensitivity theory, pointing out its strengths and limitations. We then provide an overview of the major kinds of interactions among organisms (predator-prey, competitive, mutualistic) and how those inter-actions yield patterns in nature that help us contemplate the significance of loss of particular species. Finally, we consider the problem of which species are most likely to be lost, how those probabilities affect the patterns of change we are likely to see, and the management implications of these possibilities.

Ecological Roles

For convenience, we use the term "role" to refer to the activities of a species without any of the teleological implications often associated with that word. Because of major differences in the activities they carry out, we introduce roles separately for photosynthetic plants, microorganisms, and animals.

Photosynthetic Plants. Collectively, photosynthetic plants are the earth's over-whelmingly important producers of organic carbon molecules from simple pre-cursors. They account for about 97 percent of all carbon fixed in terrestrial and aquatic communities, and their photosynthetic activity is the foundation for the energetics of all other components of ecological communities. Vascular plants carry out photosynthesis by one of three principal mechanisms—C_3, C_4, and CAM—details of which can be found in Black (1971), Bjorkman and Beery (1973), and Beery (1975).

Nearly all woody plants at all latitudes and herbaceous plants of high and mid-latitudes use the C_3 pathway—the process that performs best under moderate temperatures. C_4 plants are primarily grasses and small shrubs of hot climates and are especially well adapted for photosynthesizing under hot, moist conditions. CAM plants have relatively low photosynthetic rates, but they are unusually economical in their use of water, and thus particularly successful in hot, dry environments.

Because so many plant species use each of the three primary pathways of photosynthesis, and because all use the same nutrients, many species would have to be lost from most ecological communities before there would be a failure of

147

replacement of their roles as photosynthesizers. The exceptions occur in poor, unusual, and extreme environments such as (1) sites with high levels of toxic metals (natural or mine tailings), (2) soils of extreme pH or salinity, and (3) sites with extreme temperatures (hot springs, surfaces of glaciers and snowfields, very arid deserts). The key feature of these "extreme" environments may be that they are very different from the more prevalent ones surrounding them. Thus the "infertile" serpentine soils, which usually outcrop in small patches, generally support an impoverished vegetation, whereas in New Caledonia and parts of California, where soils derived from serpentine are widespread, those soils support very rich plant communities (Kruckeberg, 1985; Proctor and Woodell, 1975).

Although plant species are likely to be highly replaceable as photosynthesizers in most environments, with respect to other features their substitutability (replaceability) is not as extensive. Photosynthetic plants differ from one another in their physical structure, in the depth in the soil from which they tap water and mineral resources, in the characteristics of their litter, in the nature and extent of their accumulations of specific elements, in the physical and chemical characteristics of their tissues (leaves, stems, wood, roots, flowers, fruits), and in the nature and extent of their participation in coevolved mutualisms. Because of these characteristics, plants are involved in a rich variety of interactions with other organisms, and they also provide the physical structure within which these interactions take place. Therefore, losses of particular species of plants could result in measurable effects on many processes and species in ecological systems.

Animals. Animals all require preformed, high energy organic compounds for food. Therefore, unlike plants, much of their diversity relates to what they eat, how they find it, and how their foraging activities influence both their prey and other species in their environments. Animals have precise habitat requirements, are involved in diverse competitive interactions, and participate in a variety of coevolved mutualisms with plants (pollination, fruit dispersal, plant protection), other animals (mimicry, protection), and microorganisms. For the most part, although animals are diverse in structure, they do not exert the same communitywide effects via their structure as do plants, except in some rocky marine environments where animals rather than plants actually compete for space and form the dominant structural elements of the communities in which they live. Animals exert most of their influence through consumption of their prey.

Microorganisms. Unlike plants, microorganisms are very diverse in their synthetic pathways and use a wide variety of chemical substrates as sources of energy and carbon. For example, halobacteria, which live in exceptionally salty environments, use retinal, a carotenoid pigment also found in the vertebrate eye, as their photosynthetic pigment (in place of chlorophyll). Other bacteria derive energy not from light but from such simple reduced substances as hydrogen sulfide (H_2S) taken from solution around them. These special synthetic abilities,

148

although they are of rather minor importance in the overall energetics of the world ecosystem, are important in providing evidence of the possible pathways of evolution of the synthetic abilities of more complex organisms. In special, rather rare environments, such as in deep-sea hydrothermal vent ecosystems, microorganisms are the primary or only synthesizers of carbohydrates, and are therefore irreplaceable.

Microorganisms also have special abilities to synthesize compounds required by many complex plants and animals (nitrogen fixation, synthesis of vitamins), and they may be the only organisms capable of breaking down important biologically synthesized molecules (cellulose, waxes, lignins). For these reasons, they often participate in mutualisms with plants and animals. Although they are generally inconspicuous and do not attract attention, except when they spoil food or cause illness, some microorganisms would be harder to replace ecologically than the organisms to which we usually direct our attention.

Time Scales of Analysis

Species deletion effects may vary greatly depending on the time scale of the analysis. If a species is removed, quick adjustments among associated species can occur as a result of physiological and behavioral changes affecting, say, their habitat distributions or food intake rates. Over periods long enough to encompass a number of generations of associated species, functional replacement of a species might occur through population growth of other species. Interspecific differences in generation time, hence in population responses, complicate the issue in practice.

Among organisms with short life cycles, population responses to perturbations may occur very rapidly. Davidson et al. (1980) excluded rodents alone, ants alone, and both rodents and ants from experimental plots in the Sonoran Desert. Ant and rodent densities each grew quickly where the other group had been eliminated. But gradually selective seed predation by the remaining species altered the composition of the plant communities in favor of those plants less suitable as food sources for the dominant animals, showing that immediate responses are not necessarily good predictors of longer term ones (Davidson et al., 1984). Similar changes in plant communities dominated by long-lived woody species might take several centuries.

A species may be poorly replaced in one of its roles over time, yet eventually be replaced to some extent by the evolution of associated species. While this process may be important over a very long period, such responses are too slow to be of management significance, except when organisms with very short generations are involved.

DELETION SENSITIVITY OF ECOSYSTEMS

Natural ecosystems are subjected to a variety of perturbations that differ in kind, intensity, and frequency. Some perturbations change only the abundance of species. Others completely remove one or more species from the system for long periods. Common disturbances, such as fluctuations in weather, fires, and windstorms, usually affect a number of species simultaneously. Ecologists hold diverse opinions about the relation between the complexity and nature of connections among species in ecosystems and the sensitivity of those systems to disturbance. Some have asserted that simple ecosystems with few species are much less stable than more complex ecosystems (Elton, 1958; Hutchinson, 1959; MacArthur, 1955; Margalef, 1968; Watt, 1973), while others have suggested the reverse (May, 1973; Gilpin, 1975; Pimm, 1979; Goodman, 1975; but see Lawlor, 1980). This diversity of opinion reflects the rather primitive state of development of predictive ecological theory, the use of different definitions of stability, and different ways of thinking about perturbations.

We begin with a consideration of one particular type of perturbation: the loss of a species. Focusing on species deletions is both an appropriate and a theoretically tractable approach to our study of species roles. The most explicit theoretical models of deletion sensitivity have been developed and analyzed by Pimm and his associates (Pimm, 1979, 1980, 1984; Pimm and Lawton, 1977, 1980). These models have used a set of Lotka-Volterra equations with parameters randomly selected over values designed to reflect both patterns of interactions in the food webs and reasonable biological restrictions. The models analyzed varied in the number of species in the webs and in the degree of their connectance (the number of actual interspecific interactions divided by the number of possible interspecific interactions). Of all possible webs, only those with locally stable and positive equilibrium densities for all species were retained for further analysis. From this set, each species was removed in turn and the resulting model checked for feasibility (all species have positive equilibrium densities that are locally stable), singularity (predators without prey, two predators requiring the same prey, etc.), and stability.

Pimm restricted his analysis to cases in which predators are much larger than their prey, hence have much greater per capita effects on their prey than the reverse. Many trophic interactions—especially insect-parasitoid, insect-plant, and host-parasite interactions—do not have this property and may well behave differently. Moreover, Pimm's models include neither interference competition nor mutualisms. Models with these properties have not yet been developed.

Under the conditions included in Pimm's models, the following statistical results emerge: (1) The average species deletion stability decreases with increasing complexity when complexity is measured by connectance. (2) Average species deletion stability decreases with increasing species numbers when connectance is held constant. (3) Plant species removals cause least loss when the herbivores dependent on the plant feed on a variety of other species. (4) The more generalized the diet of an herbivore, the less the effect of removing that herbivore. (5) Predator removal results in more species losses when the herbivores on which the predators feed are generalized rather than specialized. (6) Whether a predator is specialized or generalized has little effect on the consequences of its removal.

Removal of predators has a much greater effect when prey densities are normally controlled primarily by their predators (predator controlled) than when the prey are close to their own environmental carrying capacities, so that most individuals would die of resource shortages whether or not they were taken by predators (donor controlled). Pimm (1980) suggested that donor-controlled communities or community components are rare in nature, except among detritivores, but in fact they may be quite widespread and important in ecosystems. Included are all forms of exudates (nutrients leaking from stressed cells, root exudates, aphid honeydew, nectar), the rewards offered in many mutualistic associations (fruits, food resources provided to ants by many plants), eggs, seeds, detritus, sunlight, and plankton for sessile animals.

EVIDENCE FOR DELETION SENSITIVITY IN ECOSYSTEMS

Deletion sensitivity can be examined from a number of perspectives. We have chosen to follow the standard divisions of ecological interactions into competition, predation (broadly conceived to include parasitism), and mutualism. We will review the current status of theory and data in each of these three areas to suggest some implications for deletion sensitivity and to note where additional research is especially needed.

Understanding the consequences of deletion of a particular species is difficult, not only because of the indirect effects the focal species may have on other components of the community but also because nonlinearities characterize many ecological interactions (e.g., density dependence, keystone relationships, and processes influenced by spatial and temporal scales of patchiness in distributions of species and habitats).

COMPETITION

Competition occurs "when a number of animals (or plants) . . . utilize common resources, the supply of which is short; or if the resources are not in short supply, competition occurs when the animals seeking that resource nevertheless harm one another in the process" (Birch, 1957).

Competition can be either *intraspecific* (the competitive effect of individuals

151

of the same species on one another) or *interspecific* (where the individuals competing are of different species). Since we concentrate here on the effect that removing a particular species has on the remaining members of its community, whenever we mention "competition" without modifiers in this section, we refer to interspecific competition.

Where only two species are in competition, the removal of one of them will generally cause the other's population to increase, and the strength of the deletion effect can be measured as the magnitude of that response. When more than two species are in competition, the issues raised are more complex. When one competitor is removed from such a system, how is the "stock" of competitive release parceled out among the various remaining species? Does one species generally take up all the slack, or do all of the competitors register some moderate gains? If one species monopolizes the gains, will it necessarily (or even usually) be the species that was most abundant? That is, does the relative success of a species in coexisting with a competitive dominant predict its ability to succeed competitively once that dominant is removed? Can a species overcompensate for the loss of a competitor, actually driving down the populations of other competitors (acting as a "keystone competitor?" e.g., Kastendiek, 1982)? While conceivably important in competitive systems, these issues remain largely unexplored, and are raised here more to stimulate interest than to promise answers.

The *direct* effect of the removal of a species on its competitors is always positive. Thus direct competitive interactions are of greater practical concern in cases of species introductions than in cases of deletions. If the *removal* of a species is to result in extinctions among its competitors, strong indirect effects must act. The species at risk are not likely to be the ones that directly compete with the species removed, but rather ones that are preyed upon by the competitors of that species, or that compete with them. Thus the removal of damselfish from a reef community allows other herbivorous fish access to previously defended territories, resulting in a reduction in algal species diversity (Hixon, 1983). Because such complex interactions have received so little study, however, we examine only the consequences of direct competition in the hope that it will at least reflect the potential for more intricate interactions.

Several excellent reviews of the competition literature have emerged in recent years, spawned largely by a heated debate about the importance of interspecific competition in natural communities. Connell (1975, 1983) and Schoener (1983) both reviewed previous field experimentation on the subject. Their papers, while similar in subject and scope, differ on certain procedural points and often arrive at noticeably different conclusions. Connell's (1983) analysis uses a narrowly defined group of studies further reduced by restrictive criteria (e.g., generally excluding from consideration studies of a single species at a single site and time), whereas Schoener (and Connell, 1975) uses a broader sample of studies. Thus the two 1983 papers may provide a crude confidence interval, giving minimal and maximal estimates of the ubiquity of competition in the current experimental literature.

Competition and Trophic Organization

Competition generally occurs only among organisms of the same trophic level (herbivores compete with herbivores, carnivores with carnivores, etc.), although exceptions exist, especially among organisms competing for space. Several authors have suggested that the likelihood of strong competition varies systematically between trophic levels. Hairston, Smith, and Slobodkin (1960; hereafter referred to as HSS) reason that, given the fact that dung, detritus, and dead bodies rarely accumulate, decomposers as a group must be close to their collective carrying capacity, and thus should commonly compete for food. Similarly, plants (or producers generally), which tend to utilize fully their limiting resources (generally space and light, but sometimes water or particular minerals), are prime candidates for competition. Herbivores, on the other hand, particularly ones that eat the plants themselves rather than "plant products" like seeds, pollen, or nectar (Slobodkin et al., 1967), seldom exhaust their resources under natural conditions, although they may do so if protected from predation. If predation keeps herbivore populations well below carrying capacity, predator populations should, in turn, be "food limited," with competitive interactions between them fairly likely (Figure 10a).

The HSS hypothesis, with its sweeping generalizations, has been widely cited and often criticized (Murdoch, 1966; Ehrlich and Birch, 1967), but only with the recent reviews by Connell (1983) and Schoener (1983) has any evaluation been possible. Schoener found strong support for the HSS hypothesis in both terrestrial and freshwater environments, but only weak support among marine studies. Competition among terrestrial carnivores was, however, less common than expected, with most of the exceptions being spiders. Schoener suggests that their small size and intermediate trophic position may be responsible.

Connell, on the other hand, found no support for HSS among the cases he analyzed. In terrestrial systems, competition was, if anything, less common among the carnivores in Connell's data set than among the herbivores. His sample, however, includes only nine terrestrial phytophagous herbivores, all of which were insects living on *Heliconia* plants (mostly species that Hairston, 1985, suggests should not be termed phytophagous at all). Schoener's sample, while it includes a wider array of taxa, contains only twelve species of terrestrial phytophages. Also underrepresented in these surveys are detritivores; Schoener lists only four studies on terrestrial detritivores, involving seven species altogether (all of which showed competition); Connell includes none.

Menge and Sutherland (1976) note the tendency for predation pressure to reduce competition among prey species (e.g., Paine, 1974) and hypothesize that the importance of competition should rise with trophic level (Figure 10b). They *a. HSS Model:*

PLANTS ——> herbivores ——> **CARNIVORES** (——> **2nd CARNIVORES?**)

b. Menge and Sutherland Model:

plants ——> **Herbivores** ——> **CARNIVORES** ——> **2nd CARNIVORES**

c. Varley Model, Short Food Chain:

PLANTS ——> herbivores ——> **CARNIVORES**

d. Varley Model, Longer Food Chain

plants ——> **HERBIVORES** ——> carnivores ——> **2nd CARNIVORES**

FIGURE 10. Diagrammatic representation of trophic patterns in competition. The larger and bolder the printing in the food chain, the stronger the hypothesized competition at that trophic level.

further suggest that competition should be especially noticeable in trophically simple communities. The first prediction is similar to that of HSS, except that plants, being low on the food chain, would be expected to show little competition under this hypothesis. Neither Connell's nor Schoener's surveys support this prediction.

Another model, proposed by Varley (1959), is based on an analysis of the success of biological control agents. He suggests that top carnivores, lacking predators of their own, should generally hold down the populations of their prey, become food limited, and consequently compete with one another. The prey's prey (the third rung down the trophic ladder) will be unlikely to be controlled by predation, because of the second group's extrinsically limited population. Thus, Varley suggests, competition should be important on alternate links of a food chain (Figure 10c, d). When food chains are three species long, as is often the case (e.g., Cohen, 1978), the predictions of this model closely resemble those of HSS. However, where predators are themselves preyed upon by second order predators (as in the spiders mentioned by Schoener), Varley's model would predict that competition among them should be eliminated, directly contrary to HSS's suggestion.

The one trophic pattern in competition that seems clearly agreed upon does not follow from any of the above models. Competition seems to be particularly light among "partial predators" such as parasites and small herbivores (Price, 1980), organisms that consume parts of their prey without necessarily killing them. The presence of one parasite species on a host may have very little detrimental effect on another species that parasitizes it in a different way.

Indeed, under some conditions, one parasite may weaken the host's defenses sufficiently to improve conditions for other parasites (Rhoades, 1985; Kareiva, 1985). Where such interspecific facilitation occurs, competition between partial predators should occur only at extremely high population levels.

Other Patterns in Competition

It has long been suggested that competition should be strongest between closely related organisms, since they are likely to use similar resources in similar fashions (Darwin, 1859, chap. 3; Lack, 1954). Many studies demonstrated patterns predicted by this generalization (e.g., closely related species cooccurring on islands less frequently than should happen by chance; Grant, 1966, but see Simberloff, 1984), but strong competitive interactions have also been demonstrated between very different organisms: ants compete with rodents (Kodric-Brown and Brown, 1979), moose compete with hares (Belovsky, 1984), and so forth. There have not, however, been good experimental tests of the notion that, among a guild of species sharing limited resources, competition should be commonest and strongest between those most closely related.

Perturbation and Competition

The size of an organism may be a reasonably good predictor of its competitive interactions. Connell (1975) reasoned that some populations may be held below competitive levels most or all of the time by "harsh physical conditions or natural enemies," and further inferred that vulnerability to perturbations is inversely related to size, suggesting that small animals should compete less commonly than large ones. Invertebrates in Connell's (1983) sample (which tended to be relatively small) showed a lower average incidence of competition than vertebrates, although the difference was statistically significant only among marine organisms. Schoener categorizes organisms by size, and finds similar trends. Social Hymenoptera (ants and bees) seem to be an exception; competition among them appears almost universal.

In Wiens's (1977) examination of the effects of environmental fluctuations on competition, competition was evident only during unusually hard times, when resource supplies are low—the reverse of Connell's (1975) prediction. The difference lies in the relative vulnerability to perturbation of consumers and their resources. Where consumers are more impervious to perturbations than their prey are (as is presumably true of Wiens's birds), bad times will bring on competition. If the reverse is true (as is probably the case for many herbivorous insects), competition is more likely during periods of favorable weather.

Asymmetries in Competition

A consistent finding of the recent reviews of competition is that competitive interactions are seldom symmetrical. Lawton and Hassell (1981) found strong

155

asymmetries in two-thirds of thirty-five studies of insect competition they examined. Connell (1983) found that in 61 percent of fifty-four field experiments
examined where competition was evident, the effects were so asymmetrical that
the removal of one species had a strong positive effect on the population of a second species, but removal of the latter had little or no effect on the first species.
Schoener's (1983) sample included fifty-one studies demonstrating strongly asymmetrical competition, three with "slight" asymmetries, and only seven where
interactions were clearly symmetrical. Thus strong asymmetries in competition
are, in Connell's words, "the rule rather than the exception."

Such a result is not surprising. Laboratory experiments have consistently demonstrated that if two species compete in a uniform environment, one generally
monopolizes the resource in question, eventually excluding the inferior competitor (Gause, 1934; Park, 1948). The real world, of course, is not perfectly
uniform, and different species may dominate competitively in different shared
roles or niche components. If, however, one species is always dominant over
another where they compete, the subordinate species may be forced to retreat
ecologically into those portions of its "fundamental niche" where it does not
compete with the dominant species. If so, removing the dominant species should
allow a major increase in the subordinate species' population, whereas removing the subordinate would leave the dominant relatively unaffected. Thus, in
a complex world, mutually competitive interactions are likely to be strongly
asymmetrical.

The only well-documented predictor of competitive dominance is body size.
Where two species of markedly different sizes compete for a resource, the larger
is almost always dominant. Schoener finds twenty-seven studies that support
this generalization and only five that contradict it. This tendency is particularly
strong among terrestrial vertebrates. Connell also finds size differences important
in predicting competitive asymmetries, but suggests that such differences matter
most among sessile organisms competing for space.

The main advantage conferred by size in these interactions probably comes
in facilitating interference competition where an individual directly excludes a
competitor from access to a resource, either by threatened or actual aggression
(in the case of motile organisms) or by poisoning or growing over it (among
sessile organisms). Where competition was described as asymmetrical in favor of
the larger of the two species, in the cases cited by Schoener, 68 percent of the
mechanisms listed fell under the heading of interference competition; among the
few interactions favoring the small species, such mechanisms accounted for only
25 percent of the citations. When the cases where a large species dominated
were compared with all other cases of asymmetrical competition examined by
Schoener, interference competition methods were significantly more common
among the former ($\chi^2 = 7.07$; $p = .004$).

While larger species may be able to exclude smaller ones from resources, smaller organisms, having lower per capita energy needs, may be better able to cope with low resource densities (although this need not hold where food acquisition and processing scale with size, as in some groups of folivores). This would provide small species with the niche refuge required for coexistence. By analogy, the larger competitor may be able to monopolize access to the cream; but if it cannot survive on milk, it can never fully extirpate its smaller rival. Recent studies of asymmetrical competition between moose and snowshoe hares (Belovsky, 1984) and among Hawaiian honeycreepers (Pimm and Pimm, 1982) describe precisely this dynamic. If this pattern proves to be common, it has some important repercussions; if large, dominant competitors are restricted in general to high quality resources, they will be the first affected by overall reductions in environmental quality. Thus, paradoxically, it could be the best competitors that are most in jeopardy of extinction. This apparently has occurred among the honeycreepers of Hawaii (Pimm and Pimm, 1982).

Overall Patterns in Competition

In summary, the following patterns emerge from the literature on interspecific competition. Some are reasonably well documented; most remain largely speculative.

1. Competition seems to be stronger in certain trophic levels than in others. Terrestrial and freshwater herbivores (and in particular phytophages) appear to exhibit competition less frequently than other groups do. Several theories have been advanced to explain this.

2. Parasites seem to compete very little except when population densities are extremely high, and they may display interspecific facilitation at intermediate densities.

3. Closely related species may compete especially strongly.

4. Small animals compete less than large ones do (or: invertebrates compete less than vertebrates). Size-related vulnerability to environmental perturbations may be involved.

5. The strength of competition commonly varies over time and space. In theory, at least, competition may peak during either particularly harsh or particularly benign periods, depending on the relative vulnerabilities of predator and prey to perturbations.

6. Competition is often extremely asymmetrical in effect. Where competitors differ markedly in size, the larger one is generally competitively dominant. Where they do not, very little is known.

Several of these points can be brought together into a single profile. To paraphrase Colinvaux (1978), *big fierce animals compete*. Large animals are relatively invulnerable to weather and predation, and they are usually behaviorally dominant over their smaller competitors, perhaps being predators on them as well. Thus the extinction of large animals, especially carnivores, should send competitive shock waves throughout the guild, causing significant changes in the populations of other species.

The contrasting extreme is the small phytophagous herbivore. If the generalizations discussed above hold true, the extinction or introduction of such a species should have virtually no competitive effect on other species. There should be relatively little risk of competitively induced extinctions of innocent bystanders if we intentionally perturb a particular herbivore population. However, this implies that biomass lost when such a species disappears is not replaced by increased growth of similar species. Thus predators of such guilds may encounter reduced food supplies as a result.

Ultimately, our concern is not simply to predict the strength of competition, but to understand the conditions under which the introduction or removal of a species from a community will cause extinctions or other adverse consequences for its competitors (or, secondarily, for other species). For the most part, we can shed very little light on this issue at present. Future research into some of the questions raised above may bring us closer to such understanding.

PREDATION

While competitive interactions are often asymmetrical, trophic ones always are. The loss of a predator has very different implications for prey populations than the prey's extinction has for the predator. Consequently, we deal with prey deletion effects and predator deletion effects separately.

Uniqueness of Prey

How will the loss of a prey species affect its predators? Assuming no competition or other complicating factors, the general answer is clear: impacts are almost certain to be negative. Even if the predator feeds on a wide range of species, the loss of one cuts the hunter's options and makes food that much harder to find. The effect may be vanishingly small, but where it exists it will be negative. Only if a predator preys on a group of species that compete in a strongly asymmetrical manner among themselves (or that compete symmetrically but that differ markedly in their value as prey) is any positive impact of prey extinction possible. The loss of a competitively dominant (or relatively inedible) species could theoretically cause an increase in the population of another prey species of such magnitude that the predator's lot would actually improve, but we know of no documented case of such an effect in a natural population.

Although the polarity of prey deletion effects is reliable, their magnitude is not. Prey deletion impacts may be as much a function of the characteristics of the predator involved as of intrinsic attributes of the prey species. The remainder of this section will examine a number of plausible but largely untested predictors of such impacts.

158

Food Limited Predators

For the loss of a prey species to have much impact, food must be a limiting commodity for the predator populations affected. Many factors may control a particular population, either simultaneously or at different times, and the lines are not generally sharp, either in theory or in nature (but see Murray, 1979). Even so, some populations are relatively insensitive to changes in their food supply while others seem to track food closely. The former are unlikely to be affected greatly by the loss of a prey species, unless reliance is so great that the loss of a species would create food limitation where there had been none previously. A monophage (a predator that eats only one type of prey) will quickly find its population to be food limited if its prey species is removed, regardless of what might have regulated its population previously. In general, though, the deletion of prey of food limited predators should have a greater impact than deletion of prey of nonfood limited predators.

If the HSS hypothesis is borne out, or those of Varley (1959) or Menge and Sutherland (1976), the trophic level of a predator (and thus, by implication, the trophic level of its prey) should help predict the deletion impact of its prey. Plant deletions in general should have a lower potential for impact than deletion of herbivores, all other things being equal (which they are not likely to be). Similarly, we expect high deletion impact for organisms eaten by large-bodied predators, and for prey species very similar in environmental vulnerability to their predators.

Specialized Predators

Almost by definition, the more a predator is dependent on one or a very few prey species, the greater the trophic impact of loss of these prey. This dependency is likely to be particularly strong: (1) when the predator is monophagous; (2) when the prey species is an important source of some necessary component in the predator's diet (Westoby, 1977); in this way, some prey may be termed "nutritionally unique"; and (3) when the prey species is important during some temporal or developmental bottleneck faced by the predator. Certain prey species may be available to predators at certain times of the year, under particular environmental conditions, or at stages of the life cycle when other sources of food are scarce or hard to procure.

Monophagy has been much discussed, but the other two cases require more study. As an initial guess, plants are probably more nutritionally unique to their predators than animals are. Plants differ strikingly in their mineral, vitamin, and toxin contents (in part because of the availability of chemical defenses), but animal tissues are fairly conservative chemically. Plants known to have unusual nutritional characteristics are prime candidates for this sort of trophic importance.

Similarly, prey species that are temporally unusual may be high in deletion impacts. A predator must be able to find food throughout its active life. In environments where seasonality imposes a degree of synchrony on prey populations, those species that are out of step with the majority and thus available during what Wiens (1977) calls "crunch" times may be of particular interest in terms of deletion impacts. As an example, Terborgh (1986) finds that only about ten "keystone" plant species support about 85 percent of the mammalian biomass of a Peruvian rain forest through the dry season. Animals are also behaviorally diverse, and predator specialization may be induced by difficulties in capturing prey with varied and unusual escape behavior.

Monophagy

The theoretical patterns of dietary specialization are discussed below.

1. Dietary specialization should be most common in stable habitats where conditions and thus resources fluctuate over narrow ranges. The available evidence on this issue is sketchy. Futuyma (1976) finds no trend toward dietary specialization among Lepidoptera of the humid tropics (commonly supposed to be climatically stable); Janzen (pers. comm.) does.

2. Monophagy should be prevalent among species whose food supplies are superabundant, presumably because their populations are limited by some factor other than food. Foraging theory suggests that, where nonfood factors are limiting, organisms should feed only on that subset of available foods that can be caught and consumed most efficiently, thereby leaving the maximum possible amount of time and energy for more pressing tasks (Tev, 1970; Estabrook and Dunham, 1976). Note that these conditions, where food is available in excess, are precisely those under which we suggested above that removing prey species should have little effect. If such conditions result in the evolution of narrow food specializations that are not labile over time, however, it could be precisely these species that would be most sensitive to loss of prey species.

3. Monophagy should be particularly common among predators of prey species that require special adaptations to capture or eat. Thus prey species that have unusual defenses, either physically, behaviorally, or chemically, should be disproportionately likely to have unique trophic roles. For plants at least, there is a growing body of evidence (e.g., Feeny, 1976; Rhoades, 1979) suggesting that "inconspicuous" species, rare or ephemeral plants, tend to have such "qualitative" defenses, and thus presumably should be high in deletion impact. In animals, unusual defenses seem to be most prevalent among sessile, communal, and slow moving species (e.g., social Hymenoptera, porcupines). Looking in particular at chemical defenses, the greater the taxonomic (or rather evolutionary) distance between predator and prey, the more feasible it may be for the prey to develop chemical defenses against predation. It would be difficult for a bird to develop chemical defenses against hawks without poisoning itself in the process, while

a plant or an insect may easily be able to evolve the capacity to store poisons against birds.

4. Levins and MacArthur (1969) point out that predators and, especially, parasites that must depend on simple criteria to distinguish between acceptable and unacceptable prey should be particularly prone to monophagy. Essentially, they argue that under many conditions it is better for the predator to adopt narrow selection criteria, thereby missing some appropriate prey, than to risk attacking prey that are inappropriate. The prey of predators with simple nervous systems may thus be expected to have rather high deletion effects.

5. There should be more species of specialists that are predators on common species than on rare ones. This argument is based on application of MacArthur and Wilson's (1967) theory of island biogeography, in which host species are viewed as resource islands to be colonized evolutionarily by predators. The larger the island, in theory, the more predators are likely to have colonized it, and the less likely those predators are to go extinct. Several recent studies of herbivorous insects (reviewed by Strong, 1979) lend this prediction a degree of empirical support. Thus common, widespread prey species may have generally lower trophic replaceability (and thus higher deletion impact) than otherwise similar species.

Before concluding our discussion of prey deletions, we return briefly to one point alluded to at the beginning of this section. Competition among species with a common predator will tend to make all the participants more trophically replaceable in their role as prey. We expect a certain degree of what Pimm (1984) has termed "biomass stability" in a community of this kind, where the loss of one prey species will result in a compensatory gain in the population of another, leaving the predator's food supply relatively unaffected. In an extreme case, the predator could even find a net benefit in the transaction. If such assemblages are, as seems likely, relatively stable in the face of extinctions, these predation-competition complexes should be well represented in natural communities. To our knowledge, little or no work has been done on the ecology of such groupings.

Predator Uniqueness

As we have defined the term, for a predator to be unique in its trophic role it is not enough for it to eat something that hardly anything else eats; it must also have an effect on the population dynamics of its prey. The removal of a predator that feeds on nonpredator-limited prey could have little or no impact on the prey's population density, but simply shift the patterns of its mortality (and perhaps its selective environment). For example, the extinctions of the passenger pigeon (*Ectopistes migratorius*) and the Carolina parakeet (*Conuropsis carolinensis*), and the virtual elimination of the wild turkey over much of its range, removed most of the dominant large seed predators from the forests of

161

eastern North America. While the levels of predation on acorns and hickory nuts must have dropped precipitously, there may be no more oaks and hickories growing in the forests today than when pigeons, parakeets, and turkeys were abundant. Tree populations seem to be limited by the availability of space, not by seed predation.

Where prey *are* predation limited, their populations should rise if a predator is removed. Populations of the prey's prey might also fall, if they, too, were predation limited, but that seems unlikely if there is an alternating trophic level pattern of population regulation. In a *noncompetitive* world, then, the most likely effect of removal of a predator would be an increase in the density of its prey. The prey's prey, in turn, is expected to decrease because its predator has become more common.

Interspecific competition between prey species makes the picture considerably more complex, however, because the removal of the predator of a predation-limited species may cause its population to increase at the expense of its competitors. Even where the prey's population was not strictly limited by predation, in a competitive world predation could serve to shift competitive balances, so that, in its absence, the former prey species would find its position improved. Having both predator limitation and competition among prey species requires that competition occur despite the availability of excess resources. This is often the case among sessile organisms, where local competition for space occurs even if there is much unoccupied space elsewhere.

Both theoretical and experimental studies show that predation may affect not only the population density of prey but their species diversity as well. To the extent that predators play such "keystone" roles in fostering coexistence, the loss of a predatory species may result in the virtual or complete elimination of other species that compete with its prey. There are several theoretical explanations of this effect:

1. When the population of a competitive dominant is predation limited, the presence of a specialized predator may allow the persistence of species that would otherwise be excluded by interspecific competition. Similar effects may occur if predation pressure is proportional to prey density, since in such cases the commonest (dominant) species always bears the brunt of the predation load (Roughgarden and Feldman, 1975). Since some features that improve competitive abilities may also increase vulnerability to predation (e.g., a fast growth rate and low investment in energy-costly defenses), specialization of predators on competitive dominants may not be uncommon. Similarly, the "island biogeography" and "conspicuousness" effects alluded to above may make competitive dominants particularly vulnerable to specialized predators.

2. Predation may reduce overall densities of lower trophic levels to the point that interspecific competition is no longer important. In this way, even a gen-

eralist predator could augment prey species diversity if predation pressure was sufficiently intense (Hanski, 1981; but see Yodzis, 1977).

3. Predation can serve as a source of disturbance, especially when it acts upon sessile or space-limited prey species, setting in motion a successional progression. Intermediate levels of such disturbance should result in greater diversity than either very high or very low levels do (Connell, 1978; Abugov, 1982).

We have examined a large but unsystematic sample of recent literature on predator removal experiments (see Appendix) to search for such patterns. The results are far from conclusive, but they do allow some tentative analysis of the following questions.

1. Is the type of predation (e.g., specialist, generalist, switching) an important factor in the occurrence of predation effects on prey species diversity? While many of the cases of predator-mediated coexistence examined involved predators with strong prey preferences, generally, for a competitively dominant species, no overall generalization holds. In several studies (e.g., Hurlbert and Mulla, 1981; Harper, 1969) grazers maintained diversity among prey despite their relatively generalized eating habits. Pimm (1980) surveyed a number of predator removal experiments and concluded that the degree of predator specialization has no effect on its keystone status.

2. Does the "keystone predator" effect occur only or primarily where the prey are sessile (thus making possible the sort of successional disturbance effect discussed above)? A survey of recent studies indicates that in the majority of cases where a keystone role was discovered, the prey were indeed sessile, although keystone effects have been found among motile organisms as well (e.g., Morin, 1983). This pattern may, however, be an artifact of the general predominance of studies involving sessile prey in the literature.

3. Does the background level of disturbance affect the result of predation? If the intermediate disturbance hypothesis applies, predation should augment diversity where other sources of disturbance are few, while it should reduce diversity when exogenous disturbance is high. Among marine environments, for instance, exposed rocky intertidal zones, subject to periodic desiccation, rain-induced osmotic stress, wave shear, and surf-propelled flotsam seem to face a higher background disturbance regime than either protected intertidal habitats or rocky subtidal habitats. Lower still in background disturbance should be benthic mud habitats. Yet a survey of marine predator manipulation studies gives little indication of the predicted trends. If anything, the pattern seems to be the reverse of that predicted: keystone effects are found in the majority of rocky intertidal studies, in roughly half of the rocky subtidal work, and hardly ever in studies of benthic mud communities. A complicating factor is that predation may also be affected by such disturbance differences: predation pressure can be significantly more intense in a protected habitat than in an exposed one. The

fact that several studies (e.g., Lubchenco, 1978; Morin, 1983) indicate key-stone effects only at low or moderate predation levels lends some support to the intermediate disturbance hypothesis.

Trophic Deletion Effects

In summary, the following traits may be correlated with high deletion effects in trophic interactions. For a prey species to have a high deletion impact its predators must be food limited; this suggests that in general plants may have less deletion impact than animals. Its predators should be specialized; this, in turn, is most likely (1) in stable habitats, (2) where prey have been abundant over evolutionary time, (3) where prey have unusual defenses (most likely in the case of "conspicuous" species: common, easily located, and perhaps communal), (4) among widespread and common prey species, and (5) where predators have simple nervous systems. Its prey should be chemically or temporally unusual (particularly likely in plants and invertebrates).

For predators to have important effects on their communities: (1) their prey should be predation limited, (2) their prey should be strongly competitive, and (3) these conditions seem particularly plausible where prey are sessile and pre-dation pressure is moderate.

No simple overall generalizations emerge from this list, and several provisions appear to be contradictory. Plants, for instance, are suggested to be both par-ticularly high and particularly low in deletion effects. The fine trade-off between predator specialization, food limitation, and prey abundance is also puzzling. Given the evident importance of predators in maintaining diversity in many systems, there is a great need for more research in this area.

MUTUALISTIC INTERACTIONS

Mutualisms occur when two species reciprocally benefit one another. Mutu-alisms are found among all major groups of living organisms (plants, animals, fungi, protists, and prokaryotes), and they range from obligate to facultative. Some of these associations are based on provisions of mutual protection, but in most of them there is some transfer of energy. For this and other reasons, the origins of many mutualisms are believed to lie in predator-prey interactions that have evolved into relationships that provide at least some benefits to the origi-nally damaged partner (Thompson, 1982). This is particularly obvious in such mutualistic relationships as those between plants and the pollinators of their flowers, where pollinators are thought to have evolved from pollen predators.

Although mutualism is defined in terms of shared benefits, the "evolutionary interests" of the interacting mutualists are rarely, if ever, identical. For example, a plant would benefit if it could attract pollinators without the investment of energy in expensive rewards. A pollinator, on the other hand, would benefit if plants provided greater rewards than they do. Thus, existing systems represent the result of compromises between conflicting selective pressures on the inter-acting partners rather than "pure solutions" favoring any one of them. In most cases the strengths of interactions are unequal, so that benefits to one partner are stronger than benefits to the other.

Fruit-Frugivore Interactions

Fruit-frugivore interactions probably had their origins in seed predation and depended on the fact that some predators did not kill all the seeds they consumed and also facilitated dispersal of seeds that survived. The evolution of traits such as fleshy nutritious rewards surrounding seeds, which increased the proportion of visits by less deleterious animals, should have been favored by conditions of very high predation rates on seeds dropped beneath the parent plant (Janzen, 1970, 1971, 1972).

For several reasons, the evolution of fruits and frugivores has not resulted in tightly coevolved relationships. To be a frugivore rather than a seed predator an animal must be large enough to swallow fruits. Such animals require many fruits each day and are, with rare exceptions, unable to hibernate or estivate when fruits are not available. Because not all frugivores move viable seeds to suitable germination sites with equal probability, fruits have evolved traits that make them attractive to some frugivores but unattractive or even toxic to others.

For most plants, there is a particular time during the year when survival of dispersed seeds is highest. As a result, they generally have short fruiting seasons and are visited by only a few species of frugivores. Conversely, most frugivores eat many different kinds of fruits over the course of a year, and most of them include animals in their diets as well, at least when young (Snow, 1971, 1981; Wheelwright, 1983; Wheelwright et al., 1984). These patterns cause an asym-metry in the sensitivity of the systems to deletion. Very few frugivores are likely to be substantially affected by the loss of one species of plant whose fruits they consume, unless that species is essential during some critical period, whereas many plants are likely to be severely affected by the loss of a single species of frugivore.

Evidence for the importance of loss of fruigivores is fragmentary but sugges-tive. Janzen and Martin (1982) have pointed out the existence of an array of plants in Central America whose fruits are not eaten by any of the native frugi-vores currently found in that region. These fruits have characteristics similar to those of African plants that are dispersed by large mammals, and they may have been dispersed by Pleistocene gomphotheres. Their ranges and abundances may have greatly decreased since the gomphotheres went extinct. If Janzen and Martin are correct, extinction or severe restrictions in the ranges of the large African mammals responsible for the dispersal of fruits there might have a serious impact on the populations of many plant species (but see Howe, 1985).

The dodo (*Raphus cucullatus*) was the only large ground frugivore on a number of islands in the Indian Ocean. Some plants with large seeds were dependent upon passage through the gut of a dodo for the chemical stimulus required to initiate germination. Since the extermination of the dodo on Mauritius in 1681, there has been no known natural germination of a seed of the tree *Calvaria major* (Sapotaceae). The species survived because of the longevity of the adults (Temple, 1977).

Plant-Pollinator Interactions

While many plants, particularly those of high latitudes and windy areas, are wind pollinated today (Whitehead, 1983), most angiosperms rely on animals—primarily bats, birds, and insects—for the dispersal of their pollen.

The number of plant species is much greater than the number of pollinator species in all areas where detailed information has been gathered. For example, in North America north of Mexico there are 3,465 species of bees (Krombein et al., 1979) but 14,575 species of flowering plants, excluding grasses, most of which are visited by bees. Similarly, North America north of Mexico contains many dozens of hummingbird-pollinated flowers, yet has only thirteen species of hummingbirds, only seven of which occur very far north of the Mexican border.

In North America bees show a high level of host-plant specialization. Two-thirds of the 960 species of bees for which there are adequate data are relatively highly specialized (Moldenke, 1979). The degree of specialization is especially well marked among bees of arid regions that visit composites (family Asteraceae). Of the 117 bee species that are specialized to a particular family, 108 are specific to the Asteraceae. Of these, 40 species visit only flowers of a single genus and 13 are confined to a single plant species. But the reverse is not true. Most plant species, including the composites, are visited by many different bee species, because there are many very generalized bee species. Thus the pattern is the opposite of that found among frugivores.

The ranges of specialized bees seldom coincide with the ranges of those plants upon which they specialize (Feinsinger, 1983), strongly suggesting that the limits of the ranges of both plants and bees are determined primarily by factors other than the availability of particular pollinators or plants. Where the "typical" pollen plant is absent, the bees use other species; and where the specialized bee is absent, other insects, sometimes from different orders, may pollinate the plants. Substitutability of species, at least with respect to their mutualistic roles, is high.

Another factor restricting the evolution of specialization is the difficulty of evolving means of excluding less desirable pollinators from floral rewards. Plant species that regularly grow in high density patches tend to have open flowers with low rewards readily available to a variety of visitors. They do not appear to have special relationships with any particular visitor, relying instead on the high density of conspecifics to move pollen to nearby conspecific flowers. Rare plants, in contrast, often have more specialized flowers and offer higher rewards (Feinsinger, 1983).

The existing patterns of specialization, or lack of it, among flowers and visitors suggest that losses of individual species of pollinators are unlikely to have serious effects on reproductive success of most plant species. Similarly, even specialized pollinators may be able to survive the loss of host plants, although competition with other pollinators may complicate the situation. However, rare plants may be very dependent on the presence of the "right" pollinator that can deliver pollen to conspecific stigmas that are far away (Rathcke, 1983).

Plant-Fungus and Plant-Microorganism Mutualisms

One commonly accepted view of the origin of mutualisms between plants or animals and microorganisms is that they began with the incorporation of free-living prokaryotes into the host cell (Margulis, 1970). Indeed, as Margulis (1976) notes, "the biosphere is conspicuous for the frequency and diversity of associations between organisms that share only remote ancestry." Here we examine some of the more common types of associations between remotely related organisms from the standpoint of the deletion sensitivity of those relationships.

Lichens. A lichen consists of two organisms, an alga (the phycobiont) and a fungus (the mycobiont) living in close association. When the initial stages of formation of lichens have been examined experimentally, the fungi are frequently observed to destroy algal cells with which they come in contact. The high frequency of these incompatible combinations suggests that some of the mycobionts are facultatively parasitic and that the symbioses may have evolved from parasitic associations (Barrett, 1983). Of the more than 70,000 species of fungi so far described, about 20,000 are known to be involved in lichen symbioses (Smith, 1975; Ahmadjian, 1970). Only about twenty genera of algae are involved in lichen symbioses and, strikingly, half of all lichens consist of algae of a single genus, *Trebouxia* (Ahmadjian, 1970). *Trebouxia* has never been described in the free-living state. Loss of members of this genus would result in the extinction of many "species" of lichens, perhaps slowing colonization of the many environments in which lichens are dominant pioneering organisms. Conversely, the richness of species of fungi involved in lichen associations suggests that the loss of any one of them would be of little consequence for the algae.

Mycorrhizae. Many plants have nonparasitic fungi associated with their root systems, and if those plants are grown without their associated fungi, they exhibit markedly inhibited growth. These nonparasitic fungus-root associations, known as mycorrhizae, have evolved independently many times. These relationships can be divided into groups based on gross structure and morphology. *Ectomycorrhizae* form a sheath of hyphae around the roots of their host plants and suppress production of root hairs. Many different kinds of fungi are involved in these associations, some of which are capable of free-living existence, while others are obligately mycorrhizal. The hyphae of *Endomycorrhizae* penetrate among the cells of the hosts and form clusters of fine branches and oil-rich vesicles which seem to be digested by the host plant. As far as is known, all fungi involved in these relationships are obligate and all attempts to culture them on artificial media have failed. Many of these fungi excrete compounds that are toxic to bacteria and other fungi, suggesting that part of the benefit to the plant derives from protection against pathogens. Like lichens, mycorrhizal associations may have a parasitic origin.

So little is known about plant-fungus interactions that speculation concerning their deletion sensitivity is hazardous. Given the fact that the fungi are obligately mycorrhizal, it is clear that loss of hosts is likely to lead to losses of fungi. More information on the specificity of host and fungal interactions is needed before the deletion of sensitivity of these relationships can be assessed. The importance of mycorrhizae for plant growth makes this a high priority area for research.

Insect-Microorganism Mutualisms

Insects, particularly those species that utilize plant tissues as food, have a rich variety of mutualistic relationships with microorganisms (Jones, 1984), but the degree of specialization and substitutability is unknown. Many plant tissues lack or are deficient in some nutrients necessary for insect growth and reproduction, or contain such nutrients in a form that cannot be used by the insects. Insects deal with these dietary inadequacies by careful selection of tissues to eat, by chemically altering components of their food, by detoxifying others, and by initiating changes that modify the chemistry of the plant tissues. Microorganisms are commonly used by insects to assist in these processes. The value of microorganisms lies in their abilities to concentrate nutrients from dilute media, their possession of synthetic abilities lacking in more complex organisms, their possession of special detoxifying abilities, and their ability to evolve rapidly.

Nitrogen Fixation

Many environments, both terrestrial and aquatic, are nitrogen deficient. The availability of nitrogen for macroorganisms in these environments depends largely on the action of microorganisms. Certain cyanobacteria assimilate N_2 during active development and liberate much of it in the form of simple soluble compounds. In laboratory cultures from 5 to 60 percent of the N_2 assimilated may appear outside the cell, in part as peptides and amino acids (Stewart, 1964).

The most thoroughly studied nitrogen-fixation mutualism is that between legumes and bacteria of the genus *Rhizobium*, the only microorganisms known to inhabit their root nodules. Similarly, *Rhizobium* is known to form mutualisms only with legumes, inhabiting nearly 90 percent of all species studied. Although but a single bacterial genus is involved, there are many strains of the bacteria and they show a high degree of host specificity. Considerable genetic work has been done on *Rhizobium*, and mutants are known which influence bacterial invasiveness, host susceptibility, and effectiveness in the acquisition of N_2 (Alexander, 1971).

Root nodules involved with nitrogen fixation also occur in members of several other plant families, but little is known about the identities of the microorganisms involved. Evidence suggests that they are mostly actinomycetes. In some tropical Cycadales, associations form with cyanobacteria and green algae (Allen and Allen, 1965; Bond, 1967; Bond and Scott, 1955). Given the extremely limited ecological and taxonomic data available for nitrogen-fixing organisms, it is difficult to determine their short- or long-term substitutability.

CONCLUSIONS

Thus far, our analyses have been directed at the probable consequences of the deletion of particular species on other members of the community, with special attention to the problem of secondary extinctions triggered by deletion of a focal species. Useful though this perspective is as a way of approaching the problem of determining the ecological value of species, it leaves many important topics unaddressed. Here we remedy, at least in part, these deficiencies by briefly considering which species are most likely to be lost, which types of ecosystems are most likely to be highly deletion sensitive, and which areas should receive high priority for future research.

Thus far we have allowed ourselves the luxury of assuming that any species could be surgically excised from a community. Real extinctions, however, are unlikely to be either so neat or so random. Some species are much more likely to go extinct than others, and factors driving one species to extinction will often have effects on many other species populations.

Extinction Proneness of Species

Both common sense and experience suggest that some species are much more prone to extinction than others (e.g., Terborgh, 1974). Decisions concerning the allocation of conservation efforts require not only information about the uniqueness quotients of species but also estimates of the probability that a particular extinction will actually occur.

Terborgh (1974; and Terborgh and Winter, 1980) lists the following groups as being particularly extinction prone: (1) species on the top trophic rung, and the largest members of guilds (generally requiring large territories, having low population densities and reproductive rates, and being desirable targets for hunters); (2) widespread species with poor dispersal and colonization ability (intolerant of habitat fragmentation); (3) endemics of both mainland areas and oceanic islands (highly vulnerable to localized disturbance and introduced predators);

(4) species with colonial nesting habits (particularly vulnerable to hunting, introduced predators and disease); (5) migratory species (double jeopardy: vulnerable to habitat destruction in both their breeding and nonbreeding ranges); (6) rare organisms (overlaps with 1 and 3, but also because of vulnerability to stochastic extinction); (7) organisms dependent on unreliable resources (especially frugivores and nectarivores) prone to great fluctuations in abundance; and (8) organisms having little evolutionary experience with disturbances.

Data to support these generalizations come almost exclusively from studies of birds and mammals. The minimum critical size experiments now being carried out by Lovejoy and his associates (Lovejoy et al., 1984) should provide us with a wider taxonomic and empirical basis for predictions.

No consistent relationship is apparent between uniqueness as we have defined it and extinction proneness. Rare organisms should be generally low in deletion impact, while they seem unusually vulnerable to loss. On the other hand, large animals (particularly predators) may rank high in both deletion impact and extinction proneness. Cases of this kind, high on both scales, should receive high priority in the allocation of conservation efforts.

Multiple Effects of Disturbance

Many of the most important causes of extinctions have effects on many species simultaneously. Habitat destruction or pollution may affect different species to different degrees, but all species in a community are touched. Many natural agents of disturbance behave in a similar fashion. When a drought or a development project drives some competitively unique species to extinction, populations of its competitors may not immediately respond, since they, too, may be stressed. Only when the agent of disturbance is narrowly focused on a particular species (as in certain forms of hunting or the introduction of a pathogen) can we safely apply the conclusions of our deletion sensitivity analysis.

If two species are eliminated simultaneously, are their impacts additive? Do effects generally overlap or interact, so that the whole shock is less or more than the sum of its parts? When responses of other species to an extinction are delayed by their having been affected directly by the disturbance as well, is the ultimate response the same, or can alternate stable solutions be reached? Our focus on single-species elimination studies is a sensible first step to understanding deletion sensitivity and uniqueness, but it is only a first step.

Ecosystems with Structures Favoring Extinction

Natural ecosystems differ in their structure (food webs) such that they are differentially sensitive to deletion of species. Highly sensitive systems are the ones toward which special conservation efforts might profitably be directed. At the most general level, ecosystems that have been subjected to repeated disturbances in the recent past, such as climatic changes, massive species introductions, and

170

large-scale human disturbances, should already have undergone most of the adjustments that might be induced by still further losses of species. If so, we should look to systems where the hand of human intervention has heretofore been light, and climatic fluctuations relatively benign during recent millennia, as systems likely to show the strongest responses to further losses of species. Such a view lends support to the considerable attention now being given to the preservation of tropical wet forests and islands.

A major conclusion emerging from both a general theoretical analysis and empirical studies is that predator-controlled systems are likely to be especially sensitive to the removal of their main predators. Examples are systems where animals as well as plants compete for space, such as hard-bottom substrates in coastal marine systems, and those terrestrial environments where large mammals are able to exert powerful control over vegetation succession, primarily arid and semiarid areas. It is likely that mammals exerted powerful control over vegetation in much of the world until recent times; but today, except for the widespread influences of domestic animals, communities with these dynamics are confined primarily to Africa. It is in Africa where the loss of large mammals is likely to exert the greatest influence on extinctions of other species. This likelihood strengthens the already strong aesthetic reasons for attempting to preserve large tracts of these ecosystems.

RESEARCH PRIORITIES

It is clear from our analysis that much research needs to be carried out if we are to develop a reasonable predictive theory of deletion sensitivity of ecological communities. At a purely theoretical level, analyses incorporating competition and mutualisms are needed to give abstract models greater reality. Also needed are models in which all members of, say, a particular food subweb or guild are removed. Empirical data of the kind found in Lovejoy's studies of Brazil are a vital component of research needs. Otherwise, theoretical speculations remain unconnected to reality. In particular, it is important to regard major environmental perturbations as experiments to be studied from the point of view of yielding valuable information on deletion sensitivity that can be obtained in no other way. The time and space dimensions of many large-scale development projects exceed those of experiments that can be deliberately performed for strictly scientific purposes. They therefore represent unique learning opportunities that are, unfortunately, usually wasted. With proper planning and care, such projects can be carried out so as to permit testing of hypotheses about the effects of the perturbation.

Finally, it is evident that even with the most favorable outcomes of the efforts of conservationists, considerable losses of species will occur during the coming century. Some of this is unavoidable, but it may be possible to develop means of

171

managing ecosystems for species "supersaturation"—that is, managing to maintain in the systems more species than would be accommodated if the systems were allowed to undergo natural processes of species relaxation that accompany habitat fragmentation. Knowledge of species deletion sensitivity may help guide efforts to determine which species should be managed most intensively because their loss would trigger other extinctions, and how those species might be protected from perturbations likely to cause their extinction in the first place.

APPENDIX

The table below summarizes thirty-four predation experiments reviewed in the preparation of this study. The list is neither exhaustive nor systematic, but represents a large proportion of the literature to date. Here we tally the effects predation was found to have on prey population densities and on three indices of community diversity: species richness (the number of species persisting), evenness (in the relative populations of those species), and H' (an aggregate index involving both richness and evenness). The symbols "+," "−," and "0" indicate the sign of the effects found; − − indicates a strongly negative effect; blank spaces indicate that no information was recorded. The designation "mod" is used where effects occurred at moderate predation levels only.

| | | | | | | | Effect of Predation | | | |
Reference	Predator	Prey	Habitat	Type of Predation	Manipulation	Prey Density	Prey Species Rich-ness	Prey Even-ness	H'	Remarks
Addicott 1974	Mosquito larvae	Protozoans Rotifers	Pitcher plant leaves		Additions and removals	−	−	+		Seems to be little competition between prey
Allan 1982	Brook trout	Inverts.	Stream	Selective	Removal	0				No diversity index examined
Bell & Coull 1978	Grass shrimp	Nematodes Rotifers	Salt marsh tanks	Generalist	Tanks with and without predators	−	0		0	Same species dominated with or without predation
Connor & Teal 1982	Mud snails	Benthic algae	Aquaria		Various pred. levels	Mod.: +				Algae density highest with moderate predation or addition of nitrogenous waste
Day 1977	Fish	Encrusting algae and animals	Shallow reefs open site	Less selective	Exclusion cages		−			Effect depends on site
			Protected site	More selective	Exclusion cages		+			
Estes et al. 1982	Sea otter	Urchin	Oceanic island		Previous hunting	− −				Much higher algae density where otters present; urchins rare

Reference	Predator	Prey	Habitat	Type of Predation	Manipulation	Effect of Predation				Remarks
						Prey Density	Prey Species Richness	Prey Evenness	H'	
Fulton 1982		Zooplankton	Estuary	Selective	Experimental tanks	–	Depends on initial condition			Predation alters relative abundances, increases diversity only if favored species predominates
Harper 1969	Rabbit	Grasses	Chalk grasslands	Selective	Enclosures disease		+?	+		A few grasses dominate in absence of predation; eventually shrubs take over
Himmelman et al. 1983	Sea urchins	Algae	Subtidal estuary		Removal	– –	–			No predation on urchins; very high predation pressure
Hixon 1983	Fish	Algae	Coral reef		Damselfish territories, exclusion cages		Mod.: +	Mod.: +	Mod.: +	Diversity highest with moderate predation pressure
Hulberg & Oliver 1980	Fish	Polychaetes	Marine sandy bottom		Various cage types	+?				Diversity higher in cages; probably cage effect rather than predation

Study	Predator	Resource	Habitat	Foraging	Manipulation				Comments
Hurlbert & Mulla 1981	Mosquito fish	Plankton	Experimental ponds		Varying densities	Varies	+	+	Some species eliminated; other increases with predators
Inouye et al. 1980	Rodents	Annual seeds	Desert	Selective	Removal	−	−?	0	Densities of particular prey species generally cut by either ants or rodents, not both
	Ants	Annual seeds	Desert	Opportunistic	Removal	−	+	+	
Kitting 1980	Limpet	Encrusting algae (2 spp.)	Marine intertidal	Diet mixing	Removals and additions	0	0	0	Algae eaten in fixed proportion regardless of relative abundance
Kneib & Stiven 1982	Mummichogs (fish)	Benthic inverts.	Salt marsh		Enclosures with varying densities, varying size	Large or med. preds.: + Small preds.: −			Large fish eat invertebrate predators and disturbers
Lubchenco 1978	Gastropod (*Littorina*)	Algae	Marine tidepools	Selective	Removals and additions	Mod.: +	Mod.: +	Mod.: +	Predator prefers competitive dominant; patchiness in tidepools prevents eradication of preferred prey
			Marine emergent substrata	Selective		−		−	

Reference	Predator	Prey	Habitat	Type of Predation	Manipulation	Effect of Predation				Remarks
						Prey Density	Prey Species Richness	Prey Evenness	H'	
Miller 1982	Fish	Algae inverts.	Intertidal reef beach		Exclusion cages	−				
Morin 1981	Newts	Tadpoles (3 spp.)	Artificial ponds	Selective	Varying densities			Mod.: +		Different species dominate with high or no predation; moderate predation allows mixture
Morin 1983	Salamanders (2 spp.)	Tadpoles (6 spp.)	Artificial ponds	Selective	Very high densities and composition	Low pred.: + High pred.: −	Mod.: +	Mod.: +		Total biomass and evenness maximized at moderately low (but not zero) predation
Oberndorfer et al. 1984	Predatory insects	Litter processing macroinverts.	Alpine stream		Enclosures with varying pred. densities	−				Predation cuts rate of litter processing
Otsuka & Dauer 1982	Misc.	Fouling community	Estuary		Exclusion cages	− −				Could be container effect
Paine 1974	Starfish (Pisaster)	Mussels	Marine intertidal		Removal		+			Preferred prey is competitive dominant

Reference	Predator	Prey	Habitat	Method					Comments
Paine & Vadas 1969	Sea urchins	Algae	Marine intertidal	Removal	—	?			Predator exclusion produces initial increase in prey species, then gradual monopolization by competitive dominant
			Marine subtidal	Cages with and without predation	—	?			
Porter 1972	Starfish (*Acanthaster*)	Corals	Reefs	Natural outbreak	—	+	+	+	
Reise 1977	Misc.	Benthic infauna, Epifauna	Marine mudflat	Exclusion cages	—	—			Container effects possible
			Marine seagrass	Exclusion cages	0	0			
Risch & Carroll 1982	Ant	Arthropods	Agriculture field semitropical	Removal (insecticide baits)	—	—			Aphids highest in presence of ants, other pests reduced greatly
Russ 1980	Fish	Epifauna	Marine subtidal	Selective; Open and closed cages	—		+		Predation prevents monopolization of space by competitive dominant
Sammarco 1982a	Sea urchins (2 spp.)	Coral spat (?)	Reefs	Removal	+		+	+	Predation and disruption of coral outweighed by removal of algal competitors
			Reefs	Removal	—		—	—	

| | | | | | | Effect of Predation | | | | |
Reference	Predator	Prey	Habitat	Type of Predation	Manipulation	Prey Density	Prey Species Richness	Prey Evenness	H'	Remarks
Sammarco 1982b	Sea urchin	Algae	Shallow subtidal (?)		Cages with varying densities	–	–	0	–	One algal species immune to predation
Thorp & Bergey 1981	Fish, turtles	Benthic inverts.	Fresh water pond		Exclusion cages	0?	+?			Probably container effect
Virstein 1977	Crabs, fish	Benthic infauna	Marine subtidal mud		Enclosures and exclosures	– –	– –	0		Prey appear strongly predation limited
Walde & Davies 1984	Stonefly	Invert. detritivores	Stream		Enclosures and exclosures	– (some)				Predation cut populations of some detritivores, not others
Waser & Price 1981	Cattle	Annuals	Desert		Exclosures		–			Similar effect in high and low rainfall years
Wicklow & Yocum 1982	Fly larvae	Fungi	Dung	Indiscriminate	Containers with varying densities		–			

REFERENCES

Abugov, R. 1982. Species diversity and phasing of disturbance. *Ecology* 63:289–293.

Addicott, J. F. 1974. Predation and prey community structure: An experimental study of the effect of mosquito larvae on the protozoan communities of pitcher plants. *Ecology* 55:475–492.

Ahmadjian, V. 1970. The lichen symbiosis: Its origin and evolution. In *Evolutionary Biology*, ed. T. Dobzhansky, M. K. Hecht, and W. C. Steere, 4:163–184. New York: Appleton-Century-Crofts.

Alexander, M. 1971. *Microbial Ecology.* New York: John Wiley and Sons.

Allan, J. D. 1982. The effects of reduction in trout density on the invertebrate community of a mountain stream. *Ecology* 63:1444–1455.

Allen, E. B., and R. T. T. Forman. 1976. Plant species removals and old field community structure and stability. *Ecology* 57:1233–1243.

Allen, E. K., and O. N. Allen. 1965. In *Microbiology and Soil Fertility*, ed. C. M. Gilmour and O. N. Allen, 77–106. Corvallis: Oregon State University Press.

Anderson, J. M., and H. Hall. 1977. Cryptostigmata species diversity and soil habitat structure. *Ecol. Bull.* 25:473–475.

Barrett, J. A. 1983. Plant-fungus symbioses. In *Coevolution*, ed. D. J. Futuyma and M. Slatkin, 137–160. Sunderland, Mass.: Sinauer.

Beery, J. A. 1975. Adaptations of photosynthetic processes to stress. *Science* 188:644–650.

Bell, S. S., and B. C. Coull. 1978. Field evidence that shrimp predation regulates meiofauna. *Oecologia* 35:141–148.

Belovsky, G. E. 1984. Moose and snowshoe hare competition and a mechanistic explanation from foraging theory. *Oecologia* 61:150–159.

Birch, L. C. 1957. The meanings of competition. *Amer. Nat.* 91:5–18.

Bjorkman, O., and J. Beery. 1973. High efficiency photosynthesis. *Sci. Am.* 229:80–93.

Black, C. C. 1971. Ecological implications of dividing plants into groups with distinct photosynthetic capacities. *Adv. Ecol. Res.* 7:87–114.

Bond, G. 1967. Fixation of nitrogen by higher plants other than legumes. *Ann. Rev. Plant Physiol.* 18:107–126.

Bond, G., and G. D. Scott. 1955. An examination of some symbiotic systems for fixation of nitrogen. *Ann. Bot.* 19:67–77.

Cohen, J. E. 1978. *Food Webs and Niche Space.* Princeton, N.J.: Princeton University Press.

Cohen, J. E., and F. Briand. 1984. Trophic links of community food webs. *Proc. Nat. Acad. Sci. (USA)* 81:4105–4107.

Colinvaux, P. 1978. *Why Big Fierce Animals Are Rare: An Ecologist's Perspective.* Princeton, N.J.: Princeton University Press.

Connell, J. H. 1975. Some mechanisms producing structure in natural communities. In *Ecology and Evolution of Communities*, ed. M. L. Cody and J. M. Diamond, 460–490. Cambridge, Mass.: Belknap Press of Harvard University Press.

———. 1978. Diversity in tropical rain forests and coral reefs. *Science* 199:1302–1310.

———. 1983. On the prevalence and relative importance of interspecific competition: Evidence from field experiments. *Amer. Nat.* 122:661–696.

Connor, M. S., and J. M. Teal. 1982. The effect of feeding by mud snails *Ilyanassa obsoleta* (say) on the structure and metabolism of a laboratory benthic algal community. *J. Exp. Mar. Biol. Ecol.* 65:29–45.

Darwin, C. 1859. *The Origin of Species by Means of Natural Selection.* London: Murray.

Davidson, D. W., J. H. Brown, and R. S. Inouye. 1980. Competition and the structure of granivore communities. *BioScience* 30:233–238.

Davidson, D. W., R. S. Inouye, and J. H. Brown. 1984. Granivory in a desert ecosystem: Experimental evidence for indirect facilitation of ants by rodents. *Ecology* 65: 1780–1786.

Day, R. W. 1977. Two contrasting effects of predation on species richness in coral reef habitats. *Mar. Biol.* 44:1–5.

Ehrlich, P. R., and L. C. Birch. 1967. The "balance of nature" and "population control." *Amer. Nat.* 101:97–107.

Elton, C. S. 1958. *The Ecology of Invasions by Animals and Plants.* London: Methuen.

Estes, J. A., R. J. Jameson, and E. B. Rhode. 1982. Activity and prey election in the sea otter: Influence of population status on community structure. *Amer. Nat.* 120:242–258.

Estabrook, G. F., and A. E. Dunham. 1976. Optimal diet as a function of absolute abundance, relative abundance, and relative value of available prey. *Amer. Nat.* 110: 401–413.

Feeny, P. P. 1976. Plant apparency and chemical defense. *Recent Adv. Phytochem.* 10: 1–40.

Feinsinger, P. 1983. Coevolution and pollination. In *Coevolution*, ed. D. J. Futuyma and M. Slatkin, 282–310. Sunderland, Mass.: Sinauer.

Fulton, R. S. III. 1982. Preliminary results of an experimental study of the effects of mysid predation on estuarine zooplankton community structure. *Hydrobiologia* 93:79–84.

Futuyma, D. J. 1976. Food plant specialization and environmental predictability in lepidoptera. *Amer. Nat.* 110:285–292.

Gause, G. F. 1934. *The Struggle for Existence.* Baltimore: Williams and Wilkins.

Gilpin, M. E. 1975. Stability of feasible predator-prey systems. *Nature* 254:137–139.

Goodman, D. 1975. The theory of diversity-stability relationships in ecology. *Quart. Rev. Biol.* 50:237–266.

Grant, P. R. 1966. Ecological compatability of bird species on islands. *Amer. Nat.* 100: 451–462.

Hairston, N. G. 1985. The interpretation of experiments on interspecific competition. *Amer. Nat.* 125:321–325.

Hairston, N. G., F. E. Smith, and L. B. Slobodkin. 1960. Community structure, population control and competition. *Amer. Nat.* 94:421–425.

Hanski, I. 1981. Coexistence of competitors in patchy environments with and without predators. *Oikos* 37:306–312.

Harper, J. L. 1969. The role of predation in vegetational diversity. In *Diversity and Stability in Ecological Systems.* Brookhaven Symposia in Biology 22.

Himmelman, J. H., A. Cardinal, and E. Bourget. 1983. Community development following removal of urchins, *Strongylocentrotus droebachiensis*, from the rocky subtidal zone of the St. Lawrence Estuary, Eastern Canada. *Oecologia* 59:27–39.

Hixon, M. A. 1983. Damselfish as keystone species in reverse: Intermediate disturbance and diversity of reef algae. *Science* 220:511–513.

Howe, H. F. 1984. Constraints on the evolution of mutualisms. *Amer. Nat.* 123:764–777.

————. 1985. Gomphothere fruits: A critique. *Amer. Nat.* 125:853–865.

Hulberg, L. W., and J. S. Oliver. 1980. Caging manipulations in marine soft-bottom communities: Importance of animal interactions on sedimentary habitat modifications. *Can. J. Fish. Aquat. Sci.* 37:1130–1139.

Hurlbert, S. H., and M. S. Mulla. 1981. Impacts of mosquitofish (*Gambusia affinis*) predation on plankton communities. *Hydrobiologia* 83:125–151.

Hutchinson, G. E. 1959. Homage to Santa Rosalia; or why are there so many kinds of animals? *Amer. Nat.* 93:145–159.

Inouye, R. S., G. S. Byers, and J. H. Brown. 1980. Effects of predation and competition on survivorship, fecundity, and community structure of desert annuals. *Ecology* 61:1344–1351.

Janzen, D. H. 1970. Herbivores and the number of tree species in tropical forests. *Amer. Nat.* 104:501–528.

————. 1971. Escape of juvenile *Dioclea megacarpa* (Leguminosae) vines from predators in a deciduous tropical forest. *Amer. Nat.* 105:97–112.

————. 1972. Escape in space by *Sterculia apetala* seeds from the bug *Dysdercus fasciatus* in a Costa Rican deciduous forest. *Ecology* 53:350–361.

Janzen, D. H., and P. Martin. 1982. Neotropical anachronisms: What the Gomphotheres ate. *Science* 215:19–27.

Jones, C. G. 1984. Microorganisms as mediators of plant resource exploitation by insect herbivores. In *A New Ecology*, ed. P. W. Price, C. N. Slobodchikoff, and W. S. Gaud, 53–99. New York: John Wiley and Sons.

Kareiva, P. 1985. Patchiness, dispersal and species interactions: Consequences for communities of herbivorous insects. In *Community Ecology*, ed. J. Diamond and T. J. Case. New York: Harper and Row.

Kastendiek, J. 1982. Competitor mediated coexistence: Interactions among three species of benthic macroalgae. *J. Exp. Mar. Biol. Ecol.* 62:201–210.

Kitting, C. L. 1980. Herbivore-plant interactions of individual limpets maintaining a mixed diet of intertidal marine algae. *Ecol. Monogr.* 50:527–550.

Kneib, R. T., and A. E. Stiven. 1982. Benthic invertebrate responses to size and density manipulation of the common mummichog *Fundulus heteroclitus* in an intertidal salt marsh. *Ecology* 63:1518–1532.

Kodric-Brown, A., and J. H. Brown. 1979. Competition between distantly related taxa in the coevolution of plants and pollinators. *Amer. Zool.* 19:1115–1127.

Krombein, K. V., P. D. Hurd, Jr., D. R. Smith, and B. D. Burks. 1979. *Catalog of Hymenoptera in America North of Mexico*. Washington, D.C.: Smithsonian Institution Press.

Kruckeberg, A. 1985. California serpentines: Flora, vegetation, geology, soils and management problems. University of California Pubs. Botany 78.

Lack, D. 1954. *The Natural Regulation of Animal Numbers*. Oxford: Clarendon Press.

Lawlor, L. R. 1980. Structure and stability in natural and randomly constructed competitive communities. *Amer. Nat.* 116:394–408.

Lawton, J. H., and M. P. Hassell. 1981. Asymmetrical competition in insects. *Nature* 289:793–795.

Levins, R., and R. H. MacArthur. 1969. An hypothesis to explain the incidence of monophagy. *Ecology* 50:910–911.

Lovejoy, T. E., J. M. Rankin, R. O. Bierregaard, Jr., K. S. Brown, Jr., L. H. Emmons, and M. E. Van der Voort. 1984. Ecosystem decay of Amazon forest remnants. In *Extinctions*, ed. M. H. Nitecki, 295–325. Chicago: University of Chicago Press.

Lubchenco, J. 1978. Plant species diversity in a marine intertidal community: Importance of herbivore food preference and algal competitive abilities. *Amer. Nat.* 112:23–39.

MacArthur, R. H. 1955. Fluctuations of animal populations and a measure of community stability. *Ecology* 36:533–536.

MacArthur, R. H., and E. O. Wilson. 1967. *The Theory of Island Biogeography.* Princeton, N.J.: Princeton University Press.

Margalef, R. 1968. *Perspectives in Ecological Theory.* Chicago: University of Chicago Press.

Margulis, L. 1970. *Origin of Eukaryotic Cells.* New Haven, Conn.: Yale University Press.

———. 1976. Genetic and evolutionary consequences of symbiosis. *Exper. Parasitol.* 39: 277–349.

May, R. M. 1973. *Stability and Complexity in Model Ecosystems.* Princeton, N.J.: Princeton University Press.

Menge, B. A., and J. P. Sutherland. 1976. Species diversity gradients: Synthesis of the roles of predation, competition, and temporal heterogeneity. *Amer. Nat.* 110:351–369.

Miller, A. C. 1982. Effects of differential fish grazing on the community structure of an intertidal reef flat at Enewetak Atoll, Marshall Islands. *Pacific Science* 36:467–482.

Moldenke, A. R. 1979. Host-plant coevolution and the diversity of bees in relation to the flora of North America. *Phytologia* 43:357–419.

Morin, P. J. 1981. Predatory salamanders reverse the outcome of competition among three species of anuran tadpoles. *Science* 212:1284–1286.

———. 1983. Predation, competition, and the composition of larval anuran guilds. *Ecol. Monogr.* 53:119–138.

Moulton, M. P., and S. L. Pimm. 1983. The introduced Hawaiian avifauna: Biogeographic evidence for competition. *Amer. Nat.* 121:669–690.

Murdoch, W. W. 1966. Community structure, population control, and competition: A critique. *Amer. Nat.* 100:219–226.

Murray, B. G. 1979. *Population Dynamics: Alternative Models.* New York: Academic Press.

Oberndorfer, R. Y., J. V. McArthur, J. R. Barnes, and J. Dixon. 1984. The effect of invertebrate predators on leaf litter processing in an alpine stream. *Ecology* 65:1325–1331.

Otsuka, C. M., and D. M. Dauer. 1982. Fouling community dynamics in Lynnhaven Bay, Virginia. *Estuaries* 5:10–22.

Paine, R. T. 1974. Intertidal community structure: Experimental studies on the relationship between a dominant competitor and its principal predator. *Oecologia* 15:93–120.

Paine, R. T., and R. L. Vadas. 1969. The effects of grazing by sea urchins, *Strongylocentrotus*, on benthic algal populations. *Limnol. Oceanogr.* 14:710–719.

Park, T. 1948. Experimental studies of interspecies competition. I. Competition between populations of the flour beetles *Tribolium confusum* Duval and *Tribolium castaneum* Herbst. *Ecol. Monogr.* 18:265–308.

Pimm, S. L. 1979. The structure of food webs. *Theoret. Pop. Biol.* 16:144–158.

———. 1980. Food web design and the effect of species deletion. *Oikos* 35:139–149.

———. 1984. The complexity and stability of ecosystems. *Nature* 307:321–326.

Pimm, S. L., and J. H. Lawton. 1977. Number of trophic levels in ecological communities. *Nature* 268:329–331.

———. 1980. Are food webs divided into compartments? *J. Anim. Ecol.* 49:879–898.

Pimm, S. L., and J. W. Pimm. 1982. Resource use, competition and resource availability in Hawaiian honey creepers. *Ecology* 63:1468–1480.

Porter, J. W. 1972. Predation by *Acanthaster* and its effects on coral species diversity. *Amer. Nat.* 106:487–492.

Price, P. W. 1980. *Evolutionary Biology of Parasites*. Princeton, N.J.: Princeton University Press.

Proctor, J., and S. R. J. Woodell. 1975. The ecology of serpentine soils. *Adv. Ecol. Res.* 9:255–365.

Rathcke, B. 1983. Competition and facilitation among plants for pollination. In *Pollination Biology*, ed. L. Real, 305–329. Orlando, Fla.: Academic Press.

Reise, K. 1977. Predation pressure and community structure of an intertidal soft bottom fauna. In *Biology of Benthic Resources*, ed. B. F. Keegan and P. O. Ceidigh. Oxford: Pergamon Press.

Rhoades, D. F. 1979. Evolution of plant chemical defense against herbivores. In *Herbivores*, ed. G. A. Rosenthal and D. H. Janzen, 3–54. New York: Academic Press.

———. 1985. Offensive-defensive interactions between herbivores and plants: Their relevance in herbivore population dynamics and ecological theory. *Amer. Nat.* 125:205–238.

Risch, S. J., and C. R. Carroll. 1982. Effects of a keystone predaceous ant, *Solenopsis geminata*, on arthropods in a tropical agroecosystem. *Ecology* 63:1979–1983.

Roughgarden, J., and M. Feldman. 1975. Species packing and predator pressure. *Ecology* 56:489–492.

Russ, G. R. 1980. Effects of predation by fishes, competition, and structural complexity of the substratum on the establishment of a marine epifaunal community. *J. Exp. Mar. Biol. Ecol.* 42:55–69.

Sammarco, P. W. 1982a. Echinoid grazing as a structuring force in coral communities: Whole reef manipulations. *J. Exp. Mar. Biol. Ecol.* 61:31–55.

———. 1982b. Effects of grazing by *Diadema antillarum* phillipi on algal diversity and community structure. *J. Exp. Mar. Biol. Ecol.* 65:83–105.

Schoener, T. W. 1983. Field experiments on interspecific competition. *Amer. Nat.* 122:240–285.

Simberloff, D. 1984. Properties of coexisting bird species in two archipelagoes. In *Ecological Communities*, ed. D. R. Strong et al. Princeton, N.J.: Princeton University Press.

Simpson, G. G. 1980. *Splendid Isolation: The Curious History of Mammals in South America*. New Haven: Yale University Press.

Slobodkin, L. B., F. E. Smith, and N. G. Hairston. 1967. Regulation in terrestrial ecosystems, and the implied balance of nature. *Amer. Nat.* 101:109–124.

Smith, D. C. 1975. Symbiosis and the biology of lichenized fungi. In *Symbiosis*, Symp. Soc. Exper. Biol. 29:373–405.

Snow, D. W. 1971. Evolutionary aspects of fruit-eating by birds. *Ibis* 113:194–202.

————. 1981. Tropical frugivorous birds and their food plants: A world survey. *Biotropica* 13:1–14.

Stewart, W. D. P. 1964. Nitrogen fixation by Myxophyceae from marine environments. *J. Gen. Microbiol.* 36:415–422.

————. 1966. *Nitrogen Fixation in Plants.* London: Athlone Press, University of London.

Strong, D. R., Jr. 1979. Biogeographic dynamics of insect-host plant communities. *Ann. Rev. Entom.* 24:89–119.

Temple, S. A. 1977. Plant-animal mutualism: Coevolution with dodo leads to near extinction of plant. *Science* 197:885–886.

Tepedino, V. J. 1979. The importance of bees and other insect pollinators in maintaining floral species composition. *Great Basin Naturalist Memoirs* 3:139–149.

Terborgh, J. 1974. Preservation of natural diversity: The problem of extinction prone species. *BioScience* 24:715–722.

————. 1986. Keystone plant resources in the tropical forest. In *Conservation Biology: The Science of Scarcity and Diversity,* ed. M. E. Soulé. Sunderland, Mass.: Sinauer.

Terborgh, J., and B. Winter. 1980. Some causes of extinction. In: *Conservation Biology: An Evolutionary-Ecological Perspective,* ed. M. E. Soulé, and B. A. Wilcox, 119–133. Sunderland, Mass.: Sinauer.

Tev, V. S. 1970. On polyphagism. *Mar. Biol.* 5:169–171.

Thompson, J. W. 1982. *Interaction and Coevolution.* New York: John Wiley and Sons.

Thorp, J. H., and E. A. Bergey. 1981. Field experiments on responses of a freshwater, benthic macroinvertebrate community to vertebrate predators. *Ecology* 62:365–375.

Varley, G. C. 1959. The biological control of agricultural pests. *J. Roy. Soc. Arts* 107:475–490.

Virnstein, R. W. 1977. The importance of predation by crabs and fishes on benthic infauna in Chesapeake Bay. *Ecology* 58:1199–1217.

Walde, S. J., and R. W. Davies. 1984. Invertebrate predation and lotic prey communities: Evaluation of *in situ* enclosure/exclosure experiments. *Ecology* 65:1206–1213.

Waser, N. M., and M. V. Price. 1981. Effects of grazing on diversity of annual plants in the Sonoran Desert. *Oecologia* 50:407–411.

Watt, K. E. F. 1973. *Principles of Environmental Science.* New York: McGraw-Hill.

Westoby, M. 1977. What are the biological bases of varied diets? *Amer. Nat.* 109:627–631.

Wheelwright, N. T. 1983. Fruits and the ecology of Resplendent Quetzals. *Auk* 100:286–301.

Wheelwright, N. T., W. A. Haber, K. G. Murray, and C. Guindon. 1984. Tropical fruit-eating birds and their food plants: A survey of a Costa Rican lower montane forest. *Biotropica* 16:173–192.

Whitehead, D. R. 1983. Wind pollination: Some ecological and evolutionary perspectives. In *Pollination Biology,* ed. L. Real, 97–108. New York: Academic Press.

Wicklow, D. T., and D. H. Yocum. 1982. Effect of larval grazing by *Lycoriella mali* on species abundance of coprophilous fungi. *Trans. Br. Mycol. Soc.* 78:29–32.

Wiens, J. A. 1977. On competition and variable environments. *Amer. Sci.* 65:590–597.

Yodzis, P. 1976. The effects of harvesting on competitive systems. *Bull. Math. Biol.* 38:97–109.

————. 1977. Harvesting and limiting similarity. *Amer. Nat.* 111:833–843.

Commentary

Douglas E. Gill

The complex issue of the conservation of natural resources, involving questions of what species or ecosystems to conserve for the future and which ones to allow to be lost, beckons opinion from experts in virtually every corner of human endeavor—biology, philosophy, economics, politics, and aesthetics. Nevertheless, the field of ecology deals explicitly with the interactions of populations of organisms with their natural environments, and should have a central position in discussions of habitat destruction, extinction of species, and potential remedies via *in situ* or *ex situ* conservation. Thus it is appropriate to review current ecological theory and experimental evidence on mechanisms of population size regulation and species interactions for guidance in the development of ecological principles related to the sensitivity of species and communities to deletions or introductions of species.

The paper by Orians and Kunin provides this conference with an impressive and comprehensive summary of the roles species play in natural ecosystems and the probable effects of deletion of particular species, given their interactions with other species. Instructive for both theoretical and applied ecologists are the in-depth reviews of current concepts of interspecific competition, predator-prey interactions, and the mutualisms of fruit and frugivore, flower and pollinator, vascular plants and fungi, and insects and their microorganisms. Nevertheless, although the paper is quite illuminating about ecology, it fails to indicate clearly how these ecological principles would or would not assist in the conservation of *genetic* resources.

A major contribution of this paper is the effort to develop a new concept, that of "ecological uniqueness." Although stimulating, the concept is troublesome. The epigraph quoting Alice and Humpty Dumpty gives warning to the reader that unconventional word usage is ahead, but that doesn't alleviate bumpy and rough going. Much of the difficulty concerns the counterintuitive definition of the term, but also there are some inconsistencies in its usage. Ecologically unique species are defined as those with roles that cannot be replaced by others. Species whose removal from the community has little impact on the population growth of those remaining are considered to have *low* uniqueness, or low deletion impact —a somewhat contradictory use of the term.

For example, consider a species, such as the giant panda, that is highly divergent both taxonomically and ecologically from other species. The fact that it currently has (and probably for many centuries has had) a very low population density is also relevant. Because there is no other animal like it, it could not be replaced if it were to go extinct. Thus it most certainly fits the first concept of uniqueness. Moreover, its demise from the bamboo forests of central China is not likely to have any significant impact on other species, including the bamboo upon which it feeds. This expectation is in part due to the giant panda's rarity (the volume of bamboo consumed must be negligible to the growth and reproduction of bamboo clones), but is also due to its extraordinary geographical, ecological, and morphological attributes. Hence, by the criterion of impact on other species, the giant panda has low ecological uniqueness; but it is unique and irreplaceable by another, equally valid, criterion.

In general, the attempt by Orians and Kunin to develop the concept of ecological uniqueness was for me distracting and unsuccessful. Some clarity is obtained by substituting more precise and less loaded terms, such as "deletion impacts" and "substitutability," in place of uniqueness. Using this terminology, the types of contradictory conclusions cited above can be resolved. Since the concept of *uniqueness* was also a major point of Antonovics's paper, it is important that the concept receive thorough discussion and perhaps constructive resolution.

Orians and Kunin discuss competition, predation, and mutualism thoroughly, but give only brief reference to parasitism as a major factor controlling population sizes. This has been a traditional approach, as illustrated by the total absence of chapters on parasitism in all standard textbooks on ecology. The inherent density dependence of host-parasite interactions, the catastrophic epidemics that are known to induce crashes if not local extinctions in many mammals, insects, and plants, and the critical role parasites play as barriers to introductions (Anderson and May, 1982) are all compelling reasons to consider parasitism seriously in discussions of preservation of species *in* or *ex situ*.

The discussion about dietary specializations of predators with simple nervous systems fails to point out that many specializations of parasites are brought about by their behavioral (and evolutionary) mechanisms of escape from immunological antagonism or physiological surveillance by the host. Humoral responses by vertebrate hosts to foreign organisms are often incapacitating, and cellular responses (cytophagous macrophages) by nearly all animals are customarily lethal to intruding parasites. Parasites are often unable to consume their hosts, usually because of powerful host defenses. Analogous interactions are doubtless rampant among genuine predators and their prey. The importance of mutual regulation of population density in hosts and parasites is unknown, has been virtually ignored in the past, and is in great need of future research.

The subject of the ecological roles species play in nutrient cycling in ecosystems was introduced early in the paper, but this interesting topic was not more fully developed later. While there is a disturbing and often misleading tendency in the ecological literature to force an "organismal" analogy onto ecosystems in the form of homeostatic mechanisms of nutrient conservation, and so forth, there are key roles upon which all else depends, such as green plants being the primary producers of biologically available energy, and nitrogen-fixing cyanobacteria (and their relatives) being the only producers of biologically active nitrogen. Were a nuclear winter to occur, or a fast-spreading antibiotic to appear, a snowballing avalanche of extinctions could ensue. Such dependencies could be highly specific: consider the size of the ecological disaster that a new antibiotic to *Rhizobium* species would have, because legumes are the third largest family of flowering plant in the world.

The development of the concept of deletion sensitivity was a valuable approach to the paper and provided many new insights. But it is not without its drawbacks. For example, the review of Pimm's theoretical work overlooked the fact that Pimm's analysis assumed the sequence (1) initial equilibrial conditions, (2) deletions, and (3) reestablishment of equilibrium conditions. Theoreticians have regularly depended on equilibrium conditions because of comfortable mathematical tractability. However, many field ecologists believe that few, if any, ecological communities are in equilibrium, but most systems are in some state of recovery from a significant perturbation in the recent past. The assumption by Orians and Kunin that reinvasion or recolonization can be ignored is, in my view, unrealistic and probably misleading. Programs of conservation that are based on the principles of nonequilibrium, and assume that organisms thrive temporarily in a recently opened habitat, but face imminent sharp population reductions because of habitat collapse or devastating epidemic, may be more effective because they are closer to reality.

Discussions about monophagy and herbivores require more clarification than provided in the paper. Rhoades and Cates (1976) proposed that conspicuous plants or plant parts should have generalized defenses, such as digestibility-inhibiting tannins, against the myriad insects that readily find them. While Feeney (1976) thought that rare and unapparent plants might depend on their isolation per se for their escape from enemies, Rhoades and Cates (1976) suggested that inconspicuous, ephemeral, or rare plants (or parts) should have highly specific toxins that are effective against the few specialized herbivores that find them. Much remains to be learned about the actual patterns found in nature, but rare or inconspicuous plants would be ecologically more unique than conspicuous plants, according to that hypothesis.

The section on mutualisms was very enlightening, but not encouraging to those who seek tight dependencies between plants and animals as a major argument for conservation efforts (see paper by Wilcox). However, given the incomplete nature of available information, some caution is in order. Better examples of facultative and obligate mutualisms are available (e.g., many of the algae and fungi of lichens do grow well alone). Current theories to predict the patterns of distribution of obligate mutualism are incomplete and often contradictory (Howe, 1984; Thompson, 1982). Orians and Kunin rather uncommittedly present Janzen and Martin's (1982) gomphotherian hypothesis, whereas Howe's (1985) recent criticism of the theory as ill founded and inherently untestable echoes the opinions of many ecologists. Orians and Kunin also overlooked Owadally's (1979) rebuttal to Temple's (1977, 1979) dodo-*Calvaria* story.

These problems all show that generalizations are difficult to come by in ecology and that our natural tendencies to seek them need to be tempered by recognition of the great variability in natural systems. We need to be especially suspicious of models that rely on unrealistic simplifying assumptions, adopted to make them mathematically tractable, however valuable those models may be in helping us approach more complex and realistic ones of ecological systems.

REFERENCES

Anderson, R. M., and R. M. May. 1982. Population Biology and Infectious Diseases. *Life Sciences Research Report 25.* Berlin: Dahlem Konferenzen, Springer-Verlag.

Feeny, P. P. 1976. Plant apparency and chemical defense. *Recent Adv. Phytochem.* 10: 1–40.

Howe, H. F. 1984. Constraints on the evolution of mutualisms. *Amer. Nat.* 123:764–777.

———. 1985. Gomphothere fruits: A critique. *Amer. Nat.* 125:853–865.

Janzen, D. H., and P. Martin. 1982. Neotropical anachronisms: What the Gomphotheres ate. *Science* 215:19–27.

Owadally, A. W. 1979. The dodo and the tambalacoque tree. *Science* 203:1363–1364.

Rhoades, D. F., and R. G. Cates. 1976. Toward a general theory of plant antiherbivore chemistry. In *Biochemical Interaction between Plants and Insects*, ed. J. W. Wallace and R. L. Mansell, 168–213. New York: Plenum Press.

Temple, S. A. 1977. Plant-animal mutualism: Coevolution with dodo leads to near extinction of plant. *Science* 197:885–886.

———. 1979. Reply to Owadally. *Science* 203:1364.

Thompson, J. W. 1982. *Interaction and Coevolution.* New York: John Wiley and Sons.

Commentary

Michael L. Rosenzweig

Gordon Orians and William Kunin have done an inestimable job revealing the state of a major segment of ecological inquiry. Again and again the questions they raise impress me as incisive and insightful. Unfortunately, the picture of ecology they convey is of an inchoate science. We use terminology inconsistently. Our theories are incoherent and our empirical justifications inadequate. Most of this the authors have noted themselves. Nonetheless, I believe that a

TABLE 7
Speciation mechanisms

Type	Special Manifestation	Deletion Sensitivity
Rate controlled extrinsically to population interactions	High species redundancy	Low
Rate also controlled by competition	Species fill niche vacuums	High

potent strategy for answering their questions, although contained in the paper, is likely to go unnoticed.

Evolution is the most promising philosophy of inquiry that one can extract from the Orians and Kunin paper. Customarily, evolution is an object of inquiry. What do I mean by calling it a philosophy of inquiry rather than an object of inquiry?

Evolution proceeds by various mechanisms. If we are to make useful predictions about extinctions, it is likely that we shall need to understand these mechanisms to understand how community structures and population dynamics evolve to be what they are. That is what I mean by evolution as a philosophy of inquiry. It is the opinion that these questions, not directly about evolution, are nevertheless best studied from an evolutionary perspective.

Orians and Kunin ask two great groups of questions. First, what are the effects of an extinction? Second, what is most likely to become extinct? Most of my comments will suggest ways of refining and organizing their questions to make them easier to approach.

Using the philosophy of evolutionary inquiry, I begin with an examination of how new species arise and what they may teach us regarding species deletion. Then I do the same thing with frameworks of community structure. The connections between it and many topics in Orians and Kunin will be more obvious than those linking speciation to their work. Finally, I shall criticize the use of imprecise and troublemaking definitions and the tendency of the authors to approach ecology with the single-variable limitation hypothesis of Hairston, Smith, and Slobodkin (1960).

MODES OF SPECIATION

Many modes of speciation have been suggested. For purposes of this discussion, however, we can lump them into two classes (Table 7). Most (Futuyma

189

and Mayer, 1980, would say all) accepted modes of speciation depend on *random events* unconnected with population interactions or ecological processes except perhaps in a totally subordinate fashion. But at least one, competitive speciation, is dependent on a *deterministic ecological process*.

Random speciation modes include all forms of polyploidy and geographical speciation. They proceed after some genetic or geological event separates a population into reproductive isolates. Then the isolates diverge. Ecological mechanisms are neither necessary nor sufficient. The most these can do is enhance the rate or certainty that the isolates diverge. For instance, an isolate nipped off a population occupying an extensive environmental gradient is much more likely to produce a new species in a short time than is one that occupies the same sort of habitat as all its conspecifics in the mother population (Mayr, 1954; Endler, 1977).

The more species there are, the higher the rates at which random modes produce new ones (Rosenzweig, 1975). This does not happen in response to an unused ecological opportunity. It happens simply because the genetic event takes place, or it happens because some mountain range or island forms. And the more species that play roulette with genetics and with geological processes, the more jackpot winners turn up during each eon.

Random processes of speciation suggest that many closely related species in the world may be ecologically redundant, at least to a large extent. Having not been evoked by ecological differences, they do not lead us to believe that their continued existence is crucial to the functioning of any ecosystem. Instead, it is quite possible, perhaps even probable, that the ecological differences that have arisen between them are sequelae of speciation and would vanish were either species to be left without the other. Of course, this depends on the processes by which natural selection forces species into different ecological niches. Some processes, as we shall see, may result in asymmetries such that only some species are easily and quickly replaced. In addition, the manner of achieving the niche separation is likely to be important. The extinction of a species separated by a morphological difference such as body size may be felt for generations compared with one separated merely by behavior. Behavior, after all, is quite plastic compared with morphology. Yet, in general, it may be fair to expect random speciation to produce rather redundant species with small deletion impacts, especially when compared with competitive speciation, which is considered next. Those species left behind should rapidly expand to do most or all of what was previously done. (They may even do better, as we shall see when we discuss interference structure.)

Yet there is a severe nonlinearity here which was overlooked by Orians and Kunin. "You may take away one species of Pacific salmon," says Mother Nature. "You may even take away two, three, or four. But please don't take away my very last one!" Species deletion is simply not a linear process; it is undoubtedly not

190

even continuous (Mares and Rosenzweig, 1978). Therefore, a set of experiments that shows no ill effects accompanying the removal of either species A or B should not convince anyone that both may be taken away without trouble. This is a direct consequence of the redundancy of random speciation modes.

Competitive speciation produces a new species only if there is a real ecological vacuum (Rosenzweig, 1978). A habitat that is not lived in, or whose resources are unutilized, promotes the expansion of the phenotypic variation required to exploit it. Then, under certain circumstances, intermediate phenotypes are rendered unfit because of the competition they face from extremes. This can result in disruptive selection leaving the extremes as different species (Maynard Smith, 1966; Rosenzweig, 1978; Rice, 1984). The most famous examples are species of *Rhagoletis (Diptera)* studied by Bush (1969).

If we could find ways to determine whether a species has been produced by competitive speciation, we would have a head start in predicting deletion impact. A species thus produced should leave behind unexploited opportunities if it becomes extinct, so its deletion impact should be high. Evolution will have to go through the whole process of speciation again in order to fill the gap.

Fuzzy clues exist to help us identify which species have been produced competitively. It seems likely that birds have not been, judging from such evidence as the one species of geospizid on Cocoa Island (Lack, 1947). Also, competitive speciation may be a rare event except during periods of adaptive radiation or new environments (Rosenzweig and Taylor, 1980). Last, habitat selectors are the most likely candidates for competitive speciation (Rice, 1984, 1985). As yet, though, this line of research is in a very primitive state.

FRAMEWORKS OF COMMUNITY ORGANIZATION

An early attempt to deduce community structure was that of MacArthur and Levins (1967). They predicted how strong competition can be before extinction begins removing losers. The parameter predicted was called the "limiting similarity of species."

But the theory of limiting similarity soon foundered. It was battered on the rocks of mathematics, which threaten any theory lacking robustness (e.g., Dayton, 1973; Abrams, 1975; Turelli, 1978). And it broke up on the reef of natural selection, which endangers any biological theory foolish enough to ignore it (Rosenzweig, 1974; Schroder and Rosenzweig, 1975; Roughgarden, 1976). Its complete absence from Orians and Kunin is both startling and refreshing.

Approaching community organization inductively, Colwell and Fuentes (1975) proposed that there are three qualitatively different patterns of niche overlap (Figure 11). Species niches could be coextensive, or each species could outperform all others in its own special region of niche space (reciprocal niche overlap), or species could have included niches, nested inside one another like

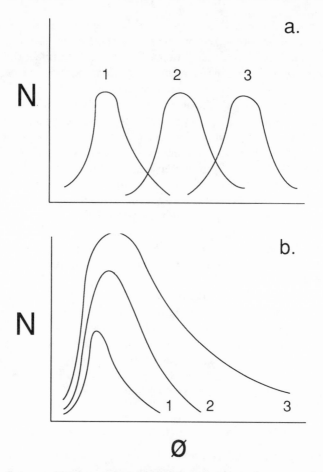

FIGURE 11. Patterns of fundamental niche overlap. N is the population size of the species when alone. Ø is the point in niche space; Ø could represent a salinity gradient on a mountainside or a body size gradiant of potential prey, or any other environmental variable that separates species. (a) Reciprical niche overlap; each species (1, 2, and 3) is most common in a different region of niche space. (b) Included niches: all species peak in the same part of niche space, but some can function in niche points where others cannot. (Coextensive niches merely have all three species with identical curves.)

a set of hollow wooden dolls. These three patterns are interesting, but to know how to apply them to any of our questions, we must first understand the mechanisms that generate them.

I tried to do this for habitat selection (Rosenzweig, 1979), by proposing three causes for it. The first cause is classic in the theory of evolutionary ecology.

It is the jack-of-all-trades principle. If we measure the ability of individuals of each species to function in various environments (i.e., at various values), we find each species has a piece of niche space in which it is the best. Second, included niches could result from interference competition if the dominant species in everyone's best habitat buys its dominance at the cost of ability to exploit poorer habitats well. The dominant would have the narrowest niche; the most subordinate would have the broadest. Pimm has called these two cases "distinct preference community structure" and "shared preference community structure" respectively (Rosenzweig and Abramsky, 1986).

The third cause for habitat selection is nonecological and leads to coextensive niches. Sometimes, species need to use different habitats as mating cues. The habitat becomes a rendezvous point for courtship (Zwolfer, 1974). Colwell (1985) argues that this mechanism is responsible for the many species of nose mite he observes. The rendezvous point mechanism seems to require preexisting separate species. Thus it is easy to conclude that any species involved in it must have been produced by geographical isolation and would be easily replaced if deleted. Colwell (1985) has actually observed the local extinction and quick replacement of a nose mite presumably involved in rendezvous point habitat selection.

Except for rendezvous point selection, these mechanisms are built on the foundation of optimal foraging theory (Rosenzweig, 1985). An additional type of community structure has recently been proposed using optimal habitat selection theory. It is called centrifugal organization (Rosenzweig and Abramsky, 1986). Species share their primary habitat type. It is an ideal mixture of two or more components, such as food and shelter. But the species specialize on which component they can manage without. Hence their secondary patch types are distinct. At least one pair of interacting gerbil species appears to be organized centrifugally (Rosenzweig and Abramsky, 1986).

Using an optimal foraging framework of habitat selection, we can begin to speculate on the deletion sensitivity and deletion impacts which they should foster (Table 8).

The one we know least about is a centrifugal organization. But already it is evident that removal of a centrifugal species is not going to be repaired fully by those left behind, except in the primary or ideal habitat type. Secondary habitat should remain depleted if the species adapted to withstanding its stress becomes extinct.

We know most about shared-preference organization. The most satisfying conclusions in the Orians and Kunin paper concern this kind of organization. It is asymmetrical and involves interference competition, but we learn a surprising thing from considering its mechanism of evolution. The dominant species exerts a powerful negative effect on its subordinates, but there may be no effect of subordinates on dominants. Consequently, the dominant will hardly be missed if

193

TABLE 8
Community structures (by optimal foraging)

Type	Special Manifestations	Deletion Sensitivity
Distinct preferences	Ghost of Competition Past	Low, especially in long term
Shared preferences	Interference; asymmetry	Low if dominant removed, high in secondary patch types if subordinate removed; dominant most likely to go
Centrifugal preferences	Ghosts impossible	High in secondary patch types
Distinct preferences with large spatial variance within a patch type	Ghosts less likely than without variance	Low even in short term
Rendezvous point foraging	Coextensive niches	Low

deleted. On the other hand, the subordinate is actually better at using the dominant's habitat than its own (Pimm et al., 1985). It may even be better than the dominant (Fellows and Heed, 1972). Therefore, it is the subordinate's absence that is likely to be felt. A missing subordinate is likely to have been the only species capable of using some secondary habitats.

Bock and Ricklefs (1983) and Brown (1984) have studied the relationship between abundance and geographical range of a species. Common species seem everywhere to be relatively common and usually have larger ranges. The explanation for this pattern might well be something like the following speculative paragraph.

In known cases of shared-preference organization, there is often a quantitative gradient that is subdivided among the species. Often this is as simple as environmental productivity. The best habitats in such cases are the richest. Since these are extreme habitats rather than average ones, it is reasonable to expect them to be scarce, and to expect the species that require them to be scarce and limited in distribution. But these same species are the dominants who monopolize the richest patches. Thus the biogeographical and abundance pattern would be a reflection of the asymmetry of shared-preference organization. If so, large areas could be left without a guild representative if a subordinate gets deleted.

On the other hand, the subordinate, if it is in fact common, is less likely to become extinct. One must therefore wonder whether there is not a natural tendency for big, fierce species (i.e., competitive dominants) to become extinct and be replaced by subordinates who would then be favored by natural selection to become big, fierce, rare, and restricted to living in the richest places. Therefore, natural selection can be expected to encourage dominance despite its making a species rare. Are shared-preference communities, in some bizarre sense, adapted to the repeated loss of their dominants? Is shared-preference organization one mechanism for driving taxon cycles?

Distinct-preference organization also should be applicable to deletion analysis. Many species involved in it should be haunted by the Ghost of Competition Past (Rosenzweig, 1981; Pimm and Rosenzweig, 1981; Joel Brown and Rosenzweig, manuscript). Dynamically they do not appear to compete at all. Their ranges may be totally nonoverlapping. Yet, if one is deleted, the other should move in. Probably the initial expansion will be weak but then accelerate as natural selection restores some generalist potential to the remaining species.

Joel Brown and I have recently added a second dimension to the optimal foraging framework of community structure by allowing spatial variation within patches of a single type. So far this has been applied to and had interesting effects on the distinct-preference case. It greatly weakens the likelihood of competitive ghosts. Deletion should therefore be followed by much more rapid compensation in such cases, because we need not wait for natural selection to act. Spatial variation is produced when foragers can sense the richness of a patch before they have invested much time in it, and can reduce that richness enough so as to be forced to leave in search of another place to forage. Distinct-preference systems in such circumstances should be particularly deletion insensitive.

Pimm and Lawton (see esp. Pimm, 1982, 1984) have been constructing another framework for community structure based on food web linkages. Like limiting similarity, its evolutionary mechanism is tolerance. It determines what sorts of linkages avoid producing dynamic instability and rigidity, both of which can lead to extinction. Unlike limiting similarity, the food web framework has successfully withstood empirical tests.

Orians and Kunin are even more aware of this framework than the optimality one. Yet they only nibble around the edges of its potential. Pimm has discussed how the symmetry of interaction coefficients appears to be a good predictor of linkage complexity. Do the various sorts of systems he envisions tend to reach a uniform extinction probability? Even if they do, may we not learn whether there is any extinction asymmetry as there is in shared-preference structure? Undoubtedly we can expect some asymmetry of deletion impact.

A term missing from the Orians and Kunin paper is biome. They give one hint that the tropics may be particularly deletion sensitive. But in general, they left this reader with a feeling that biomes may not be the best way to classify ecological systems for the purposes of deletion analysis. Is this right?

HOW NOT TO PROCEED

Fortunately, Orians and Kunin have adequately mirrored two problems ecology could do without: deficient terminology and the search for *the* limiting factor. This gives me the enviable chance to rail at both.

Terminology

Scientific terms are supposed, like university administration, to help us do our job. When those terms are imprecisely or ambiguously defined, they become troublemakers instead. Many terms in this paper relate to population dynamics. They should be defined in terms of the mathematical concepts that underlie them. Otherwise what are we to make of "benefit," "harm," "short supply," "reduce competition," not to mention "competition," "predation," and "mutualism" themselves?

"Keystone species" is a worthy term used in at least two senses in their paper. So is "competitive dominant," which, I have tried to show, has a great role to play in the context of shared-preference community structure. It has no evident meaning anywhere else.

"Unique," as the authors themselves point out, is not a comparative word. It should not be used to represent a continuous variable. "Uniqueness" is a travesty, something we should never be forgiven if we perpetuate it. What we really mean by saying "low uniqueness" is that loss of the species would need to have its effects measured by an expert. "High uniqueness" implies a cascade of apparent (not necessarily catastrophic) effects. Orians and Kunin have used a perfect term for the communities' response to the deletions: "deletion sensitivity." In this paper, I have refrained from anything but its converse, "deletion impact." I was never once bothered by my decision to forswear "uniqueness."

Single-factor limitations

Any population dynamicist knows that species are able simultaneously to be limited by any number of variables. I have discussed this for exploitative systems (Rosenzweig and MacArthur, 1963; Rosenzweig, 1973), but it really is a trivial mathematical point. Any population growth equation written with two independent variables demonstrates this point. Orians and Kunin rely much too heavily on notions of single-factor hypotheses, with resulting confusion. All of their arguments are better approached from the perspective of simultaneity of limiting factors.

196

REFERENCES

Abrams, P. 1975. Limiting similarity and the form of the competition coefficient. *Theoret. Pop. Biol.* 8:356–375.

Bock, C. E., and R. E. Ricklefs. 1983. Range size and local abundance of some North American songbirds: A positive correlation. *Amer. Nat.* 122:295–299.

Brown, J. 1984. On the relationship between abundance and distribution of species. *Amer. Nat.* 124:255–279.

Brown, J. S., and M. L. Rosenzweig. Habitat selection in slowly regenerating environments. Manuscript.

Bush, G. L. 1969. Sympatric host race formation and speciation in frugivorous flies of the genus *Rhagoletis* (Diptera, Tephritidae). *Evolution* 23:237–251.

Colwell, R. K. 1985. Community biology and sexual selection: Lessons from hummingbird flower mites. In *Community Ecology*, ed. J. Diamond and T. Case, 406–424. New York: Harper and Row.

Colwell, R. K., and E. R. Fuentes. 1975. Experimental studies of the niche. *Ann. Rev. Ecol. Syst.* 6:281–310.

Dayton, P. K. 1973. Two cases of resource partitioning in an intertidal community: Making the right prediction for the wrong reason. *Amer. Nat.* 107:662–670.

Endler, J. A. 1977. Geographic variation, speciation and clines. *Monogr. in Pop. Biol.* 10. Princeton, N.J.: Princeton University Press.

Fellows, D. P., and W. B. Heed. 1972. Factors affecting host plant selection in desert adapted cactiphilic *Drosophila*. *Ecology* 53:850–858.

Futuyma, D. J., and G. C. Mayer. 1980. Non-allopatric speciation in animals. *System Zool.* 29:254–271.

Hairston, N. G., F. E. Smith, and L. B. Slobodkin. 1960. Community structure, population control and competition. *Amer. Nat.* 94:421–425.

Lack, D. L. 1947. *Darwin's Finches.* Reprint, 1961. London: Cambridge University Press, Harper Torchbooks.

MacArthur, R. H., and R. Levins. 1967. The limiting similarity, convergence and divergence of coexisting species. *Amer. Nat.* 101:377–385.

Mares, M., and M. L. Rosenzweig. 1978. Granivory in North and South American deserts. *Ecology* 59:235–241.

Maynard Smith, J. 1966. Sympatric speciation. *Amer. Nat.* 100:637–650.

Mayr, E. 1954. Change of genetic environment and evolution. In *Evolution as a Process*, ed. J. S. Huxley, A. C. Hardy, and E. B. Ford, 157–180. London: Allen and Unwin.

Pimm, S. L. 1982. *Food Webs.* London: Chapman and Hall.

———. 1984. The complexity and stability of ecosystems. *Nature* 307:321–326.

Pimm, S. L., and M. L. Rosenzweig. 1981. Competitors and habitat use. *Oikos* 37:1–6.

Pimm, S. L., M. L. Rosenzweig, and W. Mitchell. 1985. Competition and food selection: Field tests of a theory. *Ecology* 66:798–807.

Rice, W. R. 1984. Disruptive selection on habitat preference and the evolution of reproductive isolation: A simulation study. *Evolution* 38:1251–1260.

———. 1985. Disruptive selection on habitat preference and the evolution of reproductive isolation: An exploratory experiment. *Evolution* 39:645–656.

Rosenzweig, M. L. 1973. Exploitation in three trophic levels. *Amer. Nat.* 107:275–294.

———. 1974. On the evolution of habitat selection. In *Proc. First Int. Congr. Ecol.* 401–404.

———. 1975. On continental steady states of species diversity. In *Ecology and Evolution of Communities*, ed. M. L. Cody and J. M. Diamond, 121–140. Cambridge, Mass.: Belknap Press of Harvard University Press.

———. 1978. Competitive speciation. *Biol. J. Linn. Soc. London* 10:275–289.

———. 1979. Three probable evolutionary causes for habitat selection. In *Contemporary Quantitative Ecology and Ecometrics*, ed. G. P. Patil and M. L. Rosenzweig, 49–60. Fairland, Md.: Intern. Coop. Publishing House.

———. 1981. A theory of habitat selection. *Ecology* 62:327–335.

———. 1985. Some theoretical aspects of habitat selection. In *Habitat Selection in Birds*, ed. M. L. Cody, chap. 18. London: Academic Press.

Rosenzweig, M. L., and Z. Abramsky. 1986. Centrifugal community organization. *Oikos* 46: 339–348.

Rosenzweig, M. L., and R. H. MacArthur. 1963. Graphical representation and stability conditions of predator-prey interactions. *Amer. Nat.* 97:209–223.

Rosenzweig, M. L., and J. A. Taylor. 1980. Speciation and diversity in Ordovician invertebrates: Filling niches quickly and carefully. *Oikos* 35:236–243.

Roughgarden, J. 1976. Resource partitioning among competing species: A coevolutionary approach. *Theoret. Pop. Biol.* 9:388–424.

Schroder, G. D., and M. L. Rosenzweig. 1975. Perturbation analysis of competition and overlap in habitat utilization between *Dipodomys ordii* and *Dipodomys merriami*. *Oecologia* 19:9–28.

Turelli, M. 1978. A reexamination of stability in randomly varying vs. deterministic environments with comments on the stochastic theory of limiting similarity. *Theoret. Pop. Biol.* 13:244–267.

Zwolfer, H. 1974. Das Treffpunkt-prinzip als Kommunikationsstrategie und Isolations-mechanismus bei Bohrfliegen (*Diptera: Trypetidae*). *Entom. German* 1:11–20.

Summary of the Discussion

Drew Harvell

Discussions in this session dealt with the problem of how to rank species by their ecological value. In the absence of a basis for such ordering, all species would be considered mutually substitutable, an assumption known to be untrue. The task is complicated enormously because species are a multipotential, ever-changing, and renewable resource.

Orians and Kunin defined a "unique" species to be one with a high deletion impact, whose removal would produce a cascade of changes in the community.

198

This definition of uniqueness is based solely on the functional importance of a species in a community. Discussions followed the lead of Orians and Kunin and centered largely on the consequences of species removal for population and community function. Other kinds of ecological uniqueness unrelated to community function were discussed later in the session.

There was considerable discussion about whether deletion impact was a sound measure of biological sensitivity. The conclusion was positive, with the following qualifications: (1) More attention needs to be given to the nonlinearity of impacts resulting from (a) the sequence in which removals are conducted, (b) the number of species removed, and (c) the type of species removed. (2) We need to consider the potential for generalizable results from experiments: to what extent do results obtained in tropical communities generalize to temperate ones, or aquatic to terrestrial? Contained within these qualifications are a rich array of important research issues for applied and theoretical ecologists.

What determines the rates at which surviving species fill the gaps caused by single and multiple species deletions? One hypothesis to test is that taxonomic relatedness is a good guide to suitability. Rosenzweig suggested that the nature of a speciation event would determine the potential impact of deleting the resulting species. For example, communities where two closely related species now compete could have arisen through splitting of one species into two in the same area. In such a community, deletion of one of the closely related species may have little impact, because the congener might easily fill the vacant ecological role. Likewise, in communities where species are taxonomically dissimilar, removing a species might result in a high deletion impact. Nonetheless, the significance of the nature of the speciation event should decline with time as a result of subsequent evolutionary changes in the species. Therefore, as time passes, the type of origin becomes increasingly irrelevant to present-day interaction. This may explain why closest competitors can be completely unrelated (Brown et al., 1979; Woodin and Jackson, 1979).

The impact of species deletions on community diversity was also discussed. Despite considerable research in recent years on species removals, it is still difficult to predict the effect of removal (or insertion) on species diversity. Some deletions actually result in a more diverse community. It was also pointed out that simply lowering the population density of a species may have the same effect as removing it. Often, because it is difficult to remove every single individual, individuals are removed to yield a series of graded densities. These types of studies allow identification of nonlinear threshold effects of different densities of individuals.

In addition to recommending deletion and insertion impact studies of single and multiple species, the importance of experimentally removing whole communities and chunks of habitats was discussed, because this is a type of perturbation occurring extensively in many developing and developed regions. Little is known

199

about the cumulative effects of such large-scale perturbations. While experiments on this scale are impractical, it is possible to monitor the outcome of natural or anthropogenic events that delete large geographic areas.

As originally defined (Paine, 1969), a keystone species is a species of "high trophic status" with a controlling effect on its community. For example, in the rocky intertidal zone of the Pacific Coast, removal of *Pisaster ochraceous*, a predatory seastar, leads to domination of space by the competitively superior mussels and reduction or exclusions of many other space-occupying species. Another keystone species is the sea otter, which consumes sea urchins and therefore allows kelp communities to dominate the substrate (Estes and Palmisano, 1974). The generalization emerging from these examples is that predators on dominant competitors or herbivores in space-limited systems are likely to be keystone species.

A broader definition of keystone species, unrelated to trophic status, is simply those with high deletion impact. For example, beavers are keystone species that construct and maintain a lacustrine habitat used by many other species. Removing beavers would produce a dramatic habitat change and a cascading series of local extinctions of species dependent on ponds constructed and maintained by beavers. Damselfish, by grazing on corals, play a similar habitat-constructing role in marine systems (Hixon and Brostoff, 1983). Finally, any species that forms a monoculture that provides habitat (e.g., mangroves, some corals, salt marsh), is likely to be a keystone species. Keystone effects are also expected upon deletion of any species that is the sole representative of its guild.

While strong interactors (Paine, 1980) may have clear community effects, the sum of many weak interactors may add up to the same result. Does the removal of many weak interactors produce effects similar to those resulting from removal of the dominant? A speculation was that the answer is "no" in the case of unique habitat constructors such as beavers, but probably "yes" where the keystone effect operates through a trophic web.

The group also considered how the importance of keystone species might vary geographically. Are keystone species more common in marine versus terrestrial, tropical versus temperate systems? It was suggested that keystone species are more common in species-poor than species-rich systems. This may be why keystone species appear rare in the tropics. Many separate species may fill the same role occupied exclusively by a single keystone species in a temperate system. On average this may be true, but there are clear examples of keystone species in species-rich habitats, such as the crown of thorns starfish (*Acanthaster planci*) on Australian and Indo-Pacific coral reefs (Chesher, 1969).

It is not known how widespread keystone species are, or in what kinds of communities they are most likely to be found. One point of view suggests that keystone species should not be important in nonequilibrium communities regularly perturbed by unpredictable stochastic events such as bad weather. But there are exceptions to this expectation, most notably *Pisaster* (mentioned above), which lives in a system greatly affected by irregular but strong perturbations by the physical environment (wave action, logs crashing on the rocks).

Potential keystone species may fail to have an effect because they recruit very unpredictably and often suffer massive reproductive failure. In other cases, species removals expected to have a large effect (e.g., removing trout from streams) have no effect on resident populations (Allan, 1982). The explanation may be that the trout system is donor controlled (limited by the number of insects produced, rather than by predation on the insects). If so, as suggested by Orians and Kunin, a high deletion impact is not expected if we remove the consumer.

There is evidence that species whose populations fluctuate widely under "normal" conditions are more likely to go extinct than species having more stable populations. For example, on Barro Colorado Island in Panama, many of the species that went extinct when habitats were reduced by rising water levels had highly variable and unpredictable population sizes before Lake Gatun was formed (Zaret and Paine, 1973).

Species often provide ecosystem-level services unrelated to their importance to community stability and function. For example, nutrient cycling is a critical function often accomplished by species, such as nitrogen-fixing bacteria, whose ecological role is otherwise inconsequential. In some cases the rates of critical ecosystem processes may be controlled by key species. An example, in addition to those discussed by Orians and Kunin, is the sea otter-urchin-kelp relationship. It has been suggested that on islands where sea otters live and therefore kelp abound, the food web is detritus based. Where otters are absent and macroalgae are rare, the food web may be less productive and plankton based (C. Simenstad and D. Duggins, University of Washington, pers. comm.). The origins of nutrient input are poorly known in many marine systems, but generally can be described as either detritus or plankton based. A detritus-based food web is dependent on production of detritus from kelp and other macroalgae. The community-level ramifications of these differences are not yet known.

Species contain important, unique traits that often are not directly related to their community role. Among these are genetic, taxonomic, trophic, defensive, reproductive, and architectural features. Uniqueness in any of these functions provides us with valuable information about the ecology and evolution in natural systems. Species unique in one or more of these traits may therefore be more important than species not so unique. *Sphenodon*, the last surviving representative of a reptilian lineage, for example, may be considerably more valuable than a species of iguanid lizard, because there are hundreds of other species in that family. The last *Sphenodon* would contain valuable taxonomic, evolutionary, and genetic information to aid in unraveling the evolution of reptiles and higher vertebrates.

201

Considering ways of evaluating different classes of uniqueness posed the most difficult problem for workshop attendees. Theoretical and experimental community ecology provides a rich body of knowledge with which to rank species value in relation to their effects on other species. Unfortunately, however, little is known about how these measures relate to ecosystem functioning or taxonomic uniqueness. Building a conceptual and empirical base for addressing these problems is a key objective of ecological research designed to serve purposes of preserving biological resources.

REFERENCES

Allan, J. D. 1982. The effects of reduction in trout density on the invertebrate community of a mountain stream. *Ecology* 63:1444–1455.

Brown, J. H., D. W. Davidson, and O. J. Reichman. 1979. An experimental study of competition between seed-eating desert rodents and ants. *Amer. Zool.* 19:1129–1143.

Chesher, R. H. 1969. Destruction of Pacific corals by the sea star *Acanthaster planci*. *Science* 165:280–283.

Estes, J. A., and J. F. Palmisano. 1974. Sea otters: Their role in structuring nearshore communities. *Science* 185:1058–1060.

Hixon, M. A., and W. N. Brostoff. 1983. Damselfish as keystone species in reverse: Intermediate disturbance and diversity of reef algae. *Science* 220:511–513.

Paine, R. T. 1969. A note on trophic complexity and community stability. *Amer. Nat.* 103:91–93.

———. 1980. Food webs: Linkage, interaction strength and community infrastructure. *J. Anim. Ecol.* 49:667–685.

Woodin, S. A., and J. B. C. Jackson. 1979. Interphyletic competition among marine benthos. *Amer. Zool.* 19:1029–1043.

Zaret, T. M., and R. T. Paine. 1973. Species introduction in a tropical lake. *Science* 182:449–455.

5

Valuation of Genetic Resources

Gardner M. Brown, Jr.

In writing this chapter I am awed by how little, it seems, economics can contribute at present to the valuation of genetic resources. A natural explanation is that since most of the genetic resources of interest do not trade in markets, there are no prices. And it is unlikely that price data will soon appear.

In the absence of prices, we can draw on a body of economic theory to guide evaluation of some nonmarket values, if there are adequate observations of the behavior of consumers or firms related directly or indirectly to genetic resource use. Such investigations would, in principle, provide lower boundary estimates of the value attributed to being able to view or hunt some species. However, these are glancing blows and miss the center of the problem or the potential value of genetic resources.

To use a frequent analogy, the problem is not in valuing the books we have read, but in deciding which books to preserve from the vast array of unread books. The books we have read make up a small share of all books. From the unread books we must choose not only those that appeal to us but also those valuable for future generations. Their tastes will differ from ours in unknown ways because those tastes are a function of knowledge, and we do not know where the path of knowledge will lead. At least, our understanding of epistemology is too imprecise to guide us in a yes/no decision about species preservation.

It may be that the role of economists in the arena of genetic resources will be limited largely to urging noneconomists to utilize some simple but useful principles in their deliberations, such as substitution or trade-offs. Money spent listing more endangered species is money that cannot be spent to preserve those already listed. Or, self-evident budget constraints necessitate a choice about which species to save and which to sacrifice. We are not rich enough to save all species and enjoy other goods and services thought to be important. Other things being equal—they rarely are—species with a greater likelihood of usefulness in the next decade are more valuable than species with a comparable likelihood of usefulness in the more distant future. Time preference—the discount rate—makes this so. Other things being equal, genetic resources that have the least cost of preservation ought to be ranked above those with greater costs. When other things are not equal—say, some species are believed to be more beneficial than others—then some decision about ranking rules, at least of an ordinal nature, is in order. If this is not done, the opportunity cost of saving an *important* species may be the loss of two or more *significant* species, which, together, are valued more highly.

The paper proceeds as follows. I will point out some salient economic features of genetic resources that make their valuation an important issue. I will remind the reader of the consequences of budget constraints. The kinds of benefits genetic resources provide will be discussed, and ways of estimating them. I will also summarize some past attempts to estimate values.

Many suggest that a species is worth preserving if it yields products of commercial worth. It is not clear what to make of this suggestion. Armed with an order of magnitude value for the rosy periwinkle, or Madagascar periwinkle (*Catharanthus roseus*), we surely cannot conclude that all plants containing alkaloids (such as the rosy periwinkle) should be preserved, or plants containing more than four or five alkaloids should be preserved, even if the rosy periwinkle has valuable medicinal uses. The results of the plant screening program of the National Cancer Institute are not encouraging. Approximately 35,000 plants have been screened since 1956, and none have produced results in cancer treatment that would make them commercially successful. To be sure, evaluation was limited, and exhaustive tests were not used. Not all parts of a plant were studied thoroughly. According to Myers (1983), the rosy periwinkle would have been missed by the screening procedures.

The recognized value of a species does not mean that it needs to be saved wherever it grows. The rosy periwinkle grows in many parts of the tropics, but there may be some habitats that have alternative uses with a higher economic value. The rosy periwinkle can still be harvested on the remaining habitats.

Many seem to conclude from the demonstrated value of so many species that there is much more wealth just waiting to be discovered by unlocking nature's secrets. Perhaps the future will bear out this belief. One could argue, however, that most of the valuable germ plasm has been discovered and we are facing diminishing returns. I know of no theory that provides a convincing argument one way or the other.

Norman Myers argues that 70 percent of the plant species that possess anticancer properties occur in the tropics. An acre in the tropics or in an arid zone is thought to be relatively more valuable because of a greater abundance and diversity of botanochemicals in these regions, arising from pronounced ecological competition. Should we save more or less habitat in Pakistan because 17.6 percent of the plants screened show anticancer activity and the next highest fraction is 12.4 percent in Sri Lanka? (Myers, 1983:114).

On an individual species basis, one could argue that once a cultivar has contributed to an improvement in a plant, or its natural produce has been synthesized, it need not be preserved. Or, put less dramatically, the plant should be ranked lower and extinguished sooner than other plants. Past value does not guarantee future value.

The valuation of genetic resources is considered important because of the belief that genetic resources are extremely valuable, so much so that we cannot afford the predicted rate of extinction during the next century. Why the concern? If individuals or corporations believe that oil, minerals, and other natural resources will be more valuable in the future, they redistribute use to the future by reducing the rate of extraction today. Intertemporal arbitrage takes place until, at the margin, there are no apparent remaining rewards for making further trades. As long as buyers and sellers are able to trade in the resource markets, the rates of extraction reflect the collective preferences of society. The preferences of future generations are looked out for by the profit-minded today. Profit tomorrow comes from preserving today. Either the past and expected rate of extinction is optimal, given the tastes, endowment, and resource constraints of society, or the argument above is flawed.

The profit motive, combined with free entry into and exit from markets, works well when markets exist. In fact, futures markets don't extend beyond a year or two for any commodities. Thus the nonexistence of a market view of the price of lumber a decade from now (on average or for any given state of nature) leaves people on their own, unable to benefit from buying or selling risk connected to future lumber prices at the competitive market rate. This is the important service a well-functioning futures market provides. Many would conclude that failure of the futures market results in people acting as though future risk is more costly than a futures market would reveal. The higher risk premium tilts the extraction path more toward the present and leads to "too fast" rates of extinction.

Contemporary markets don't exist for any genetic resources because it is difficult or impossible to exclude users of the resource. Migrating animals were referred to in the early laws as things of the negative community, because they could not readily be bent to private property. Salmon, blue whales, and migratory waterfowl are examples. One can imagine artful institutional arrangements, designed to bend public into private property (such as the adjudication of groundwater aquifers), which seek to allocate a depletable resource in some quantitative way to all those having an acknowledged access to it. Jon Goldstein's paper in this volume addresses this issue more carefully.

Even if contemporary markets for genetic resources existed, the undepletable characteristic of many would bring about a nonoptimal rate of preservation. This is well illustrated by viewing genetic resources as a source of knowledge. (See Baumol and Oates [1975] for an excellent discussion of depletable and nondepletable externalities.) Since genetic resources are the linchpins in research and development programs, the literature on market failure in the context of R&D has immediate bearing on the issue at hand. Dasgupta and Stiglitz (1980) review many features about knowledge which drive a wedge between market performance and efficiency. Knowledge is a public good of a nondepletable variety. Once a new seed variety or a new drug has been discovered, it costs nothing to produce the knowledge for someone else, although it does cost something to produce the seed, medicine, or industrial product in which the new knowledge adheres. Since no resources are employed to produce the knowledge for the second and subsequent users, no price for the information should be charged. Clearly the incentive to discover new ideas is diminished as financial rewards decrease; there would be no discovery unless firms could acquire patent rights or find other means of rewarding their efforts. But with patents comes market power and its associated market distortions. Notice that when there is market power, the larger the market, the greater the production over which the costs of R&D can be spread. Put another way, there are increasing returns to the use of information, and this, in turn, results in fewer firms. The process encourages the acquisition of more information and further exacerbates the ills associated with market power.

Genetic resources labor under another disadvantage in a market economy. Even with perfect patent rights to established ideas, access to the pool of genetic resources allows competitors to create substitute products, such as drugs or seeds, which corrode the profitability of the earlier discoveries.

Brown and Swierzbinski (1982, 1983) have shown that no market structure preserves an optimal number of species when knowledge, once discovered, becomes freely available to all. The idea in this paper is that genetic resources are a store of knowledge of uncertain value. As more species are preserved (fewer extinguished) and greater preservation costs are incurred, the distribution of knowledge is enriched. The expected value of a draw from a richer distribution is greater, as is its variance. The purpose of a draw or experiment could be to improve the quality of product (greater resistance to leaf rust, a wider spectrum of antibiotics), but we assume that a favorable discovery results in lowering the cost of production of a good such as a drug or seed variety.

Three separate economic discrepancies arise between the social and private optimum. First, firms do not take the consumer's net benefits into account. That is, the firm's demand function is not society's, except in the limit (unless one firm perfectly price discriminates or the number of firms is very large). A common nostrum for externality problems is to privatize the property. We know this solution won't work perfectly, because the discrepancies cited above remain. It doesn't even seem to work well in the cases we examined. Welfare losses were between 16 and 47 percent when the monopolist was given an optimal subsidy equal to 92 percent of the investment cost in order to induce him to preserve the optimal level of species preservation. (The range for the loss stems from different assumptions about the cost of R&D. The proportion of research costs in relation to expected net benefits ranges between 10 and 70 percent. The loss rises to 29 to 86 percent as demand elasticity changes from 1.1 to 3; see Brown and Swierzbinski [1983]). Having done this, the firm fails to produce the optimal output, because it loses monopoly profit in so doing.

Second, the public goods aspect of genetic capital is reflected in the fact that each firm only considers $1/n$ of the cost reduction that the discovery occasions; there are n like firms in this industry. Third, a firm's ability to profit from cost reduction is eroded by similar responses from like firms.

The structure of our model is such that any assumption about the commonness of knowledge, short of complete privateness, produces the same qualitative results. Moreover, we have also shown that the same three economic discrepancies arise when there is one firm in each of many markets producing substitute goods but commonly benefiting from a new discovery.

LIMITED BUDGETS NECESSITATE RANKING

The model just discussed took for granted that probably not all species should be saved, just those that could pay for themselves. If we can't save all species, we need a ranking based on one or more criteria, from which we select the highest ranked for preservation. However, scientists, with infrequent exceptions, display an almost universal antipathy to ranking species. The aversion arises perhaps from moral reasons, but it may also spring from the biological idea that all species are equally important, since evolution is a random walk. Such a view would make more sense in a world with a zero discount rate. The positive interest (discount) rate signifies that a good event has more value today than the same good event in the future. Thus compound interest forces us to take a more myopic view of matters. Hard reality does too, because the fact is that species are becoming extinct at an increasing rate. Opler (1971) has an illustration showing that thirteen or more species have become extinct in the United States in each decade since 1880, except for five in the 1970s. The Council on Environmental Quality (1981:31) believes that species may be becoming extinct as frequently as daily.

Our collective failure to save all species suggests either that public institutions designed to implement public preferences for preservation are flawed or that other goods and services are thought to be more important than the species that are allowed to be snuffed out.

If species truly cannot be ranked, yet are thought to be important, a rational prescription is to maximize the number of species saved per period for a specific outlay. We therefore should rank species by their cost of preservation and systematically move up the list, starting with the least cost, until the available budget is exhausted. Several comments are in order. First, the results of following a least cost preservation rule almost certainly are inconsistent with current preservation activities. The California condor has cost more than $5 million to preserve. The cost of land acquisition alone was more than $500,000, according to the U.S. General Accounting Office (GAO). Over $3 million worth of land has been purchased to preserve the Key deer (GAO, 1979). The operation and maintenance costs alone for Siberian tigers in zoos are over $2.5 million annually, according to William Conway, the director of the Bronx Zoo (Hahn, 1980); whereas 500 okapis cost less than $1 million. This money alone probably would save more than two species, say a couple of tropical insects, even if we netted out the marginal viewing benefits of the tigers and okapis. The legal costs of the battle to save the snail darter arguably would have saved an insect or two, and one could have known this in advance.

Second, people's preferences very likely are inconsistent with the results of the least cost preservation rule. For example, professional employees in the U.S. Office of Endangered Species were asked to rank individual endangered species for their economic, aesthetic, social, and ecological value on a scale of one to ten. Not only do the values for individual species vary, but the means for the grouped species were not the same (at either the 1 or 5 percent significance level) (Brown, 1982). Thus, within the class of endangered species, some are believed by experts to be more valuable than others.

Third, although the least cost prescription has dubious value in general, it makes sense in some cases, such as when valuation studies cannot be done because of inadequate time, money, or knowledge. I found the most dramatic expression of this principle at the National Zoological Park Conservation Center, Front Royal, Virginia. With limited budgets, according to Guy Greenwell, a bird curator there, birds with relatively superior breeding ability in captivity were chosen for preservation. Self-feeding birds were chosen over hand-fed birds, and birds that would survive in open yards were chosen first. The park had inherited from the U.S. Cavalry roofless yards used to enclose horses. As the facility became established—that is, budget constraints relaxed—roofed pens were added. Gradually, more exotic, expensive to care for endangered species are being added. Similarly, for the hoof stock and small mammal program, the relatively hardy animals with modest demands and enclosure requirements were chosen first, according to Larry Collins, the manager of this section of the preservation program.

Fourth, elements of biogeography in their naive form are suspicious in light of the discussion above. Island biogeographers such as MacArthur and Wilson (1967) postulate a relation between species (S) and area (A), of the Cobb-Douglas form (an explanation of this form is found in Ricklefs, 1973:709–714), familiar to economists: $S = cA^z$ (where c and z are parameters whose values are fitted to each set of data). The relationship is empirically supported by data for alpine mammals on mountain tops, birds in woodlots, fish in lakes, marine worms in artificial sponges, and so forth. As ecosystems decrease in size, the

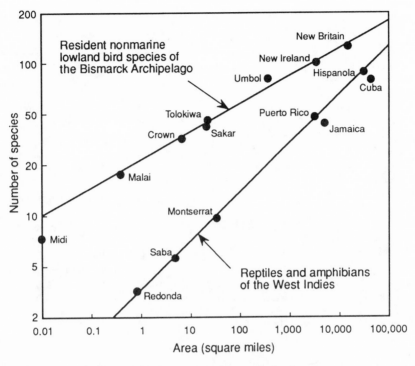

FIGURE 12. Relationship between number of species and island size.

number of species always decreases (Council on Environmental Quality, 1981). The point is illustrated in Figure 12.

MacArthur and Wilson (1967) illustrate a menu of values for the elasticity of species with respect to area, as seen in Table 9. There is a surprisingly narrow range of $0.24 < z < 0.49$.

A model of this sort makes economists a bit uncomfortable. To see why, assume that the unit price of A is constant. Then it is easy to show that we ought to try as hard as we can to save as many *very* small ecosystems as we can because an areal expansion in percentage terms has a less than proportionate favorable effect on the added species saved. Thus there are uniform diminishing physical returns, since the cost of saving species rises continuously as the area purchased increases.

One can fiddle with the constant price assumption, but my guess is that it would be more promising to revise the production function relation. Probably there are increasing returns over some range for many habitats—that is, there are ecosystems sufficiently complex that the marginal cost of preserving species

209

TABLE 9
Species-area relationships among various species groups on islands.

Fauna or Flora	Island Group	z	Authority
Carabid beetles	West Indies	0.34	Darlington (1943)
Ponerine ants	Melanesia	0.30	Wilson (Orig.)
Amphibians and reptiles	West Indies	0.301	Preston (1962)
Breeding land and freshwater birds	West Indies	0.237	Hamilton et al. (1964)
Breeding land and freshwater birds	East Indies	0.280	Hamilton et al. (1964)
Breeding land and freshwater birds	East Central Pacific	0.303	Hamilton et al. (1964)
Breeding land and freshwater birds	Islands of Gulf of Guinea	0.489	Hamilton and Armstrong (1965)
Land vertebrates	Islands of Lake Michigan	0.239	Preston (1962)
Land plants	Galapagos Islands	0.325	Preston (1962)

SOURCE: MacArthur and Wilson (1967).

decreases over some range. Increasing returns are critically important when we are agnostic about the value of any species saved and increasing returns are attractive in the more realistic instances when species can be differentiated by worth.

There is an important message here even if one deprecates island biogeography in its simplest form. It costs different amounts of resources, such as money, to save different species. These costs are necessary ingredients in future preservation decisions. If they are disregarded, we shall save one species at the expense of more than one without knowing the cost of what we have sacrificed.

Society does rank some species higher than others. Generally, more familiar mammals are preferred relative to insects. Lovejoy reflected this view, according to Hahn (1980), when he said "there is probably considerable argument for attempting to save all forms of anthropoids, on the basis of what we can learn of use to ourselves from studying our closest relations."

It is also true that some species are believed to have more scientific value than others. Jon Goldstein and I (1983) argued that scientists entertain beliefs about what drugs, seeds, and livestock are weak and will be most vulnerable to breakdown in the future. There is a continuous sampling of known gene pools to find plants and animals superior in one or more characteristics (phenotypes). The species that are expected to be the repository of the next period's products are, other things being equal, more valuable than species that are not, in some way, candidates for experiments.

We developed some simple models showing how to value species. The most important knowledge necessary for these primitive models included density functions for failure times for each relevant phenotype, density functions for successful experiments providing superior phenotypes, costs of preservation, cost functions for experiments, and demand functions for the final product(s). Much of that knowledge is not in a codified form, but what is codified is kept quite secret in order to maintain one's competitive advantage. Under these circumstances, empirical economic analysis in the matter of valuing species can be expected to proceed quite slowly.

Other preservation criteria have been set forth that have economic content and should be mentioned. Many years ago Ciriacy-Wantrup (1952) argued that we should follow the principle of the safe minimum standard. Since the extinction of a species is irreversible, we might discover that society had forgone immense benefits by allowing this to happen. Thus we should think in terms of saving minimum viable populations and the necessary supporting habitat of all species. Ciriacy-Wantrup was not an absolutist. He argued that the cost of this strategy would be small insofar as there are close substitutes for the goods and services of the threatened species, especially if there is a technological bias in favor of these substitute goods.

Bishop (1982) reiterated this theme and likened it to the strategy of minimizing the maximum possible loss. He added: unless the social costs are unacceptably high. Neither Bishop nor Ciriacy-Wantrup gave any operational hints about when the social costs are prohibitive, or how the expected benefits could be computed so we can calculate the social costs. Bishop also might have warned us that the mini-max strategy typically violates the usual postulates of rational behavior. It puts too much weight on the extremes and not enough on the middle outcomes.

Direct Productive Value

Many species have an easily ascertainable commercial value. It is not surprising that harvest pressure will greatly reduce the stock of these species and can drive them to extinction unless harvest is regulated or the harvester's cost of searching out members in a diminished population exceeds the harvest price. Harvest can be regulated with a charge system or with various rationing rules, such as time and area closures and minimum and maximum harvest sizes. Regulation can be set by the public agency or the private owner, if one exists. Three instructive examples of resources having direct productive value are the passenger pigeon, the Pribilof fur seal, and the blue whale.

The passenger pigeon is a marvelous case study for the economist. The passenger pigeon was valuable for its meat and, in part, for its feathers. Four factors contributed to the bird's extinction. First, as hunters reduced the population from several billion in the late nineteenth century to zero in the early twentieth century (1914), the cost of search did not rise, because the pigeons had the unfortunate tendency to flock rather than disperse after being shot at. Second, technological progress permitted the unit cost (not the economic cost) of harvest to fall or remain stable. Third, the market price of passenger pigeons did not rise, because chicken was regarded as a close market substitute and the price of chicken remained stable during the passenger pigeon's demise. Since price did not rise to signal growing scarcity, there was no economic force inducing entrepreneurs to attempt to save the pigeon, because there was no evident economic scarcity rent to be earned. I have no explanation for the fourth element, the failure of the public agency to intervene on behalf of the passenger pigeon. (See Tober [1981] for an excellent but brief history of the passenger pigeon.

Stocks of the Pribilof fur seal fell to precariously low levels as a result of hunting pressure from the United States, Canada, Russia, and Japan just after the turn of the century. A treaty was struck that gave the United States the right to harvest the seals, using primitive methods, but guaranteed each of the four countries a specific market share of the harvested pelts. With property rights restored or acquired, it was possible to build the stocks back up to acceptable levels (U.S. Department of Commerce, 1985).

The case of the blue whale is similar to that of the Pribilof fur seal in the sense that hunters from many countries had drawn down the stocks to precariously low levels during the first half of this century. The blue whale is an exceptionally large mammal, weighing well over a quarter of a million pounds as an adult (Spence, 1973). It is a commercially valuable harvest because the meat has a ready market in Japan. The fact that the current catch is largely incidental to the catch of other whales makes harvesting the blue whale profitable. Although the International Whaling Commission has been successful in reducing the number of harvesters, Japan and the Soviet Union continue to harvest the blue whale, and its preservation for the future remains uncertain.

In the examples above, a member of the species is harvested. But there are other important harvests that need not be destructive, as when fruits are picked, latex is drawn from rubber plants, or lac is used to produce shellac. Lac is secreted by insects, *Laccifer* spp. (Oldfield, 1984; Prescott-Allen and Prescott-Allen, 1980).

At this point the categories of value I have chosen become a bit strained. Except in unusual cases, the natural resources harvested are combined with other inputs, typically purchased, to produce an intermediate or final product which is then sold on a market. The demand for, or value of, the resources is derived from the demand for the product. Strictly speaking, the value of the resource *in situ* is indirect, if the resource is simply moved in time or place. I am taking liberties with my categorization, using indirect to refer to a natural production function separating a commercially valued output from a noncommercially valued input. Natural materials harvested for use in drugs and other industrial products qualify as directly productive in my scheme. Here the examples are myriad, since it is believed that half of the world's medicine and nearly half of the prescriptions written in the United States contain ingredients of natural origin (Schery, 1972). Foxglove is used to produce digitalis, the rosy periwinkle is used to obtain vincristine and vinblastine to treat leukemia in children, and so forth.

Direct Consumptive Value

Many species are valuable because they provide either food or recreation directly to the consumer who is the hunter or fisherman. For example, it is reported that Steller's sea cow became extinct in 1768, twenty-seven years after discovery, because ship crews killed them to provide variation in their diet; and big game animals are particularly attractive to sport hunters.

In the first case (direct productive value), economic values accrued directly to firms (ship crews); in the second, economic value accrues directly to the consumer (the sport hunter). In the first case, input prices, the production function, and output prices are needed to discover value. In the second case, the same information, along with the household's utility function, or its equivalent, is required, since the product price typically is not observed directly, because the market for the right to harvest the species usually does not exist in most countries, although there are exceptions. In Alaska one can purchase rights to harvest salmon. The permit is region specific and gear specific. In some cases the permits are worth over $100,000. In some streams in Canada the right to harvest salmon is auctioned off periodically. In Scotland one can purchase access to salmon by buying a particular length of shoreline along a river, rod days, and other specialized forms of property rights. Indians in the state of Washington provide access to salmon with tie-in sales: when one rents the services of a guide, one is allowed to catch salmon on rivers within an Indian reservation.

Indirect Productive Value

Species can also be viewed as making a valuable contribution to the ecosystem of which they are an integral part. A particular species may not have commercial value but it may be the predator or the prey of a species that has estimable economic value. It is of no conceptual consequence if the predator-prey relation is direct or indirect. Commercially valuable bottom fish can graze on the necks of noncommercial clams, or a commercially valuable species can eat bottom fish, located in nonharvestable areas, which in turn graze on the noncommercial clams. In either case the existence of the natural interspecies production relationship between output and input gives rise to the economic values, through natural demand of the worthless clams.

Much has been written about the importance of species in augmenting the genetic diversity of an ecosystem. It has been argued that more species are better than less because there is greater stability (Ehrlich and Ehrlich, 1981). The economic equivalent of this consequence, the marginal increase in stability, if true, is a difference in the risk premium associated with different classes of risks or states of the world. Orians and Kunin further analyze the biological foundations of the stability and complexity debate in their paper.

Nonconsumptive Values

When species have harvest value, they are twice vulnerable. Economic value is derived from drawing down the stock for today's use, and competitive demands for the habitat can threaten the species. There are several nonconsumptive sources of value genetic resources provide that don't depend on harvest. They are conceptually interesting because they fall into the class of undepletable externalities. Your visiting or photographing rare species does not keep me from viewing them. Knowledge obtained from genetic resources and embodied in better medicines and seeds is another example. Studies carried out in natural marine or terrestrial laboratories by one group does not deny the opportunity to another. (As always, caveats are in order. When congestion comes in the form of too many viewers or too many experimenters, one's use has spillover effects on others, and the aggregate activity of the viewers can change the viewed. According to Western and Henry [1979], the probability of viewing a lion or a cheetah in its habitat depends on the number of vehicles in the local area.)

Future Uncertain Nonconsumptive Value

One of the classic articles in natural resources literature is John Krutilla's "Conservation Reconsidered" (1967), in which he brought the idea of option demand into the conservation literature. Previously, Weisbroad (1964) had argued, in effect, that in the absence of contingent future markets, conventional site valuation studies would underestimate demand; they would capture only the behavior of people who visit a site. What about the people who are uncertain about visiting a site in the future, but who, in fact, never will purchase (visit) it? They should be willing to pay to preserve the option of purchasing the site in the future. There is, therefore, an option demand and a corresponding option value. (Note that Weisbroad was concerned about the value for those who never exercise the option. Schmalensee [1972:813] later defined option value as the maximum amount a consumer with certain preferences—and possibly income —would be willing to pay for the option to purchase a particular commodity at a specified price.) Krutilla elaborated on the determinants of an option demand

214

function. In the process, he integrated heretofore extraneous themes in the con-servation literature which earlier had been obscurely presented to economists by Ciriacy-Wantrup.

Krutilla identifies three sources of nonconsumptive value: existence, bequest, and scientific. Some people may derive value from knowing that certain species exist although they do not intend to view them. Or some may contribute to species preservation because they believe in the efficacy of species diversity. For example, contributions are made to the Nature Conservancy, which purchases habitats around the nation, or to the World Wildlife Fund, whose expenditures are typically used to preserve exotic species in remote corners of the world, where contributors are not likely to go. This is what is meant by *existence* value.

Some may not want to see a species or habitat but want their children to have the opportunity to do so. They would have an interest in purchasing an option in order to *bequeath* the possibility of seeing this species to a future generation. Future availability is a collective consumption good (the thrust of Weisbroad's paper), in that making something available to one makes it available to all. The vicarious pleasure I get out of knowing that my children will have an opportunity to view eagles along the Skagit River is different from the value of the eagles' future use to my children. My value enters my utility function. My children's value enters their utility function. In some cases, future availability may depend on the purchase of enough options to elevate preservation values high enough to dominate the value of extinction.

The third element is the *scientific* value of public knowledge that a species might provide in the future. This has been discussed elsewhere in this paper. An interesting example of private action to preserve species motivated by scientific values is the case of thirteen scientists who put up $1,000 each to buy 1,500 acres of a Costa Rican tropical rain forest for long-term ecological research. The Organization for Tropical Studies contributed another $50,000, and an appeal for funds published in a open letter in *Science* brought in another $22,000. The National Academy of Sciences indicated that the site purchased was one of the foremost promising sites worldwide for research on the fragility of tropical ecosystems (Norman, 1981).

A common feature of most genetic resources that are endangered is the irre-versible nature of some choices. A momentous decision in conservation history illustrates the point. John Muir spoke in a moving way about irreversibility when he argued against damming California's Hetch Hetchy Valley in Yosemite Park. It would "forever" remove the uniqueness of the valley for park use (Weisbroad's example) even if the dam were eventually removed. The dam would "irrevers-ibly despoil the great granite walls" of one of "Nature's rarest and most precious mountain temples, . . . a majestic wonder fashioned by the hand of the creator" (Muir, 1962).

One theme in this literature is whether option price is greater than, equal to, or less than the expected consumer's surplus. Noneconomists should think of consumer's surplus as the satisfaction from purchases made at less than the maximum price one is willing to pay. Long (1967) argued that option value was the same as consumer's surplus. Lindsay (1969), Cicchetti and Freeman (1971), Krutilla (1967), and others suggested that option price was greater than consumer's surplus. Others, including Schmalensee (1972), Graham (1981), and Hartman and Plummer (1982) have concluded that option value may be greater than, equal to, or less than consumer's surplus. The answer, of course, depends on how the problem is formulated. One discrepancy between option price and expected consumer's surplus is a risk premium that can arise if people are risk averse. In recent theoretical formulations, option price can be less than expected consumer's surplus on the assumption that the good is normal although the difference is likely to be small. Unlike a financial option, in these formulations, if the valuation of preservation falls relative to development, one must continue to pay the option price rather than choose not to exercise the option. Consumer's surplus varies throughout the world. However, the option price is constant over states of the world. When income is low and demand for a normal good is low, option price requires relatively greater sacrifice (the marginal utility of money is high) compared to when income is high and option price is a bargain. Because the utility function is concave, the pain of the constant price when there is a negative income perturbation is greater than the corresponding pleasure of a positive perturbation.

Irreversibility combines with imperfect information to provide another definition of option value, developed in papers by Arrow and Fisher (1974), Henry (1974), Hanemann (1983), and Fisher and Hanemann (1984). (Note that future use is compatible with the line of thought developed below, so this view of option value departs from Weisbroad's definition.) Suppose in the context of genetic resources that today's value of preserving each of n species is known with certainty. However, the information about tomorrow's value of preservation relative to extinction is known imperfectly now and will be known much better tomorrow. There is no expenditure we can make today to remove or reduce the veil of ignorance about tomorrow's values.

Irreversibility introduces an important asymmetry. Tomorrow, we can always extinguish a species preserved today if the new knowledge acquired warrants it. But we cannot preserve a species tomorrow that was extinguished today if the new knowledge informs us that we should have done so. Option value in these models is associated with the benefits captured by marginally postponing development today in order to better capitalize on new knowledge in the future. In the models developed to date, the amount we learn does not depend on the level of development in the first period. Of course, the benefit of the learning does depend on the level of development in the first period. Option value in these models is the (conditional) value of better information. (Conrad [1980] calls it the value of information; Fisher and Hanemann [1984] emphasize that it is the conditional value of information.) It is the difference between payoffs arising from two solution strategies. One can find today the *best values* for each period using, say, expected values of the random variables. (Miller's [1981] model substitutes growth in derived demand for the irreplaceable resource for knowledge and removes uncertainty.) Alternatively, one can choose today a *best policy rule* to follow which determines the best values for preservation to choose, conditional on choosing the best values in each subsequent period. This dynamic policy has to produce a bigger payoff, because choices each time are made with the better information.

It is important to know that solving the problem correctly, using dynamic programming, leads to preserving more species, as opposed to finding an incorrect solution to the problem. Mendelsohn (1984) has made the useful point that as the amount of information to be gained decreases and as the arrival time of the information lengthens, the option value decreases.

ALTERNATIVE PROCEDURES FOR VALUING GENETIC RESOURCES

Before illustrating estimates of the value of genetic resources, it is useful to discuss the five principal ways in which these resources can be valued. They are the travel cost procedure, hedonic analysis, discrete choice techniques, willingness to pay or contingent valuation methods, and indicative values such as the price of a genetic resource-intensive product. These major approaches are used to discover the value of consumptive or nonconsumptive goods or services in the absence of markets.

Travel Cost Procedure

Originally the travel cost procedure was used, exploiting the simple fact that people from different locations incurred different costs to visit a site, for example to view an endangered species. A demand curve for trips can be estimated, because different costs give rise to different rates of participation. Observing the frequency of trips taken by someone facing high costs tells us about the consumer's surplus, or *net* benefit, enjoyed by those who incur low costs, after controlling for other differences in viewing demand determinants among the population. This technique works best when viewing the species in question is the only purpose for taking a trip, as when trips are taken to view humpback whales in Hawaii or gray whales off the coast of California. When trips have a multiple purpose or when one sees numerous species at a site, as when trips through African game parks traverse many sites and many animals are viewed, the travel cost technique does not work well. When there are multiple characteristics, one must employ a generalization of the travel cost procedure.

217

Hedonic Method

The second approach has two versions. The hedonic travel cost method, developed by Brown and Mendelsohn (1984), views the consumer as incurring increasing costs to acquire increasing benefits. Presumably the consumer's "cost function" of viewing varies with his place of origin. For example, a package tour in which one sees tigers plus a vector of other experiences might cost $50 more than a tour offering only the vector. An individual tiger viewing then has the implicit value of $50, which must be magnified by the forecasted stream of visitors. (Probably the data will require ingenuity to extract values, such as developing the probability of sightings and the duration of sightings.)

Presumably the consumer's "cost function" for viewing varies with his place of origin. In principle, then, observations across places of origin and across destinations provide sufficient information to estimate the viewing demand for an endangered species.

The second version, typically called the "household production function approach," focuses in a detailed way on how a viewer produces a viewing. These versions differ in details not essential to this discussion. Each has its own set of strong assumptions in order for the theory to work well. Each requires a body of data of sufficient size that few if any existing data sets are useful.

Brown and Mendelsohn (1984) applied their hedonic travel cost model with a modicum of success to more than fifty streams in Washington to estimate the value of salmon. Since the values obtained are specific to a site or set of sites, one could, in principle, estimate the loss of a particular variety of salmon. Strong and Hueth (forthcoming) applied the household production function approach to estimating the value of fish in Oregon.

Discrete Choice

For many decades, investigators concentrated on nonmarket valuation of single sites. In response to an increasing public demand for valuing the characteristics of sites, researchers have broadened their analysis to cover multiple sites in order to capture variation in site characteristics or qualities.

Discrete choice models are well established in the economics literature (see Amemiya [1981] for a useful survey), but have been applied in limited ways to value natural resources. There are no applications in the area of genetic resource valuation. However, the potential usefulness of these models warrants their mention. For example, if rare birds are relatively valuable economically, we would expect to see relatively more visitation at the habitats of the rare birds. Estimates of the viewing value of a particular rare bird can be ascertained from discrete choice models.

Hanemann has the best account of discrete choice models applied to natural resources (see Bockstael et al., 1984). In each decision period, individuals are assumed to make a discrete decision about whether or not to participate in an activity, such as viewing a rare bird. The choice of what to do—which bird or set of birds to watch—is studied in the second stage of analysis. The benefit of maintaining or improving any qualitative characteristic, if it is quantifiable (and sought out), is then computed from the two-stage analysis.

Other Approaches

The fourth approach used to measure consumptive and nonconsumptive values is to ask consumers how much they are willing to pay (WTP) for the right to bag a species or how much they are willing to sell (WTS) their implicit or hypothetical right to hunt or bag or view a species. Davis (1963) first developed the method, and it has been used subsequently by many.

The fifth approach is to use indicator or proxy values as a substitute for measuring the true values of species. The most common proxy indicator is the price of a product of which the genetic-based resource is only one part. The sales value of a drug containing natural components is an example. Valuing species lost by oil spill, toxic discharge, or other human acts is often accomplished by using the price or cost of replacement. Price lists of firms supplying organisms for experiments or supplying animals to zoos are one source of replacement costs. There are serious drawbacks to this approach. The price of an organism to a zoo or laboratory differs greatly from the price of the organism *in situ*, which is the value we are after. Embedded in the market price are the costs of collection, storage, and transportation to storage sites. These costs are not losses to society. Even the normal profit component is not a cost to society if that profit can be earned by collecting the same organisms elsewhere, where they were not destroyed by the spill. Some might want to argue that since some biomass was lost, it ought to be replaced. If it is not replaced, the damaged parties ought to be given the costs of the replacement. This argument might hold up in court, but the economic flaw is that it does not take into consideration the notion of opportunity costs. (In the August 11, 1980, appeal to *Commonwealth of Puerto Rico* v. *S.S. Zoe Colocotroni*, the judge rejected the argument for replacement costs since, as a practical matter, the organisms were not to be purchased and would not survive the return to their damaged habitat. See *Commonwealth of P.R.* v. *S.S. Colocotroni*, U.S. Court of Appeals, First Circuit, No. 78-1543, 79-1468, August 12, 1980.) Suppose that replacing a given number of amphipods takes labor, capital, and other resources that could have produced goods valued at $1 million by consumers. If, indeed the amphipods are replaced, society will have given up $1 million worth of goods for the replaced amphipods. Society will have made a poor trade if the replaced amphipods are not worth at least $1 million. Other resource development agencies in the United States have long recognized this idea and have adopted the following standard used in project analysis: benefits (losses) are limited by alternative costs (the value of damage costs in alternative use). Replacement cost certainly is an overestimate of benefits in most instances.

A second valuation technique used in practice also draws on the concept of alternative costs. A swamp might generate natural methane gas. If the swamp is drained, the methane gas is lost. The value of the gas is considered to be the cost of the alternative: purchasing the gas elsewhere. Of course, one must be careful to deduct the cost of developing the product (methane gas, in this example) to obtain the net value of the natural resource. Arguably this is but another case of using price, which was discussed above. At the risk of redundancy, I include it here in order to weave together valuation of genetic resources with the literature on valuing water resources, using benefit-cost analysis. Two provisos have to be made explicit. First, the least cost alternative has to be chosen, otherwise one could justify any project by choosing high cost alternatives. Second, the alternative itself has to be economically feasible.

A third example of indicator values is to find a focal point: something that has a price and is believed to be at least equivalent in some acceptable sense to the object being valued. For example, a rare bird might be evaluated at the market price of a lower valued species such as a canary. The market price of a canary in a pet shop is not the economic value of a bird *in situ* but that value plus the cost of harvesting resources that have a value elsewhere. Including harvesting costs as a component of price produces an upward bias in the estimated value. The other drawback of this approach is, of course, the leap of faith it takes to measure one species by the value of another. Few would argue seriously that the value of goldenrod or witchgrass should be estimated by the market price of alfalfa. Almost certainly this technique creates an upwardly biased estimate of unknown magnitude.

Another method is to choose some arbitrary unit value and apply it to the number of organisms destroyed or damaged, as the judge did in the Colocotroni decision when he used a value of six cents per organism and applied it across all organisms to reach a total cost.

PAST VALUATION ATTEMPTS

Direct Productive Value

Few species or habitats have been evaluated in economic terms in a complete way or using proper estimation techniques. If this negative finding can be sustained, it will point the way, I think, to a useful piece of future research.

I know of no accurate estimates of the directly productive value of species. Michael Spence (1973) could find no trustworthy data to use in his analysis of the blue whale. According to Spence, the *in situ* commercial value of a blue whale is calculated to be about $644, which seems low to me.

Each year the pelts of the Pribilof fur seal are auctioned off. In 1983, a pelt was valued at about $15. This is nearly the *in situ* commercial value of the Pribilof fur seal, since the harvest costs are low.

Margery Oldfield (1984) and Norman Myers (1983) provide the most complete catalogue of commercial values arising from genetic resources. For the most part, the numbers are suggestive or speculative, and in many instances the estimates are based on sales value. Nevertheless, some values are instructive. In the 1970s a leaf blight eliminated about 20 percent of the U.S. corn crop, resulting in a price increase of about 20 percent and an estimated loss of about $2 billion. Plant breeders reacted quickly, drawing successfully on several blight-resistant strains. This gives us an order of magnitude estimate of the value of all the genetic material used, not of separate parts of it.

More generally, Myers (1983) says that the U.S. Department of Agriculture estimates the value of genetic improvement to be about 1 percent, about $1 billion, of sales of U.S. crops (about $100 billion). Again, this benefit is not keyed to a particular or specific species. Myers refers to seven biological control projects in California in the last fifty years that have reduced crop losses and produced input savings of about $1 billion, and in Australia four current biological control programs are estimated to have a net present value of about a quarter of a billion dollars. Vincristine and vinblastine derived from the rosy periwinkle have a sales value of approximately $100 million annually, from which development and production costs, of course, need to be netted out.

Margery Oldfield (1984) presented a wide array of estimates for the values of harvesting animals—small monkey carcasses were sold at $7 each, the harvest value of the slender-billed shearwaters was $70,000 annually, and internationally traded game meat amounted to $140 million annually. She also summarized pelt and snakeskin prices along with values for turtle flippers, gorilla heads and hands, and other parts valued by people. Most of these values, perhaps with the exception of those coming from the biological control projects, rely on using product prices as a proxy for the value of the resource.

Many of the values are annual figures, and we are left to imagine how long the stream of value might last. Most of the values are inframarginal. Perhaps they tell us in broad terms what some species are worth commercially, but they rarely accurately reflect the value of a threatened species. Oldfield reports that sea turtle skins have been sold for a couple of dollars a pound. Sea turtle harvesting is prohibited by CITES.

Direct Consumption Values

All the studies on valuing direct consumption with which I am familiar arise from the value game hunters place on species that are not very rare. Hammack and Brown (1974) used willingness to pay (WTP) studies to estimate the value of waterfowl at about $20 in early 1970s dollars. Several studies using hedonic

221

analysis, WTP, or willingness to sell (WTS) techniques have found that the value of sport salmon varies from $50 to $100 each in current prices (Mathews and Brown, 1970; Brown and Mendelsohn, 1984). Deer have been valued at $100 to $200 using the techniques described earlier (Brown and Gamponia, 1984) and by converting acreage leasing costs to a price per buck harvested (Teer et al., 1965). Oldfield (1984) concludes that the African oryx is worth around $233 as a game species and notes that the Arabian oryx is endangered.

The most ingenious experiment undertaken to date is that of Bishop and Heberlein (1979), who studied the value of the right to shoot a Canadian goose in the Horicon zone in Wisconsin. Individuals, responding to hypothetical questions, said that they were willing to pay about $20 on average for one permit and that they were willing to sell their permit for about $100.

In an annual lottery conducted by the Wisconsin Department of Natural Resources, the winners receive a permit to shoot a Canadian goose. A sample of hunters whose names had been drawn were offered checks of different amounts if they would give up their permits. Based on actual sales of the permits, the estimated average value was $60—halfway between the hypothetical willingness to sell ($100) and willingness to pay ($20) responses in the survey. There are sound theoretical grounds why WTS responses are greater than WTP, but the difference should be small—say a discrepancy of less than 5 percent. On the other hand, it is not uncommon for surveys to come up with differences between WTP and WTS of factors of five or more, as in the Bishop and Heberlein experience. Part of the problem resides in questionnaire design, which has been discussed elsewhere in the literature. Another difficulty is in creating the proper competitive market setting in a questionnaire or during a brief interview. The respondent must pretend to be a bidder whose bid will be binding only if the market settles on a price not higher (lower) than his or her sell (buy) order. (Bishop and Heberlein also estimated the value of a goose permit using the travel cost method. The value varied between $9 and $32 depending on whether the time was valued at 0 or one-half the wage rate.)

Indirect Productive Value

Most of the species destroyed by people probably are linked to species having commercial value. However, the links are so poorly known that there are, to my knowledge, no reputable estimates of indirect value.

Nonconsumptive Values

All of the estimation procedures discussed earlier can be used to value nonconsumptive uses.

Stoll and Johnson (1984) applied the hypothetical WTP method to value whooping cranes at the Aransas National Wildlife Refuge and found that the addition of whooping cranes on the refuge would be worth $1.40 per visit, or an estimated expected contemporary annual value of over $100,000.

Brookshire et al. (1983) survey the WTP/WTS literature, particularly the work done in valuing clean air and visibility in the Four Corners (northwestern) region of New Mexico and in the Los Angeles metropolitan area of California. The subject of these studies is a bit removed from the genetic resources area, so the results will be summarized briefly. However, they are mentioned because similar valuing for species is neither more difficult nor easier than placing value on the visual benefits of clean air, for example.

Most of the early travel cost studies could be cited here because they typically were used to value park sites, which have a strong nonconsumptive element. One of the most interesting applications is that of Cicchetti et al. (1976), who used a sophisticated version of the generalized travel cost analysis, involving several sites, to value wilderness threatened by a proposal to build a road through Sequoia National Park in order to reach a ski area to be developed by Walt Disney Productions. Cicchetti et al. concluded that the social costs of the project were greater than the social benefits.

Brown, in unpublished research, used an indicative value approach to estimate the social cost of waterfowl destroyed by an oil spill in Chesapeake Bay. A statistically significant regression equation using time series data on visitor days and the waterfowl population at a nearby refuge indicated that the number of viewing days decreased as the waterfowl population declined. The estimated lost days were then valued at the midrange of the Water Resource Council's published values for visitor days. The values are consensus (indicative) values policy makers have agreed to use in water project benefit-cost analyses.

ESTIMATING FUTURE VALUES

Future values cannot readily be captured by observing behavior. They are always based on hypothetical situations. There have been several empirical studies of option value, the kind that ask people if, hypothetically, they are willing to pay something for preserving a resource. Not surprisingly, there have been no studies based on the arrival of new information, although Fisher and Hanemann (1984) have worked out an example. In order to empiricize their model, it is necessary to know or estimate the mean and variance of the distribution of future net benefits. Their example is placed in the context of a new corn variety. Using their example, we need to know the future demand and supply function of corn, with and without the existence of a perennial variety of seed corn obtained by, say, a successful cross of the related wild grass (Zea diploperennis). The random variable can be thought of as governing the level of the unit cost of corn production. This is the case that Fisher and Hanemann consider.

TABLE 10
Value of option bids (in dollars).

TABLE 10
Value of option bids (in dollars).

Survey Sample	Grizzly Bear	Bighorn (Wild Sheep)
Hunters	21.50	22.90
Expect to observe but not hunt	21.80	23.00
Expect not to observe but value existence	24.00	7.40

SOURCE: Brookshire et al. (1983).

The first empirical study of option value, conventionally defined, was the one I did with Steve Mathews in 1967 valuing sport salmon fishing sites in Washington (Mathews and Brown, 1970). We found that more than half of each fishing area's estimated value stemmed from those who had not fished in the given area but might like to in the future. These numbers seemed unreasonably high and were not published for that reason. I still think the numbers unreasonably high. Curiously, option value is typically 50 percent or more of the total value of a resource in more recent studies. They are mentioned only briefly because they do not bear very directly on genetic resources. Greenley et al. (1981) questioned people about preserving water quality in the South Platte River basin. Total value is made up of bequest (14%), existence (19%), option value (19%), and recreation (48%). (See Brookshire et al. [1983] for a criticism of the procedure used by Greenley et al.)

Schulze et al. (1983) found that direct use values were insignificant in relation to the existence values of improved visibility resulting from clean air in the national parks of the Southwest.

Walsh et al. (1985) found that actual recreational use value is less than 20 percent of the total value that residents of Colorado place on their three, seven, or fifteen more valuable rivers. The remaining value of rivers is made up of nonconsumptive use values.

Desvousges et al. (1983) sampled users and nonusers of rivers. The sample population valued a marginal improvement in river quality to permit boating. There was no statistically significant difference between user and nonuser valuing of the river.

Finally, Brookshire et al. (1983) surveyed several hundred grizzly bear and bighorn sheep hunters in Wyoming and found that the existence value of the bears was greater for those who did not want to hunt bear than the hunting value was for those who wanted to hunt, as is illustrated in Table 10.

These and other studies that could be cited reach the empirical conclusion that nonuse value is *very* high in relation to use value. I would like to believe these studies, because it would give me much more confidence about the prospects for saving endangered species. But they are difficult to accept. Can we really believe that a representative nonbird-watcher values some randomly chosen bird to the same degree a bird-watcher does? Do nonclimbers and non-hikers value mountains and back country to the same degree that climbers and hikers do? (Would a fisherman value a stream he never uses as much as his favorite stream?) If one wanted to obtain donations to preserve a wild river and could draw 100 names from a user group, such as a kayak club, or 100 names from a nonuser group—both groups having the same income, education, and so forth —is there any doubt which group would be more likely to provide donations?

CONCLUSIONS

There are four attractive areas for future research. A discussion of them is appropriately placed in the final chapter of this book, so they are only mentioned here. First, the need to refine existence value studies is apparent and will require closer collaborations between survey researchers and other social scientists, such as social psychologists and sociologists. Second, bringing the existing finance literature to bear on the topic of option value is a natural research opportunity. Concepts shared between this paper and Goldstein's (this volume) give rise to the next two topics. Third, past and present attempts to manage common property need much more careful scrutiny. From these studies we will learn empirically the relative importance of the determinants of success and failure. Fourth, in the final analysis, distributive questions are paramount. Correct answers to questions of efficiency are not useful unless we can devise operational schemes for exchange between the future beneficiaries and those bearing the opportunity cost of preservation. Not only is there the complication of different generations, so to speak, but typically the two groups reside in disparate political jurisdictions. On the international scale, the problem is further complicated by the very uncertain nature of the present value of the future benefits resulting from particular preservation decisions today.

REFERENCES

Amemiya, T. 1981. Qualitative response models: A survey. *J. Economic Lit.* 19:1483–1536.

Arrow, K. J., and A. C. Fisher. 1974. Environmental preservation, uncertainty, and irreversibility. *Quart. J. Econ.* 88(2):312–319.

Bachmura, F. 1971. The economics of vanishing species. *J. Natural Resources* 11:647–692.

Baumol, W., and W. Oates. 1975. *The Theory of Environmental Policy.* Englewood Cliffs, N.J.: Prentice-Hall.

Binkley, C. S., and W. M. Hanemann. 1978. *The Recreation Benefits of Water Quality Im-*

provements: Analysis of Day Trips in an Urban Setting. EPA-600/5-78-010. Washington, D.C.: Environmental Protection Agency.

Bishop, R. C. 1982. Option values: An exposition and extension. *Land Econ.* 58:1–15.

Bishop, R. C., and T. A. Heberlein. 1979. Measuring values of extra-market goods: Are indirect measures biased? *Amer. J. Agric. Econ.* 61: 926–930.

Bockstael, N. E., W. M. Hanemann, and I. E. Strand, Jr. 1984. *Measuring the Benefits of Water Quality Improvements Using Recreation Demand Models.* CR-811043-01-0. Washington, D.C.: Environmental Protection Agency.

Brookshire, D., L. Eubanks, and A. Randall. 1983. Estimating option prices and existence values for wildlife resources. *Land Econ.* 59:1–15.

Brookshire, D., B. Ives, and W. Schulze. 1976. The valuation of aesthetic preferences. *J. Envir. Econ. Manage.* 3:775–782.

Brown, G., Jr. 1982. Survey of methodologies for ranking species. Report prepared for Department of Interior, Office of Policy Analysis.

———. 1983. Tourists and residents. In *Assessing the Economic Damage of Oil Spills: The AMOCO CADIZ Case Study*, chap. 4. Washington, D.C.: National Oceanic and Atmospheric Administration (NOAA).

Brown, G., Jr., and V. Gamponia. 1984. Investigation of the application of a willingness-to-pay approach to the 1980 National Survey of Fishing, Hunting and Wildlife Associated Recreation. Report prepared for Division of Program Plans, Department of Interior.

Brown, G., Jr., and J. H. Goldstein. 1983. *A Model for Valuing Endangered Species.* Washington, D.C.: U.S. Department of the Interior.

Brown, G., and R. Mendelsohn. 1984. The hedonic travel cost method. *Rev. Econ. Statistics* 66:427–433.

Brown, G., Jr., and D. Ragozin. 1985. Harvest policies and non-market valuation in a predator-prey system. *J. Envir. Econ. Manage.* 12:155–168.

Brown, G., Jr., and J. Swierzbinski. 1982. Endangered species, genetic capital and cost reducing R and D. Discussion Paper 82-9. Seattle: Department of Economics, University of Washington.

———. 1983. Cost reducing R and D and public knowledge with an application to the conservation of genetic capital. Discussion Paper 83-8. Seattle: Department of Economics, University of Washington.

———. 1985. Endangered species, genetic capital and cost-reducing R&D. In *Economics of Ecosystem Management*, ed. D. O. Hall, N. Myers, and N. S. Margaris. Dordrecht, Netherlands: Dr. W. Junk Publishers.

Cicchetti, C. J., A. C. Fisher, and V. K. Smith. 1976. An econometric evaluation of a generalized consumer surplus measure: The Mineral King controversy. *Econometrica* 44:1259–1276.

Cicchetti, C. J., and A. M. Freeman III. 1971. Option demand and consumer surplus: Further comment. *Quart. J. Econ.* 85:528–539.

Ciriacy-Wantrup, S. V. 1952. *Resources Conservation.* Berkeley: University of California Press.

Conrad, J. M. 1980. Quasi-option value and the expected value of information. *Quart. J. Econ.* 94:813–820.

226

Council on Environmental Quality. 1981. *Environmental Quality: 1980.* Eleventh Annual Report. Washington, D.C.: Government Printing Office.

Cropper, M., D. Lee, and S. Pannu. 1979. The option extinction of a renewable natural resource. *J. Environ. Econ. Manage.* 6:341–349.

Dasgupta, P., and J. Stiglitz. 1980. Industrial structure and the nature of innovative activity. *Economic Journal* 90:266–293.

Davis, R. 1963. The value of outdoor recreation: An economic study of the Maine woods. Ph.D. thesis, Harvard University.

Desvousges, W. H., V. K. Smith, and M. P. McGivney. 1983. *A Comparison of Alternative Approaches for Estimating Recreation and Related Benefits of Water Quality Improvements.* Washington, D.C.: U.S. Environmental Protection Agency.

Dorfman, R. 1962. Basic economic and technologic concepts: A general statement. In *Design of Water Resource Systems,* ed. A. Maass et al., 88–158. Cambridge, Mass.: Harvard University Press.

Ehrlich, P. R., and A. H. Ehrlich. 1981. *Extinction: The Causes and Consequences of the Disappearance of Species.* New York: Random House.

Evensen, R. E., and Y. Kisley. 1976. A stochastic model of applied research. *J. Pol. Econ.* 84:265–281.

Feenberg, D., and E. S. Mills. 1980. *Measuring the Benefits of Water Pollution Abatement.* New York: Academic Press.

Fisher, A. C., and W. M. Hanemann. 1984. Option value and the extinction of species. Working Paper 269. Berkeley: Giannini Foundation of Agricultural Economics, University of California.

GAO. Report to the Congress of the United States. 1979. Endangered Species: A controversial issue needing resolution. By the Comptroller General (July 2).

Goodman, D. 1975. The theory of diversity-stability relationships in ecology. *Quart. Rev. Biol.* 50:237–266.

Graham, D. A. 1981. Cost-benefit analysis under uncertainty. *Amer. Econ. Rev.* 71: 715–725.

Greenley, D. A., R. G. Walsh, and R. A. Young. 1981. Option value, empirical evidence from a case study of recreation and water quality. *Quart. J. Econ.* 96:657–673.

Hahn, E. 1980. A reporter at large: Eleventh hour. *The New Yorker,* September 1, pp. 37–69.

Hammack, J., and G. Brown, Jr. 1974. *Waterfowl and Wetlands: Toward Bioeconomic Analysis.* Baltimore: Johns Hopkins University Press, for Resources for the Future.

Hanemann, W. M. 1983. Information and the concept of option value. Working Paper 228. Berkeley: Department of Agricultural and Resource Economics, University of California.

Harrington, W. 1981. The Endangered Species Act and the search for balance. *J. Nat. Resources* 21:71–92.

Hartman, R., and M. L. Plummer. 1982. Option value under income and price uncertainty. Seattle: Institute for Economic Research, University of Washington. Mimeographed.

Henry, C. 1974. Investment decisions under uncertainty: The irreversibility effect. *Amer. Econ. Rev.* 64:1006–1012.

Henry, W. R. 1979. Management planning for tourism in Amboseli: Incorporating behavioral information on park users. Nairobi: Institute for Development Studies, University of Nairobi.

———. 1976. A preliminary report on visitor use in Amboseli National Park. Working Paper 263. Nairobi: Institute for Development Studies, University of Nairobi.

Krutilla, J. V. 1967. Conservation reconsidered. *Amer. Econ. Rev.* 57(4):777–86.

Lindsay, C. M. 1969. Option demand and consumer's surplus. *Quart. J. Econ.* 83:344–346.

Long, M. F. 1967. Collective-consumption services of individual consumption goods: Comment. *Quart. J. Econ.* 81:351–352.

MacArthur, R. H., and E. O. Wilson. 1967. *The Theory of Island Biogeography.* Princeton, N.J.: Princeton University Press.

Mathews, S., and G. M. Brown, Jr. 1970. Economic valuation of the 1967 sport salmon fishery of Washington. Technical Report 2. *Washington Department of Fisheries Research Bulletin.*

May, R. M. 1973. *Stability and Complexity in Model Ecosystems.* Princeton, N.J.: Princeton University Press.

Mendelsohn, R. O. 1984. The benefits of preserving endangered species: With special attention to the humpback whale. Honolulu: Southwest Fisheries Center, National Marine Fisheries Service.

Miller, J. 1978. A simple economic model of endangered species protection in the United States. *J. Envir. Econ. Manage.* 8:292–300.

———. 1981. Irreversible land use and the preservation of endangered species. *J. Envir. Econ. Manage.* 8:19–26.

Morey, E. R. 1981. The demand for site-specific recreational activities: A characteristics approach. *J. Envir. Econ. Manage.* 8:345–371.

Muir, J. 1962. *The Yosemite.* New York: Doubleday and Company.

Myers, N. 1983. *A Wealth of Wild Species.* Boulder, Colo.: Westview Press.

National Academy of Sciences. 1972. *Genetic Vulnerability of Crops.* Washington, D.C.: National Academy of Sciences and American Association for the Advancement of Science.

Norman, C. 1981. Biologists buy a piece of the tropics. *Science* 214:1106.

Oldfield, M. L. 1984. *The Value of Conserving Genetic Resources.* Washington, D.C.: U.S. Department of the Interior, National Park Service.

Opler, P. A. 1971. The parade of passing species: A survey of extinction in the U.S. *Science Teacher* 44:1–5.

Prescott-Allen, R., and C. Prescott-Allen. 1980. The contribution of wild plants and animals to industrialized societies. Mimeographed.

Ricklefs, R. E. 1973. *Ecology.* Portland, Ore.: Chiron Press, Inc.

Roush, G. J. 1982. On saving diversity. *Nature Conservancy News* 32:4–10.

Russell, C. S., and W. J. Vaughan. 1982. The national recreational fishing benefits of water pollution control. *J. Envir. Econ. Manage.* 9(4):328–354.

Schery, R. W. 1972. *Plants for Man.* Englewood Cliffs, N.J.: Prentice-Hall.

Schmalensee, R. 1972. Option demand and consumer's surplus: Valuing price changes under uncertainty. *Amer. Econ. Rev.* 62: 813–824.

Schulze, W., D. Brookshire, E. Walter, K. MacFarland, M. Thayer, R. Whitworth, S. Dean-David, W. Main, and J. Molenar. 1983. The economic benefits of preserving visibility in the national parklands of the Southwest. *J. Nat. Resources* 23:149–173.

Spence, A. M. 1973. Blue whales and applied control theory. In *Systems Approaches and Environmental Problems*, ed. H. W. Gottinger. Gottingen, Federal Republic of Germany: Vandenhoeck and Ruprecht.

Stoll, J., and L. A. Johnson. 1984. Concepts of value, nonmarket valuation, and the care of the whooping crane. Technical Article 19360. College Station, Tex.: Texas Agricultural Experiment Station.

Strong, E., and D. Hueth. Valuing Oregon steelhead using a household production approach. In *A Comparison of Alternative, Indirect Valuation Techniques for Estimating Recreational Values of the Oregon Salmon and Steelhead Fisheries*, ed. D. Hueth, E. Strong, and R. Fight. Portland: Northwest Forest Experiment Station. Forthcoming.

Teer, J. G., J. W. Thomas, and E. A. Walker. 1965. Ecology and management of white-tailed deer in the Llano basin of Texas. *Wildlife Monographs* 15:1–62. Washington, D.C.: Wildlife Society.

Tober, J. A. 1981. *Who Owns the Wildlife? The Political Economy of Conservation in Nineteenth Century America*. Westport, Conn.: Greenwood Press.

U.S. Department of Commerce. 1985. *Environmental Impact Statement on the Interim Convention on Conservation of North Pacific Fur Seals*. Washington, D.C.: Government Printing Office.

Walsh, R., J. Loomis, and R. Gillman. 1984. Valuing option existence, and bequest demand for wilderness. *Land Econ.* 60:14–29.

Walsh, R., L. Sanders, and J. Loomis. 1985. *Wild and Scenic River Economics: Recreation Use and Preservation Values*. Englewood, Colo.: American Wilderness Alliance.

Weisbroad, B. 1964. Collective consumption services of individual consumption goods. *Quart. J. Econ.* 77:71–77.

Western, D., and W. Henry. 1979. Economics and conservation in Third World national parks. *BioScience* 29:414–418.

Commentary

Partha Dasgupta

Gardner Brown's article contains, to the best of my knowledge, the most comprehensive discussion to date of the problems that one faces in placing social values on (genetic) resources. This assertion may seem odd, especially since Professor Brown spends the better part of his essay explaining alternative procedures for estimating the *private* values that people place on environmental resources. But I take it we would all agree that social choice ought to be founded on social values, and I guess most would suppose that social values in turn ought to be based on private values. Thus it should not be surprising that Brown spends most of his time in this essay discussing a critical ingredient for social choice.

The essay begins with the right question: why do genetic resources pose a social problem; that is, why might we think that genetic resources are "misallocated" in market environments? Or to put it another way, why do markets fail to ensure the intelligent use and adequate preservation of genetic resources? (I am distinguishing "preservation" from "use" because a resource may be simultaneously preserved and misused.) It won't do to argue merely that genetic resources are natural resources, for the question is then merely shifted to a larger category of goods. The answer Brown favors lies in the *absence* of markets for genetic resources ("Contemporary markets don't exist for any genetic resources because it is difficult or impossible to exclude users of the resource"). Here, Brown is basing his argument on a fundamental theorem in economics, that under certain conditions a decentralized allocation, in which every commodity—now and in the future—has a market, is efficient (efficiency is, of course, a very weak virtue of allocation, but at the level of generality of Professor Brown's analysis it is very reasonable to concentrate on efficiency). Genetic resources are often common property. Counterexamples notwithstanding, there is a strong presumption that such resources are overused. Environmentalists will know of this as the problem (or, as Hardin would say, "tragedy") of the commons. Now this may not agree with what comes next in Brown's article, which is that society most likely *under*values genetic resources because they "are a store of knowledge of uncertain value," and because "knowledge is a public good of a nondepletable variety." For one may rightly ask why there would be *excessive* use of a commodity that is undervalued. The answer is that the "locations" of genetic resources have alternative uses. The point is not that genetic resources are overused; quite the contrary, being undervalued they are neglected, and thus unprotected from the competing uses to which their sites are put. This is the problem.

The general line of the argument seems correct to me, but there are problems (not because I think it is a shade misleading to regard genetic resources as a "store of knowledge"; ask if you think it appropriate to regard coal as a store of knowledge, unlocked with the success of the Henry Corts process). Such resources are a possible input in the production of useful goods and services (contingent upon the acquisition of new knowledge). No, the problem is elsewhere. If we undervalue genetic resources, they will be in danger of neglect even if private property rights are assigned to their use. In the extreme, if people place no value on them, commercial or otherwise, there will be no incentive to protect them, no matter what the pattern of ownership and control. It seems, therefore, that we don't really need the common property feature to argue that these resources are in danger. They would be in danger if knowledge were to accumulate too slowly or accumulate in areas that don't enhance the use of genetic resources, and they would be in danger if private discount rates were high. Brown presents the classic argument that there is probably an underinvestment in the production of knowledge because of its having the characteristics of a public good. But this is only one side of the picture. The other side is that if there is free entry into the production of knowledge, whatever rents there are to be collected are all likely to be whittled down to zero as firms race to get there first. But this provides a bias in the opposite direction: too fast a growth in knowledge, not too little! I am not suggesting by this that genetic resources are not under threat, they probably are. I am only arguing that a priori arguments don't seem to nail down the bias.

I began by remarking that Professor Brown is ultimately concerned with the social valuation of genetic resources, and I suggested that it may seem eminently reasonable to base social values on individual values and therefore that for this reason the author has concentrated on the latter in his essay. In fact, there are three problems here. First, individual valuation of resources is conditioned by information, because information influences beliefs (e.g., about the potential technological use of a particular resource). The processing of information, however, is costly, and there is evidence that people consistently underestimate low probability events. So then perhaps social values ought not to be based exclusively on individual values. Second, even if as a first approximation one begins by basing social values on individual values, there is a problem in assessing individual values accurately. There is a hallowed tradition in econometrics of inferring *values* (or *preferences*) from *choice*, and much of the literature addresses techniques for making this kind of inference. Professor Brown's review of this literature is excellent (but note the sleight of hand econometricians have been practicing). Originally, econometricians intended to explain individual choice on the basis of individual preferences. The literature that Brown discusses reverses the process. It tries to *recover* preferences from choice. This reversal isn't entirely devoid of problems even though it might satisfy the hard-nosed empiricist. The conditions under which one can recover underlying preferences from actual choice are stringent. For example, one can't successfully carry out this enterprise in circumstances where individuals, or groups, can behave strategically. Third, individual preferences can be influenced by education, exhortation, and so forth; that is, individual values aren't mere raw data on which one may base public policy regarding the use of genetic resources.

If economists rely on the expected utility hypothesis of von Neumann, Morgenstern, and Savage to discuss choice under uncertainty, ecologists emphasize the need to *preserve* choice. They favor decisions that maintain flexibility, leaving future options open, arguing that information will become available with the passage of time. The twin presence of uncertainty about the future value of genetic resources and the fact that their habitat destruction is irrevocable suggests that there is a social value in keeping options open by not destroying the

231

habitats. The expected utility hypothesis doesn't appear to capture this concern. Brown discusses option values and argues, rightly, that the economist's prescription is in some sense consonant with the ecologist's concern, but I don't think he quite brings out the intimate connection. He formalizes the expected utility hypothesis but does not formalize the ecologist's concern. I want to conclude these comments by trying to draw the formal connection.

Let Z be the set of feasible objects of choice (e.g., amounts of biological diversity that are preserved). Let X be the set of all nonempty subsets of Z. An element of X is thus a set of objects of choice, a set of options. The expected utility analysis is based on preference orderings defined directly on Z. The ecologist's desire for flexibility (i.e., for keeping one's *options* open) suggests that he wants to consider preference orderings defined directly on X. Now, one may trivially obtain an induced preference ordering on X from a preference ordering on Z. This one does by saying that of two members x_1 and x_2 of X, x_1 is at least as good as x_2 if (and only if) the best element of x_1 is at least as good as the best element of x_2 (notice at once that this means that if x_2 is a subset of x_1, the induced ordering regards x_1 to be at least as good as x_2).

What might one mean by flexibility? Presumably that whenever x_1 and x_2 are elements of X, and x_1 is at least as good as x_2, then $x_1 \cup x_2$ (i.e., the union of x_1 and x_2) is not merely as good as x_1 but may well be strictly preferred to x_1. Now, choice based on expected utility is consistent with a desire for flexibility if utility is state dependent. To see this, suppose Z contains precisely two elements, z_1 (preserve site A containing one set of organisms) and z_2 (preserve site B containing another set of organisms). Next suppose there are two states of nature, s_1 and s_2, the probabilities of the occurrence of which are 0.9 and 0.1 respectively. Assume finally that the *realized* social utilities, net of preservation costs, at s_1 and s_2 from z_1 are 1 and 0 respectively (thus utility is state dependent) and that the *realized* social utilities, again net of preservation costs, at s_1 and s_2 from z_2 are 0 and 1 respectively. Expected social utility of option z_1 is clearly 0.9, and that of option z_2 is 0.1. Therefore, z_1 is at least as good as z_2 (in fact it is strictly better). But if both sites are preserved [$(z_1) \cup (z_2)$], expected social utility is 1, and this is still better than z_1. There is thus a gain in keeping one's options open.

Therefore, we see that the desire for flexibility, for keeping one's options open, is consonant with the expected utility analysis. If one starts from an ordering on Z, provided the utility function is state dependent, the induced ordering on X can readily satisfy the desire for flexibility. But does the argument run the other way? Suppose, following the ecologist's instinct, one begins with a preference ordering on X, and suppose that it reflects a desire for flexibility (that is, to repeat, that if x_1 and x_2 belong to X, and if x_1 is at least as good as x_2, then $x_1 \cup x_2$ is at least as good as x_1 and possibly better; and that if x_2 is a subset of x_1, then x_1 is at least as good as x_2). In an interesting article, Kreps (1979) has shown that if one imposes a mild additional condition on this ordering, there is a corresponding ordering on Z that implies the expected utility hypothesis (with a state-dependent utility function) such that the ordering on X can be induced from the ordering on Z. One concludes from this that the ecologist's concern and the economist's prescription can be made entirely consonant.

REFERENCES

Kreps, D. 1979. A representation theorem for "preference for flexibility." *Econometrica* 47:565–578.

Commentary

W. Michael Hanemann

Since I agree with most of Gardner Brown's excellent paper, I will use this space to provide a gloss on his text. Before proceeding, I must confess that I keep thinking of Oscar Wilde's definition of a cynic, which may also apply, I fear, to the economist: a man who knows the price of everything and the value of nothing.

Accepting that economics has the answer, it may be useful to begin by asking: What is the question? One can distinguish three types of questions discussed in Brown's paper:

1. Questions in which species preservation is the central concern: Should this species be saved? Should nature preserves be established? How should nature preserves be designed? Which species should be included? How much money should be spent on nature preserves?

2. Questions in which species preservation is one of several issues: Should this dam be built? Do the hydroelectric benefits outweigh the environmental costs? How should the dam be designed?

3. Metaquestions: Will agents make the correct decision in answering questions such as those under headings 1 and 2? Is there reason to believe that they will make a wrong decision? Should there be public intervention? What form should the intervention take?

Questions of type 3 are raised in Jon Goldstein's paper as well as in Brown's, and in Brown and Swierzbinski (1983). I will focus, however, on questions of type 1 and 2. Here, economics can play two roles. First, it can help to structure the decision, such as by reminding the decision maker of the distinction between objective function and constraints; by critiquing alternative objective functions ("maximize the number of species saved," say, versus "maximizing the value of species saved"); and by elucidating the choice set ("Just what alternatives are available?"). Second, it can generate numbers (benefit estimates, cost estimates, discount rates, etc.) which are inputs in the decision-making process.

Although the title of Brown's paper is "Valuation of Genetic Resources," it is important to remember that valuation and optimal investment decisions are two sides of the same coin: the appropriate valuation procedure usually depends on the context of the decision to which the valuation will contribute. As an example, it may not be necessary to measure benefits exactly as long as one can show that the minimum benefit exceeds the maximum cost. My point is that the chief question is not "What is the correct valuation procedure?" but "What is the correct decision?" The tendency for economists to emphasize the first question in isolation is partly explained by institutional factors; much of the time economists are not actually involved in helping to make species preservation decisions.

With this caveat, I will discuss some valuation issues raised by Brown's paper. Since he has already provided a taxonomy of the benefits involved in species preservation, I will focus on trying to answer the following two questions: (1) what is special about species preservation, and (2) in what way are the valuation problems different from more conventional applications of cost-benefit analysis?

In some cases, genetic and other natural resources may be construed as final consumption goods—that is, providing some service directly to individuals. In other cases they are inputs, valued not in their own right, but as contributors to the production of something else that is valued in its own right. Consider the first case and questions such as: What is the value of a wildflower in Tuolumne Meadow or a gorgeous sunset over San Francisco Bay? In principle there are two ways to answer such a question. One is to ask people directly, and the other is to observe choices that they make on which the genetic resources somehow impinge, and infer their values from these choices. The latter method is often based on the following scenario. An individual has preferences for various commodities whose consumption is denoted by the vector x, and for environmental or genetic resources by q (e.g., wildflowers in Tuolumne Meadow). These preferences may be represented by a utility function $u(x, q)$. The individual chooses the level of the x variables by maximizing his utility subject to a budget constraint, $\Sigma p \, x = y$. He does not determine the level of the q variables. These are in the nature of public goods for him, and he takes them as given. The utility maximization generates a pattern of consumption behavior represented by the ordinary demand functions, $x_i = h^i(p, q, y)$ $i = 1, \ldots, N$. Substituting these demand functions for the utility function yields the indirect utility function $v(p, q, y) \equiv u[h(p, q, y), q]$. The economic valuation of a change in q from q'

to q'' is based on either the direct or the indirect utility function. Thus, choosing the Nth commodity as a numeraire, say, we may measure the welfare effect of the change by the quantity S where

$$u(x_1, \ldots, x_{N-1}, x_N - S, q'') = u(x_1, \ldots, x_{N-1}, x_N, q') \tag{1}$$

which is the compensating or equivalent surplus; or we may measure it by the quantity C where

$$v(p, q'', y - C) = v(p, q', y) \tag{2}$$

which is the compensating or equivalent variation. Of course, we do not observe $u(x, q)$ or $v(p, q, y)$ directly, but we *do* observe the demand functions, $h^i(p, q, y)$. From these, it is possible (with a caveat noted below) to recover the direct and indirect utility functions and measure S or C.

I have said nothing yet about what type of commodities the x's are; the only requirement is that the consumer's choice among them is somehow affected by the q's. Thus, at this level of generality, there is no distinction between cases for which the individual's enjoyment of q would be classified as a consumptive value and cases for which it would be classified as a nonconsumptive value. Moreover, notions of altruism and John Krutilla's (1967) bequest value can be fitted into this framework; the q's could represent other people's consumption of natural resources or their utility from them—both of which are exogenous to this individual.

Krutilla's concept of existence value can also be incorporated into the above model. There may be some commodity, say, x_1, with the property that, when $x_1 = 0$, $\partial u / \partial q = 0$ (i.e., when there is no consumption of good 1, the individual gets no benefit from q). In Maler's (1971) terminology, x_1 is said to be weakly complementary with q. Alternatively, there may be no goods with the weak complementarity property; in that case, the quantity C can be expressed as the sum of two components—one representing the value of the change in q when positive quantities of the x's are consumed, and the other representing the value when the x's are not consumed. One might choose to call the latter existence value. Not surprisingly, it is possible to determine whether weak complementarity holds only if one has data on both cases for which $x_i = 0$ and cases for which $x_i > 0$.

There is a second concept which might, perhaps, be called existence value but which is quite distinct. Suppose that the direct utility function has the separable form

$$u(x, q) = f[g(x, q), q]. \tag{3}$$

235

In this case the quantity C can be expressed as the sum of two components—one representing a valuation of the change in q based on the $g(\cdot)$ component of the individual's preferences and the other representing a valuation based on the $f(\cdot)$ component. However, here the valuation procedure runs into an obstacle: the ordinary demand functions yield information only on the utility subfunction $g(\cdot)$ not on $f(\cdot)$. Thus, from observed consumption behavior, we can recover only the first component of C, not the second. Note that the presence of weak complementarity is neither necessary nor sufficient to rule out (3). Thus this is a different issue. An extreme form of the problem occurs when

$$u(x, q) = f[g(x), q]. \tag{4}$$

In that case, the first component of C is zero; the entire value of the change in q derives from the $f(\cdot)$ function, about which *no* information can be obtained from observed market choices.

By definition, the second component of C in the case of (3) and (4) is a non-consumption value. It cannot be measured from observed market choices—only from direct interviews. To be more precise, direct interviews yield an overall estimate of C rather than its individual components.

My original question—What is special about the valuation of genetic resources?—can now be answered. First, when one sets out to estimate demand functions $h^i(p, q, y)$, as a practical matter, it is often hard to know how to measure the q's. What aspects of ecosystems matter to people? How do people perceive and care for ecosystems? These are questions that economists cannot answer alone, and will require collaboration with biologists or psychologists. Indeed, there may be a divergence between "objective" measures of ecosystem performance or quality and people's subjective perceptions, so that it becomes necessary to model perception formation—that is, the mapping from objective to subjective quality—in addition to demand behavior (the latter a function of subjective measures of the q's).

Second, for many genetic and environmental resources, it is likely that people's preferences have the structure given in (3) or (4). Their marginal rate of substitution between market commodities is partly or totally independent of these resources. This leads to the valuation problem mentioned above. The third problem applies to the direct interview approach. Most people are not used to thinking about genetic resources and are simply unable to say (at least accurately) how much these resources are worth to them. I have suggested (Hanemann, 1985) that people generally come to know their preferences for objects by observing their own past choices. If they have not had to make trade-offs between wildflowers, say, and money in the past, they may not be able to give a realistic response to a hypothetical question in a contingent valuation survey.

This raises a fundamental question: Who are the people whose valuation the

236

economist is attempting to elicit? The utilitarian tradition in economics puts the emphasis on consumer sovereignty, allowing private individuals to be their own judge of what is good for them and of changes in their welfare; but what if their preferences are unstable, fickle, or readily subject to manipulation; or (perhaps more relevant) what if they do not know their own preferences? In that case, should one rely on the preferences (valuation) of experts; if so, which experts?

I have focused, so far, on genetic resources as consumption "goods." What about their role as inputs? Some microorganism may be valued not for its own sake but because it may lead to a cure for cancer or a better species of corn. Given some knowledge of the production relation (production, cost or profit functions, input demand, or output supply functions), economists have a well-developed tool kit for valuing changes in the quantity, quality, or cost of inputs. What is special about genetic resources, however, is the relative lack of knowledge about the production relation. This must come from natural sciences. Among development economists, there used to be a joke about the crass economist who was invited to visit a developing country as a consultant. He flies to the country, is taken in a limousine from the airport to the president's residence, and ushered into a meeting with the president. He takes a pad from his briefcase, turns to the president, and says, "Please state your social welfare function." In the present context, the crass resource economist turns to the biologists and says, "Please give me the biological production function."

There are three other types of valuation problems which, although they arise in other areas, apply with special force to genetic resources. These are the problems of dealing with uncertainty, time, and learning and information. They are likely to appear simultaneously in practice, but it is useful to discuss them separately.

The stochastic nature of many biological systems, the variation in peoples' preferences, the uncertainty concerning economic variables, data limitations, and the need to rely on statistical estimates all imply that the valuation exercise must be conducted in a probabilistic framework. We are really choosing among probability distributions of outcomes. How do people value uncertainty when they choose among probability distributions? How should they? Here, economics can make both a normative and a positive contribution. For example, economics can show how you ought to value uncertainty *if* you accept certain choice axioms (e.g., the von Neumann and Morgenstern axioms on expected utility maximization). It can also show the logical implications of a given choice model, such as the Arrow and Lind (1970) theorem that, even if all individuals are risk averse, a marginal project should be evaluated in a risk-neutral manner if its consequences are widely dispersed over many people and are uncorrelated with other sources of uncertainty affecting their welfare. With some important exceptions, such as the work of Machina (1982), much of the recent work on the economics of uncertainty has tended to focus on normative analysis and the expected utility maximization paradigm. Greater attention needs to be given to positive analysis. How do people value uncertainty? What aspects of uncertainty do they care about? Do they care only about the extremes of the distribution, so that something like a safe minimum standard might be justified? Indeed, do they have preferences over probabilities as well as outcomes?

There are two specific issues in the valuation of uncertain consequences to which I would like to draw attention. The first is the question whether people have a single set of preferences for uncertainty which apply in all contexts. In other words, when one infers individuals' risk preferences by observing their behavior in some kind of choice situation involving uncertainty (e.g., occupational choice and investment in costly safety equipment), how easily can these preferences be applied to the valuation of other types of risks? How widely transferable are they? The second issue is related, but it concerns the transferability of multivariate preferences (utility functions) elicited or inferred in circumstances of certainty to the valuation of risky choices. Specifically, several recent valuation exercises have assumed that the direct or indirect utility functions, $u(x, q)$ and $v(p, q, y)$, obtained by observing deterministic choices can be treated as von Neumann and Morgenstern utility functions and can be combined with probability distributions on x, p, q, or y to form expected utility functions. From a technical point of view, this is dubious since $u(\cdot)$ and $v(\cdot)$ are ordinal whereas the expected utility hypothesis works with cardinal utility. Beyond this, there is a real psychological issue: Do people have the same "mind-set" when they switch from making deterministic to stochastic choices? If not, the valuation problem is harder, because then we have to work with two sets of utility functions.

A word on the Cicchetti, Freeman, Schmalensee, Bohm, Graham concept of option value which Brown correctly characterizes as something akin to a risk premium in the face of uncertainty. As Ulph (1982) has pointed out, another way of characterizing this concept is as a correction factor for converting from one to the other of two methods of obtaining money measures of expected welfare in the face of uncertainty corresponding to ex ante and ex post welfare measurement. Therefore, rather than worrying about the magnitude of this correction factor, the crucial question is: Which is the correct welfare criterion—the ex ante or the ex post? (In Hanemann [1984] I have shown that the same issue arises in the context of the Arrow, Fisher, Henry option value discussed below.)

As with uncertainty, the valuation of outcomes occurring at different times raises both positive and normative questions. How do people value time? Do they discount? What is their discount rate? How should they value time? Moreover, the questions of multiple preferences and the transferability of intertemporal preferences elicited or inferred from one choice context to another also arise. Do people, in fact, apply the same valuation to different types of investment regardless of context (e.g., taking out a three-year auto loan versus setting aside a wilderness area for the benefit of future generations), or does the context matter?

Time and uncertainty are often combined in practical valuation problems. From one point of view, the combination raises relatively few new conceptual issues except for the normative/positive question of how risk and temporal preferences are merged into a single preference function. However, the interplay of time, uncertainty, and the temporal resolution of uncertainty does create a genuinely new issue: the value of information and of flexibility that provides a setting in which the arrival of new information over time can be effectively exploited. This is captured by the Arrow, Fisher, Henry concept of option value which measures the value of the future information that may be gained about the consequences of an investment decision if one retains sufficient flexibility now by not taking any action whose consequences are irreversible. It should be emphasized that irreversibility is not an absolute; it is better to think in terms of lagged ecosystem responses to some stimulus with a continuum of lags from zero lag (i.e., immediately reversible) to a lag of several hundred years (i.e., irreversible for all practical purposes). It can be shown that the concept of option value does not require absolute irreversibility, but merely that at any point in the future, the consequences of current actions cannot be instantaneously offset so that today's actions and tomorrow's actions are not perfect substitutes with respect to tomorrow's benefit function. Combined with some uncertainty (e.g., about the presence or length of the lag) which will be partly, if not totally, resolved with the passage of time, this is sufficient to generate a positive option value.

REFERENCES

Arrow, K. J., and R. Lind. 1970. Uncertainty and the evaluation of public investment decisions. *Amer. Econ. Rev.* 60:364–378.

Brown, G. M., and J. Swierzbinski. 1983. Cost reducing R and D and public knowledge with an application to the conservation of genetic capital. Discussion Paper 83–8. Seattle: Department of Economics, University of Washington.

Hanemann, W. M. 1984. On reconciling different concepts of option value. Working Paper. Berkeley: Department of Agricultural and Resource Economics, University of California.

————. 1985. Some issues in continuous-discrete response contingent valuation studies. *Northeastern J. Agric. Econ* 14:5–13.

Krutilla, J. V. 1967. Conservation reconsidered. *Amer. Econ. Rev.* 57:777–86.

Machina, M. 1982. Expected utility analysis without the independence axiom. *Econometrica* 50:277–323.

Maler, K. 1971. A method of estimating social benefits from pollution control. *Swedish J. Econ.* 73:121–133.

Ulph, A. 1982. The role of ex ante and ex post decisions in the valuation of life. *J. Public Econ.* 18:265–276.

Discussion

William J. Strang

Gordon Orians began the discussion by questioning the relevance of the types of trade-offs between species which the panelists had emphasized in their remarks, such as "Given a fixed budget, should the money be spent on captive breeding of the California condor or on some other endangered species?" It was precisely this sort of "thought experiment" that was being criticized, because not spending the money on the condor does not necessarily mean it will be spent on other genetic resources. Indeed, the widespread awareness of the endangered status of the condor has contributed to increased spending on genetic resources. Although there was general agreement on this, it was pointed out that trade-offs must still be made, for example between genetic resources and other goods, as well as different types of genetic resources.

David Woodruff noted that many concerns cross state and international boundaries. For example, there are people in Africa and Asia looking to see how we in the United States deal with the condor. Perhaps we can convince Africans to save their wildlife if we save ours. Jon Goldstein responded by saying that Congress will accept the argument that we are buying more than the condor by saving the condor. After all, these indirect benefits from the condor should carry weight in Congress since they accrue to U.S. citizens, and there is a strong lobby for environmental protection. He also mentioned that within the Department of Interior, decisions are made as to how much money to allocate to endangered species, and how much to spend on habitat protection and other things. Administrators within the endangered species division need not assume that the budget for preservation of endangered species is fixed until the yearly budget has been set. Further, the question was posed: What will happen to the funding of endangered species protection if the condors go extinct? Will it rise or fall?

Margery Oldfield expressed concern about using the *in situ* price of a resource as a measure of its value. She felt that the valuation procedures used by economists attributed too much of the wholesale and retail prices of resource-using outputs to capital and labor rather than to the resource itself. For example, labor and capital currently used in the whaling industry would not be as productive elsewhere, but this surplus value of capital and labor in whaling is dependent on the existence of whales. When a rare and valuable resource like a scarlet macaw brings $10 to the person who captures it, $50 to the exporter, $250 to the importer, and $2,500 to the retailer, most of the profit in the industry is going to specialized labor inputs. She felt that more of the surplus should be attributed to the macaws themselves. In fact, the *in situ* value of many endangered species is zero, because they are open-access resources.

Some of the economists responded by asking what the conflict was between the approach taken by economists and what Margery had prescribed. In fact, economists don't make the mistake of attributing a zero value to open-access resources by using actual prices paid. The *true* value (or shadow value) of a genetic resource *in situ* will be very large if the species is rare and valuable. The problem is that it is often very costly to define and enforce property rights in mobile genetic resources, so that exploiters of that resource behave *as if* its value were zero (i.e., as if it were a free good). Economists use the *shadow* value of the resource *in situ* rather than the *actual* value *in situ* as a measure of the contribution of the resource to the welfare of society. More generally, market prices are not good measures of the net benefits (rents) provided to society by such markets. They may be even worse measures of the shadow values of various inputs used in the production of the market output.

Peter Ashton noted that the discussion so far was based on the valuation of species rather than ecosystems or species within ecosystems. He then suggested a project that might provide information on the value of currently underexploited (or even unknown) species. If the last 200 years were surveyed and all the commercial resources that have been available from a species (or a well-exploited ecosystem, or even all ecosystems) were examined, this information could be used to value the commercial resources likely to be newly exploited from underexploited genetic resources. There was general agreement that such information could be useful if sufficient data are available.

Janis Antonovics then raised the issue of how to value variances (e.g., number of genotypes) as well as means (existence or nonexistence of a genotypic race or species). For example, sexual reproduction increases the genotypic variance within a species, which can in turn improve fitness. Few suggestions were forthcoming, and the discussion returned to the valuation of uncertain outcomes.

Michael Hanemann noted that accepted models of preferences and behavior under risk may imply that variation may be good on the production side but bad on the consumption side. In any case, economists *do* have theories of how to value uncertain outcomes, although there is evidence that people perceive risk incorrectly.

As an example of the use of economic theory in public decisions under uncertainty, Partha Dasgupta asked the group to consider a mutually exclusive choice between an irrigation project and a fertilizer factory with the same *expected* benefits. The irrigation project would be the better choice because it helps us in drought years while fertilizer yields benefits only in wet years (when we are fairly well off anyway). This example illustrates that high variance in benefits can be valuable in consumption if the benefits counteract natural risk.

John Krutilla then introduced Hal Salwasser of the U.S. Forest Service to discuss the case of the spotted owl, in which the Service is currently involved.

241

This case is especially rich because it has to do with the integration of genetic, ecological, and economic issues in a public policy decision.

The case deals with five relationships that are important in making a decision on how much owl habitat to preserve. The first is the relationship between structural and functional attributes of an area and what is called "old growthness," or high natural diversity. Two simple measures of old growthness in a forest are the diameter at breast height of the dominant tree species and the amount of layering in the canopy. Fallen trees leave gaps in the canopy within which an understory develops. The fallen trees support mosses, lichens, and (indirectly) arboreal rodents upon which the owls prey. The age of timber at which this occurs depends on the quality of the site and the species composition of the vegetation on the site, but is generally several hundred years for a Douglas-fir forest. The spotted owl appears to be the vertebrate species that most requires an old-growth environment for survival.

The second relationship is the probability of survival per pair of owls and their descendants as a function of the acreage of old growth each pair has in its home range. This function appears to increase at a decreasing rate for all acreage levels, with acreages above 4,000 (about two-thirds of a typical home range) resulting in almost certain survival. Closely related to this is the third relationship, which shows the probability that a population of owls will persist for 100 years as a function of the number of individuals in the population (with density constant). This function is monotonically increasing, eventually at a decreasing rate, with several hundred individuals necessary for almost certain persistence. As the population size falls, the environmental catastrophes lower the probability. The fourth relationship concerns the effect on the probability of persistence of the distance between patches of suitable owl habitat. As this distance increases, the probability that owls can disperse themselves to colonize new habitats decreases, and extinction rates for individual patches can outweigh rates of recolonization. As a result, the probability of persistence falls.

Economic considerations enter with the fifth relationship, which is the opportunity cost of owl habitat as a function of the quantity of habitat. This opportunity cost involves the value of the timber within owl habitats that is lost when such habitats are preserved. A number of these owl habitats have zero opportunity cost for the purposes of the Forest Service planner, because some are already included in national parks, Bureau of Land Management lands, wilderness areas, state parks and forests, and private holdings. Unfortunately, most of these are at high elevations, whereas the owl prefers low elevations. There are also some Forest Service lands not suitable for timber harvesting (e.g., steep land and stream banks) that contain owl habitat. These lands of zero opportunity cost provide habitat for perhaps 300 owls. The value of timber on additional patches of land depends on the size and density of timber and ease of access; the value of timber on a patch of 1,000 acres considered to have high value and convenient access is about $16 million.

But the total amount of land is not the only consideration. One must also know the size of individual patches and the distance between them. For example, if the patches are fairly scattered, it is less likely that a large number of individuals will be destroyed by a catastrophe such as the Mount St. Helens eruptions or the Tillamook Burn. On the other hand, patches must be close enough to allow genetic exchange between them and thus avoid problems of inbreeding. Also, smaller patches are easier to obtain: few large tracts of old-growth land are available.

Salwasser emphasized that he had merely presented the framework of the analysis, and that economists, geneticists, ecologists, and natural historians were still needed to help estimate the functions and review the decision criteria. He also agreed with Roger Noll's comment that old-growth owl habitats are good for supporting other things besides owls (e.g., other vertebrate species, fungi, plants) and that the decision framework should reflect this. Other things being equal, the owl habitats that best support other species should be the first to be protected. To emphasize that owl habitats yield multiple benefits, he noted that recent research shows that terrestrial voles transfer fungi and nutrients from old-growth stands to nearby clearcuts, so that if old-growth stands are left in an area, inoculation of seedlings and fertilization are not necessary.

Gardner Brown and others noted that we still hadn't answered the question of how many owls (or how much land) to save. To do so, we would need to calculate the benefits as well as the costs of saving various amounts of land or owls. Some land may be preferable for purposes of hiking because it is closer to population centers, or allows more spectacular views, and so forth. This gives us a total benefit function of acreage devoted to preservation which is subject to diminishing marginal benefits.

It is not enough to know the present benefits and costs of preservation, however, since both benefits and costs will probably change over time. For example, assuming that incomes and population continue to rise and that the demand for environmental enjoyment rises with income, the total benefits of preserved owl habitat will rise over time, perhaps even faster than the discount rate.

This is true a fortiori if wilderness experiences are less valuable when natural areas are congested. As long as the growth in benefits exceeds the growth in costs, the optimal quantity of owl habitat will rise over time. This might seem to imply that we should increase owl habitat over time, but owl habitat cannot be created overnight. Unlike the preservation decision, the decision to harvest timber is not immediately reversible. Although recruitment to old growth can take place, there may be a long lag between the maximum age of timber in a rotating forest and the minimum age for a tract to be considered old growth. As a result, if the desired rate of increase in old growth exceeds recruitment rates, the decision of how much to preserve now depends on expected future benefits and costs as well as present ones. Choice of a discount rate is also necessary to aggregate these net benefits over time.

Brown emphasized that even if a legislative decree requires that we save the spotted owl, or save a certain amount of its habitat, economists can calculate how large the difficult-to-measure benefits must be in order to outweigh the easy-to-measure costs of owl habitat and therefore justify such a policy. Krutilla then raised the question as to who should testify about the merits of saving additional owls. He felt that it should be those who can best explain the contribution that additional numbers makes to survival of the owls, and that these people aren't economists.

Eric Charnov was skeptical about the assertion that the value of preservation relative to timber would be likely to rise over time. After all, income and population growth may increase demand for timber as well as for natural environments. Ron Johnson and Hal Salwasser noted, however, that if the desired land use pattern changes over time, decisions can and should be adaptive. Also, in the face of uncertainty about whether better information will become available over time, it pays to keep options open by leaving more old growth standing. A decision to preserve is immediately reversible, whereas a decision to cut may require hundreds of years to reverse. Salwasser also said that probably enough owl habitat exists to consider survival of the owls as highly probable for several decades even if old growth on Forest Service lands is cut at present rates.

Noll emphasized that the discount rate should not be thought of as something that gives less weight to the desires of future generations than to those of the present one. It also tells us *how* to provide for future generations if that is what we want to do. It does so because it tells us which assets are worth saving for future generations and which should be sold and the funds used to earn a higher return.

Ruth Shaw noted that the valuation procedures discussed in Brown's paper involve trade-offs between genetic resources and other goods rather than between different types of genetic resources. The "thought experiment" we had been using which asked how much of one genetic resource we would trade for another was thus artificial. The budget for genetic resources is not fixed, although the assumption of such a constraint may be useful as a first cut to keep things simple.

Orians responded by saying that bureaucratic constraints often cause such budgets to be de facto fixed. For example, the endangered species office can decide whether to spend money on condors or on some other endangered species, yet it can't buy land to protect nonendangered species. Several individuals responded that the flexibility to allocate funds between such activities is there in higher levels of decision making, like the Office of Management and Budget or congressional committees.

Brown felt that the fact of scarcity implies that there must be trade-offs, but which trade-offs are relevant is not immediately apparent. It depends very much on political processes. He also felt that one can always find an economist to defend a particular viewpoint, because there are enough areas of controversy to make the range of supportable benefit estimates quite large. Perhaps the best examples are arguments that say that the capital market is imperfect, and this makes the choice of discount rate flexible. Another example is the value of a Canadian catch of fish that originate in U.S. hatcheries. Should they be included in the measure of benefits of U.S. hatcheries?

Hanemann supported this by reminding us of the importance of sour grapes as a motivation for human behavior, and that economists have two possible roles. They can sit at the feet of the mighty and tell them what to do or they can be prophets outcast in the wilderness and say, "You're making a big mistake." Both are motivations or justifications for economic analysis.

Michael Strauss liked the economists' message of compromise. He said that in his job at the Office of Technology Assessment, he only heard one-sided views like, "We should get rid of all the U.S. agricultural lands and bring back the natural diversity of the prairie," or "We should plant crops on all the wildlands." In this context, a prescription of balance was refreshing. In a similar vein, Lawrence Riggs felt that maintaining natural areas inviolate was not necessary to maintain genetic resources. For example, a rotating forest with occasional stands of old growth can contain as much genetic diversity as a homogeneous old-growth forest.

Margery Oldfield ended the session by saying she felt that the alternative valuation procedures that had been discussed were dependent on the socioeconomic status of individuals, and they shouldn't be. For example, in a depression, individuals might not go to national parks as frequently, but that doesn't mean that they value them any less. She also felt that individuals don't have perfect knowledge and thus behave in ways that cause genetic resources to be systematically undervalued.

6

The Prospects for Using Market Incentives
for Conservation of Biological Diversity

Jon H. Goldstein

The problem of maintaining genetic diversity, preserving endangered plants and animals, and protecting sufficient habitat is both critical and growing. Arguably, many of the large, aesthetically attractive species may be saved in zoos, preserves, and safari parks. But to preserve these species in the wild, and to save the untold number of plants, invertebrates and the smaller fish, amphibians, reptiles, birds, and so forth, we will need to preserve their habitats.

In their efforts to stem the loss of wildlife, the nations of the world have relied primarily on trade restrictions, publicly maintained habitat, and governmental control of the rate of exploitation. The results have been mixed at best. Illegal harvesting and trade have severely taxed enforcement resources, and public budgets for habitat protection have been inadequate to withstand the pressure of development. Many observers are resigned to the belief that the best that can be expected from the current preservationist strategy is a postponement of the inevitable depletion of species, habitat, and ecosystems.

There are two important aspects to the loss problem. First, the burgeoning illicit trade in wildlife products (which has encouraged overharvesting of the more valuable species), together with the poaching of endangered species, even in the most protected parks and preserves, is rapidly outstripping the financial resources of most Third World (and many developed) countries to cope with the problem. Second, and more important, the vast majority of species losses are no longer due to human harvesting in the wild. Instead, the rapid growth of indigenous populations with their concomitant demands for food, fiber, housing, and energy, together with the development of natural resources for export revenues, is eliminating many wild places altogether.

For example, the Brazilian government estimates that between 1966 and 1975, 38 percent of all deforestation in the Brazilian Amazon was attributable to large-scale cattle ranching (primarily for export), 27 percent to highway construction, and 31 percent to agricultural colonization. At the current rate of clearing, 63 million acres will be lost by the end of the century. But that rate is likely to increase: 60 percent of all deforestation in the Brazilian Amazon occurred in the three years between 1975 and 1978 (Caufield, 1985). In southern Kenya, adjacent to some of the largest and most important national parks, the human population (including in-migration) is growing at 8 to 10 percent annually (Myers, 1982). With pressure of that magnitude, enforcement alone is unlikely to prevent the conversion of wildlife habitat to farmland and grazing land, nor is it likely to deter people from poaching elephants and rhinos, slaughtering wildlife that compete with cattle for grass and water, and killing off predators. In spite of Kenya's well-publicized, able, and well-financed efforts at protectionism, over the past decade its elephant population dropped from 165,000 to 50,000 and the rhino population from 15,000 to less than 1,000 (Myers, 1981).

One of the principal reasons for the discouraging results of the preservationist efforts emanates from the status of wildlife as a common property resource. Wildlife are considered to be in the public domain, to be owned by all and hence by none. This common property status puts almost the entire burden for preserving wildlife on the public sector. It does nothing to motivate the self-interest of individuals and create private incentives to husband wildlife or preserve habitat.

The urgency of the current crisis has led a number of conservationists, scientists, and economists to argue that unless the living resources and habitat can generate some economic benefits, protectionism will be overwhelmed in the long run. The pressures for development will cause the wildlife to be eliminated and much of the habitat to be converted to farmland or grazing pastures, cut down for fuel and housing materials, or, at best, converted to monoculture woodlots or plantations. This group has called for greater reliance on commercialization, economic incentives, and private property rights to encourage people to harvest wildlife on a sustained-yield basis and to preserve habitat.

The use of market incentives and property rights is not the only approach to managing the commons. The principal nonmarket techniques are limitations on factors of production (e.g., restrictions on equipment or on harvesting methods, limitations on season lengths) and direct output controls (e.g., harvesting quotas). There are also several mixed strategies: taxes, royalties, fees, licenses, and franchises. These may have more or less economic content, depending on the criteria used for determining the magnitude of the taxes, royalties, or fees or the allocation of the licenses or franchises. (For an excellent survey of techniques used in attempting to manage natural resources that are common property see National Oceanic and Atmospheric Administration [1984].)

In this paper I examine the prospects for market solutions to the problem of the commons—that is, the prospects for conserving wildlife and their habitats by establishing markets in wildlife where an unmanaged commons existed previously. Several cases of markets in wild species (seabirds, crocodiles, and butterflies) are analyzed to determine whether trade successfully avoided the tragedy of the commons; what conditions—political, social, economic, and biological —led to success or failure; and whether these conditions are unique or likely to be encountered or replicable elsewhere. (Earlier I coauthored a paper that examined trade in several species and attempted to determine whether commercialization relieved pressure on the wild populations [Smith et al., 1983]. When I revisited these examples for the current effort, I found that the evidence simply did not support the earlier findings. I have extended the research and corrected the analysis below.) To establish a context for examining the case studies, I briefly review the supporting arguments for commercialization as a conservation strategy and some of the difficulties likely to be encountered in implementing such a strategy.

TRADE AS A CONSERVATION DEVICE

The essence of the argument in support of commercialization is that if property rights and incentives for private sector management and maintenance of species and their habitat are established, trade in wild species (either wild or captive reared) can contribute to conservation. If the wildlife and habitat have a marketable value, landowners or lessees vested with enforceable property rights have an incentive to nurture and protect these resources. If trade is prohibited or property rights are vague or unenforceable, no such private incentive exists, and virtually the entire responsibility and cost of husbanding wild resources falls upon the public sector.

Indeed, public costs are likely to be exacerbated by the absence of appropriately designed private incentives to preserve wildlife. Habitat frequently has competing commercial uses, which make conversion attractive and increase the price that public entities must pay for acquisition. Wildlife includes pests and predators, whose control or elimination may be desired, creating an incentive that is not mitigated if there is no capturable commercial value associated with wildlife. The incentive for private efforts to control wildlife that are pests and predators makes it more difficult and more expensive for public wildlife managers to maintain any given wildlife population level. Where there are penalties associated with the taking or trading of wildlife, but no private property rights, the cost of controlling illicit activity is borne entirely by the public sector. If rights to the wildlife were vested in private entities, there would be considerably more identification, prevention, and reporting of illicit activity by the private sector.

Market mechanisms are remarkably efficient devices for achieving socially desirable objectives. In many instances they are self-enforcing, requiring few supporting public resources and limited governmental oversight and regulation to function properly. Where such mechanisms can be designed for the wildlife sector, they can ease the burden of conserving wild species and their ecosystems.

Laissez-faire commercialization is not a panacea for the problems of preserving wildlife. The concept cannot be applied to many species, because they have no apparent economic value. Even for species that have obvious utility, a market-oriented solution may be difficult or impossible to design. For migratory species it may prove difficult for habitat owners to capture the contribution of their resource. In less developed countries (LDCs), political instability, unreliable biological information, and primitive permitting procedures, complete with mountains of numbing documentation, frequently complicate the problem of implementing a conservation-oriented commercialization strategy. In many instances biological constraints or economic or social institutions make it difficult to establish property rights. Where property rights can be established, the cost of enforcing them may be prohibitive. Even where enforceable property rights can be established, management of the species on a sustained-yield or profit-maximizing basis in perpetuity is not assured. There are conditions under which harvesting of the entire stock (liquidation of the firm and extinction for the species) is financially preferable for the resource owner (Clark, 1974; Fife, 1971). The list of potential problems is lengthy, well broadcast, and often legitimate.

There are, however, conditions that favor commercialization and the establishment of property rights. If the product is such that a reliable supply is important to maintain commerce, opportunities are likely to exist for undermining illicit trade and poaching. If the range of a species does not cross political boundaries, the problems of establishing a market and enforcing property rights may be simplified. If the conditions under which a species can be raised or marketed lend themselves to self-enforcement, public funds now devoted to protection or enforcement can be saved, and commercialization may prove attractive. Conversely, a market solution may not be appropriate for species that require extensive support from law enforcement in order to establish a market structure that contributes to conservation.

Trade in species for which the cost of farming or ranching is less than the cost of taking from the wild can be used to relieve pressure on wild populations. Although this is often difficult to effect, parakeets, canaries, and freshwater tropical fish are cases in point. Extensive trade occurs in these species, all of it with specimens bred and reared in captivity. If a uniform product is desirable, captive breeding has an advantage. Obvious examples are species used for research and testing and some animals used in the fur trade. Indeed, either because of diet modification or because the captive population is shielded from the hazards of the wild, captive breeding and raising can be used to generate not just a uniform product, but an enhanced, more valuable product. But captive breeding and rearing has its limitations as a conservation device. Although it can be used to reduce removal from the wild, captive breeding and rearing does not create property rights in wild populations and their habitats, and in and of itself establishes no direct incentive to maintain these populations and their habitats. Thus, where development threatens, captive breeding and rearing has minimal benefits as a strategy for maintaining the wild.

Given the number of species, populations, and habitats in need of attention, the problem of identifying species for which market incentives are likely to work takes on discouraging dimensions. But the current protectionist strategy has failed to halt the precipitous decline in wildlife populations, and without a massive increase in funding (which does not seem to be forthcoming) the prospects for protectionism are dim. Although the decline in wildlife populations is generally unwanted and potentially detrimental to the function of ecosystems, unless a way can be developed to make it in people's self-interest to preserve wildlife populations and habitats, extensive depletion is inevitable during this century. It is not noble to have to consider and engage in the exploitation of the earth's remaining species in order to create conditions under which they are valued enough to be preserved. In light of the precarious status of numerous species, it would be even less honorable, however, to avoid consideration of commercialization because of its offensive nature.

We move now to examine the case studies and some of the difficulties encountered in marrying conservation and commercialization.

SEABIRDS

In Iceland, eiders, murres, razorbills, kittiwakes, and puffins have been exploited for food and feathers for over 1,000 years. Since the eleventh century a variety of efforts have been instituted to protect the birds, control overharvesting, and supplant the commons. Indeed, as early as 1281, the civil and ecclesiastical codes stated that the birds belonged to the occupiers of the lands where they occurred (Doughty, 1979). Protective efforts were largely ineffective, however, and overexploitation continued until the middle of the nineteenth century, when enforcement became more vigorous. Currently, the management schemes for eiders, murres, razorbills, and kittiwakes make the most use of property rights, while that for puffins is more like a regulated fishery (Smith et al., 1983). In all of this, however, government enforcement, protection, and penalties play an important role.

Eiders

Eiders serve as a source of meat, eggs, and especially down. The landowners conduct fairly classic husbandry operations: creating artificial nesting sites, gathering down, collecting eggs, thereby inducing the birds to lay another clutch, and protecting the nesting colonies from poachers and natural predators.

The number of such farming operations grew during the nineteenth century and peaked during the 1920s at 250 farms. (In the official terminology of the Convention on International Trade in Endangered Species of Wild Fauna and Flora, "farming" involves close-cycled, captive breeding of wildlife, while "ranching" involves taking eggs or young from the wild and rearing them in captivity. In this paper "farming" will be used for all controlled-rearing activities.) The movement of many farmers into towns, together with the development of synthetic substitutes, has reduced the number of farms to about 200, and eider-down production is now only about half of its peak. One result of this decline in farms has been an increase in predation by gulls, ravens, and feral mink and foxes.

Despite the protections afforded by the remaining farmers and the government, Iceland's eider population has declined in recent years, while eider colonies elsewhere have grown. The fall in population has been attributed to a combination of unusually severe weather conditions, pollution attendant to development (oil, industrial wastes, and agricultural chemicals have invaded coastal waters), the increased predation (also due to human activities, since gulls are camp followers, and have thrived on municipal dumps), and the reduced number of farms (Doughty, 1979).

Among the examples of commercialization of wildlife, eider farming is one of the closest to the pure property rights paradigm. A key element, however, is that eiders are fully protected in Iceland. The government enforces the property rights, allowing no taking on public nesting grounds and imposing sizable penalties for poaching. Thus the "commons" has been replaced by a combination of private farms and government protection. The latter is central to managing the species, as well as managing the industry. The publicly protected, nonprivately owned nesting sites add to the population; and, protected from competition from poachers on public lands, the farmers are encouraged to incur costs in husbanding the nests on their property. Further supporting the management scheme is the long tradition of eider conservation, the availability of other employment and income sources, and the small, slowly growing human population. The close-knit nature of the community probably serves to further discourage poaching. Competition from synthetics and continued development in Iceland could undermine the arrangement, however, and might reduce the incentive to protect the colonies. There is a symbiotic relationship here: protection of the eiders encourages the industry, and a healthy industry results in a constituency that supports government protection. Should the industry continue to decline, interest in protection could erode.

Murres, Razorbills, and Kittiwakes

Adult birds and eggs are harvested from the sea-cliff breeding colonies of the murres, razorbills, and kittiwakes. The management and farming technique is somewhat different from that for eiders, but property rights continue to play an important role. Artificial habitat is not created to attract the birds; rather

251

property rights have been established by farmers on the natural breeding sites (Hyman, 1982). No common area remains. The entire breeding area has been subdivided into private farms, and owners frequently rent the harvesting rights to hunters. Although the law allows the taking of full-grown birds, eggs, and young, breeding stock is conserved through a combination of government controls and self-imposed husbandry practices.

Most hunting practices directed toward adults are banned by law. This is in marked contrast to the days when Iceland's nesting cliffs were known as the country's breadbasket and as many seabirds as possible were slaughtered. Taking was uncontrolled, hunting being conducted at sea and on land with nets, hooks, snares, and guns. Severe population declines occurred, and one species, the great auk, became extinct.

Eggs are collected early in the season, allowing most birds to relay. This, together with concentrating the hunting on young birds and nestlings that have a high natural mortality rate, tends to preserve the populations. All of these conservation, management, and farming practices might be for naught, however, if it weren't for the fact that the birds are extremely faithful to their nesting sites. It is this natural tendency (together with the birds' relative lack of vulnerability while dispersed at sea in the off-season) which makes it possible to establish fairly dependable "ownership rights" to the breeding population and allow substitution of a property rights structure consistent with conservation for a commons that encouraged overharvesting. If the birds tended to pioneer easily, each farmer would have an incentive to harvest as many birds as possible each year, assured that the nesting population on his land in year t was independent of his activities in year $t - 1$. Thus, in the absence of the birds' site-fidelity characteristics, a commons would reemerge despite enforceable property rights to the breeding habitat.

Puffins

The management arrangements that have evolved for puffins are similar in many respects to those for murres et al. Property rights—at least tacit property rights—play an important role in the management and conservation of puffins, but government regulation of harvesting methods has certainly been critical to the recovery of the populations from the decimated levels of the mid-1800s.

Puffins are hunted for their meat and feathers. Like the murres, razorbills, and kittiwakes, puffins are cliff dwellers, making their nests in burrows. Until the middle of the nineteenth century, hunters netted the birds at their burrow entrances, a practice that had a devastating effect on the populations. In 1875 a long-handled net called a hafur was introduced to Iceland (Hyman, 1982). It is used to capture the birds in midflight. Despite its unwieldy nature (12 feet long), no substitute has emerged for the hafur during the intervening century, and Icelandic law designates it as the only instrument that can be used to harvest puffins. Guns are specifically outlawed, because the noise frightens the birds away from the colonies, leaving the young uncared for and the eggs unprotected.

The largest puffin colonies are found in the Westmann Islands off Iceland's southern coast. The Town of Westmann Islands (TWI) owns the islands, and leases each harvesting site (island) to local hunters. Rental fees are nominal, and do not serve an allocative function. Rather, by tradition, the same group of hunters leases the same island year after year, and tacit ownership rights have evolved for each harvesting site (A. Peterson, Icelandic Museum of Natural History, Reykjavik, pers. comm., 1985). This is important, because in conjunction with the island-faithfulness of the birds, it provides an incentive not to over-harvest an island. Puffins will pioneer and recolonize an overharvested site, but recolonization takes time, and in the interim, harvest sizes will suffer.

In order to conserve the breeding population and avoid overexploitation, Westmann Islanders hunt only on the periphery of the breeding colony, station-ing themselves on the cliffs and attempting to net birds as they fly by. Hunting is conducted only during the summer months, when the bird population milling about the cliffs is composed largely of immatures. Adults rarely fly along the cliffs during the summer, traveling instead straight out to sea from their burrows in search of food for their young. Catch statistics reflect the selective nature of these hunting practices: 93 percent of the take is composed of nonbreeding juveniles (Hyman, 1982).

These restrictive hunting practices are all self-imposed. The government sets no bag limits or season lengths, requiring only that hunting be conducted with the hafur. Hunters who deviate from the accepted harvesting methods are quickly sanctioned by the small community. Further, the commitment of the Westmann Island hunters to harvesting practices that conserve the breeding population is almost certainly related to the long-term leases and tacit ownership rights that they enjoy to the harvesting sites. On islands to the north, where leases are determined annually and hunters have no long-run claim to the har-vesting sites, hunting is conducted from the middle of the colonies, rather than on the periphery. Puffin populations on these islands have declined significantly (Peterson, pers. comm., 1985).

Assessment

The arrangements for exploiting seabirds in Iceland used to be in the form of a commons, and, as expected, overharvesting occurred to the detriment of the species. Clearly, the seabirds are conserved better under the current harvesting structures than they were under a commons.

Are the bird populations larger than they would have been if commerce had been prohibited? The eider population may be, because farmers create artificial habitat for them and the birds' commercial value encourages both private and public efforts to protect them from natural predators. This is probably not the case for the other species. The management arrangements for these species do not involve habitat enhancement or predator control. Thus, if commerce were banned, the populations of cliff dwelling seabirds might well be larger. However, the implicit assumption here is that even if commerce were prohibited, the protective efforts would continue; and that is probably not the case, at least not at the level now enjoyed by the birds. On the whole, seabirds seem to have been well served by commercialization and the institutional arrangements devised for harvesting them—far better than under a commons and perhaps better than if commerce were prohibited.

It is noteworthy that this was accomplished without having to extend protective controls or property rights to the birds' entire ranges. The birds are primarily pelagic, coming to land only during the breeding season. Since the cost of harvesting them at sea is prohibitive and the rest of their habitat (the open ocean) is not terribly vulnerable, it is only necessary to control their breeding grounds to protect them (this would not have been sufficient for Norway, where overfishing has reduced the food supply for seabirds and caused the populations to decline). It is important to observe that when the portion of the commons to which protection or property rights need to be extended can be confined, commercialization consistent with conservation becomes much easier to effect.

Another principle evident from the Icelandic experience with seabirds is that in some cases (like the eiders) privatization of the commons with minimal government support of property rights may be sufficient for commercialization to be consistent with conservation. In other cases (like the puffins) more government involvement, scrutiny, and protective actions may be required. For financially pressed and overburdened governments in LDCs, providing the latter may be difficult.

There is one important condition operating in Iceland which unfortunately is not duplicated in much of the tropics. In Iceland the breeding habitat is not under pressure from alternative forms of development. The human population is stable, and there does not appear to be much use for the land except as habitat. Clearly, this is often not the case in the tropics, where the commercial opportunity costs associated with retention of the land as habitat are often high and may swamp the revenues that can be extracted from the land in its natural state.

CROCODILES

Crocodiles have been on earth for about 200 million years, but all of the twenty-six extant species have undergone such rapid depletion during the past thirty years that most of them are now listed as either threatened or endangered. The prospects for their survival in the wild are tenuous at best. In the face of a continuing worldwide demand for crocodilian leather products, there is little future for crocodiles if they remain only a common property resource.

Papua New Guinea (PNG) saw its wild populations of both fresh and saltwater crocodiles crash in a short period following the wide-scale development of commercial shooting beginning in the mid-1950s. The trade in both species peaked in 1965–66 when $1 million in skins were exported. After 1966, exports plummeted as both species disappeared from accessible areas. By 1967 populations of both species were depleted, and in 1968, even with increased hunting effort, the yield dropped by half. By 1969 the saltwater crocodile was rare throughout its range. In 1970 the PNG government estimated that all breeding-age crocodiles would be gone within five years if harvesting trends continued. In 1971–72 the total value of exports of both species had been reduced to $198,000 (National Research Council, 1983b).

In the late 1960s the government began to develop a radical national plan to incorporate all of the nation's wildlife as part of a constitutionally protected, sustainable national resource. Aware of the effects of development on other tropical regions, the government sought ways to utilize the nation's natural resources to improve the standard of living, while maintaining the traditional life-style of the indigenous peoples and the integrity of the ecosystem. PNG embarked upon a program of wildlife farming, encouraging individuals to establish enhanced habitat areas on the fringe of natural environments. The farming program included cassowaries, megapodes, wallabies, deer, butterflies, and crocodiles. PNG's butterfly and crocodile farms were designed as export industries. The other wildlife farms were conceived of as a source of domestic protein, and as of 1982 were still in the developmental stage (National Research Council, 1983a, b).

In 1968 the government formulated its first crocodile regulations, including licensing at all stages of the trade, data collection, and a ban on the sale and export of all skins exceeding 20 inches in belly width (about 8 feet long). The belly-width ban was supposed to conserve the most critical portions of the crocodile populations, the breeding adults. For saltwater crocodiles it was a well-conceived constraint, but the majority of breeding freshwater crocodiles fall below the 20 inch belly-width limit, and as a result the ban did nothing to promote their conservation (C. A. Ross, former member, UN/FAO Crocodile Project, PNG; National Museum of Natural History; project leader, World Wildlife Fund Philippine Crocodile Project; pers. comm., 1985). Further, there was no attempt to institute a ban on the killing of crocodiles for domestic consumption, because the government believed the people would not accept such a ban. Traditionally, the taking of crocodiles by natives was minimal and confined to young crocodiles and eggs for food, but harvesting increased when the international skin trade developed (National Research Council, 1983b).

The national plan culminated with the enactment of the 1969 Crocodile Trade Protection Act, which was aimed at controlling the export trade, eliminating pressure on breeding adults, and developing a nationwide program to rear crocodiles to marketable size. The government set up a three-tiered system of farms with loans, training, and informational booklets and guides. In addition, the government began establishing training centers and research farms in 1969. By 1975 there were eight such operations.

The program went into operation in 1972 with the creation of a first tier of small village farms, essentially holding pens for the young crocodiles captured by the natives. Each village had a holding capacity of 300 to 500 animals. Since the young have a high natural mortality rate, it was expected that taking of young would have a minimal impact on the size of the wild population. The natives soon discovered, however, that they could not keep hatchlings alive. They gave up taking hatchlings, and concentrated instead on juveniles (one to four year olds, 18 to 48 inches long). Juveniles have a good chance of surviving in the wild, and taking them has a distinctly negative impact on the wild population.

The second level consisted of small business farms holding up to 3,000 animals. They were located near bush airstrips, and supplied the third level of larger company farms, which had holding capacities of up to 20,000 animals. These last operations were substantial, requiring about $250,000 to establish; they were located adjacent to major population centers, and required a reliable supply of food (generally trash fish, or offal from nearby poultry farms) and fresh water. Each tier sold stock to the higher level as the crocodiles grew to market size and moved toward the export facilities. Each higher level was better financed and better equipped, and was expected to absorb the stock of the lower levels in case of food shortages, droughts, or periods of depressed market prices (National Research Council, 1983b). However, often when such pressures occurred, the crocodiles in village farms were in such a deteriorated condition that the company farms would not accept them and the crocodiles perished (Ross, pers. comm., 1985).

In 1981 the nationwide farming system held a total of about 30,000 crocodiles. The ultimate goal was to maintain a population of 100,000 crocodiles, producing about 30,000 skins per year for export. At full operation the government expected to be able to provide all its exports from captive-reared stock. By 1981, however, the system was still performing far below expectations: 10,000 skins worth $1 million to $2 million were exported, but only about 1,000 of these came from farm-reared stock. The rest were culled from the wild (Rose, 1983).

Assessment

Although still functioning, the system has been restructured and operates at a much reduced scale. The restructuring is due largely to operating problems at the village level. The retrenchment emanates from budgetary problems incurred by the government (as a result of the worldwide recession in the early 1980s, which severely reduced PNG's export earnings) and the replacement of the conservation-oriented government with a conservative government focused on development of PNG's mining and timber resources.

Village farms were largely abandoned in PNG in 1982, primarily because of the problems of instilling proper husbandry practices at the village level. With only one or two exceptions, the village farms reverted to being collection centers for the large commercial rearing operations, holding juvenile crocodiles for a brief time. Reports from UN/FAO advisers and observers in 1979 and 1980 indicated that very few of the 180 village farms were operating successfully. There were reports of shortages of food and water, crowding, a lack of hygienic conditions, leading to stress and disease. In general the stock were neglected, and the number of animals being raised had declined. All of this suggests waning interest or ineptitude on behalf of the villagers in the farming program (Rose, 1983).

This is particularly discouraging in light of the extent to which the government promoted and subsidized the program. The entire operation required sizable start-up costs, financed by the central government with development bank and UN backing. Subsidies were provided in the form of loans, training, frequent extension services, and centrally organized and controlled marketing.

Much of the deterioration in the village farm system was undoubtedly due to the technical difficulties associated with attempting to impose a rearing operation on villagers who were used to harvesting wild crocodiles for the skin trade. The villagers are primarily hunter-gatherers, unfamiliar with the requirements of operating a farm that is a supply source for an integrated commercial industry. Establishing a rearing operation in a less developed culture may require more time, supervision, and funding than PNG and UN/FAO were able to provide.

However, casual management by the government probably contributed to the problems at the village level. During the height of the skin trade, the expatriate traders maintained a regular schedule of visits to the villages. Under the integrated federal system, visits by purchasing agents were much more erratic (Ross, pers. comm., 1985). Villagers became discouraged once they realized that they were involved in a project that did not generate cash or barter rewards at reliable intervals (Rose, 1983).

Several other factors may have undermined the morale of village crocodile farmers, but the structure of property rights was definitely not one of them. In a society that is primarily communal (for example, such basic activities as food gathering and distribution are conducted collectively), the crocodile farms were organized so as to provide every incentive to operate them efficiently. Village farmers were responsible for harvesting their own crocodiles (or purchasing them from other hunters) and for maintaining their captive stock in salable condition. In return, any revenues from sales accrued to the pen operators. Since a stock pen is a fairly simple thing to construct, and since the government guaranteed the farmers a market, there was no lack of incentives. Indeed, in any one village there might be as many as a dozen crocodile farms (Ross, pers. comm.,

1985). What the property rights structure did not do was create an incentive to conserve wild crocodiles and their habitat.

The lack of such an incentive is a product of village social structure. Each tribe owns nearby forest (including any crocodile habitat) communally, with no portion of the forest going unclaimed. This territory is actively protected. Territorial boundaries have been established and are regularly defended through intertribal conflicts (King, member, National Research Council Panel on Crocodile Farming; director, Florida State Museum, Gainesville, pers. comm., 1985). Although a tribe will defend its territory against raids by alien hunters, it does nothing to restrict taking by tribal members. Crocodiles and their nests belong to the individual who locates them. This is precisely the recipe for a commons: multiple, unregulated entities all trying to capture and sell as much of the commonly owned crocodile population as is privately profitable. Although the central government encouraged the tribes to manage the harvesting by establishing rules such as no taking of nesting females or by rotating harvesting areas, few tribes initiated management practices, and implementation was even more limited.

Thus, although the farming system could have been structured to complement the conservation of crocodiles, there is nothing inherent in the design as implemented that does this. The farmers concentrate on collecting juveniles rather than the naturally vulnerable hatchlings, harvesting of freshwater adult breeders remains legal, and the entire gathering operation was and continues to be organized as a commons. Despite censusing efforts by the central government, no one really knows how the crocodile populations are faring. They may be doing better than when the system was operated by expatriate skin traders, or there may be no significant change except for the actors and the distribution of the revenues. If the crocodile populations have recovered or stabilized, it is not because management of the commons has improved dramatically.

With the election of 1980, the entire crocodile program sustained a severe blow. A development-oriented, conservative government interested primarily in promoting PNG's timber and mineral resources came to power. In 1981 the budget for the Department of Environment and Conservation was cut from 3 million (kina) to 0.5 million and the staff from 300 to 50. Extension services were severely curtailed, and UN financial and manpower support was terminated. As of 1984 many of the wildlife projects were closed or drastically reduced. The insect and crocodile projects continued but were struggling, even though the conservation-oriented government was reinstated in 1983 and has been attempting to resuscitate the wildlife programs (King, pers. comm., 1985). Clearly, political change places wildlife conservation programs in a precarious position, and political instability is common among the LDCs.

Although some increase in the pace of development occurred under the conservative government, it probably had little impact on the crocodiles and their habitat. The crocodile habitat is located in the swampy lowlands, and most of the development occurs elsewhere. Further, it would be misleading to imply that only the conservative government promoted the development of extractive industries. With the change in regimes there was merely a change in emphasis. Timbering and mining have been conducted for some time in PNG. They represent an important source of revenue and employment, and the returns are such that this type of activity will continue no matter which government is in power (Bruning, curator of ornithology, New York Zoological Society, pers. comm., 1985). If economically exploitable mineral reserves are discovered in the lowlands, the crocodile habitat will be threatened. Remoteness is the primary savior of this habitat.

Habitat modification has taken its toll on butterflies throughout the world. Deforestation in the tropics, and wetland conversion and environmental degradation in more developed areas, have resulted in threats to several species and extinction for several others.

Although PNG has been spared these by-products of development and remains largely an undisturbed rain forest populated by forest farmers, its butterflies have nonetheless come under stress from collectors, commercial hunters, and traders. By the mid-1960s butterfly collecting and commercial harvesting, which had been going on since the turn of the century, had reached such proportions that several of the most striking and valuable species (the birdwings) were endangered.

PNG moved to protect its butterflies, going so far as to stipulate in its constitution that insect conservation is a national objective. The government took its first protective steps toward butterflies in 1966. Collecting and trading in seven species of birdwing butterflies were prohibited, large fines were instituted, and penalties ranging to deportation were implemented. Soon thereafter, preserves were established, and research was undertaken to promote the recovery of depleted species. In 1974 PNG initiated its butterfly farming program. In 1975, in an effort to retain a monopoly in PNG's unique varieties of butterflies and reserve any economic returns for the country's natives, the government prohibited foreigners from engaging in commercial trade in butterflies, and forbade the export of live specimens (National Research Council, 1983a).

Nature of the Butterfly Trade

Although butterflies pollinate crops and wild plants and serve an important role in some food chains, commercially they are valued only for their aesthetic qualities. Taiwan, Korea, and Malaysia maintain factorylike operations where specimens are mounted in plastic and glass, in trays, coasters, tabletops, decorative screens, and even in clear plastic toilet seats. Similar but somewhat smaller enterprises flourish in Honduras, Hong Kong, the Philippines, and several African nations. Taiwan alone sells an estimated 15 to 500 million specimens annually; the take of blue morphos from the South American tropics is roughly 50 million per year (National Research Council, 1983a). Despite the magnitude of the harvest, wild population levels in Taiwan seem to have been unaffected, and biologists estimate that reproduction rates are such that the harvest from South America could be doubled without detriment.

Although the vast majority of the specimens in the butterfly trade are collected directly from the wild, increasing numbers are being supplied by breeding farms. A limited amount of farming is done in Australia, Malaysia, and the American tropics, but prior to the general reduction of PNG's wildlife programs, butterfly farming in PNG was conducted on an ambitious scale, with 500 farms operating in 1978.

Butterfly farms are operated by villagers who plant flowering hibiscus to attract adult butterflies and leafy plants for caterpillars to feed on. After mating, the females lay their eggs on the plants, which support the larvae. In turn, the larvae attach chrysalises to the underside of the leaves. Harvesting, preserving, and packaging the specimens for shipment to the government marketing agency —Insect Farming and Trading Agency (IFTA)—in Bulolo are delicate matters, and timing is critical. That some technical skill is required is apparent from the fact that in 1978 only 50 of PNG's 500 farmers were consistently able to supply marketable specimens.

Although the farms are run independently, the government maintains control over all other aspects of butterfly conservation and trafficking. IFTA trains village farmers, does research designed to increase productivity, and operates a marketing co-op, taking and filling all orders and maintaining quality control, for which it retains 25 percent of the profits.

Despite the scale of the farming program, as of 1982 cultivated butterflies accounted for only 30 percent of the specimens marketed, the rest having been field collected. Farmed specimens generated half of the revenue, however; this probably reflects management of the habitat for more valuable species. As a revenue raiser, the program is modest: $180,000 was distributed to the villages during the first seven years of operation (1974–81). There is significant potential for growth, however, since IFTA was able to fill less than 10 percent of the orders it received. In 1981, PNG's aspirations were to export 5,000 to 10,000 specimens a month and generate $120,000 income per year for the villagers.

Assessment

There are several reasons for PNG's success in stopping the depletion of wild butterfly populations, but none of them have to do with the establishment of private property rights over what was formerly a common property resource. The farms are satellite entities dependent on the wild. They take from the wild and may, by augmenting habitat, stimulating reproduction, and releasing unharvested specimens, even add to the wild. (Even if no overharvesting occurs, the farms may not be neutral in their impact on the wild. Because they are managed for production of the most profitable species, released specimens may change the relative population levels in the wild. The long-run implications of such changes are unknown.) In and of themselves, however, the farms do nothing to prevent overharvesting of wild butterflies (which except for birdwings remain an unregulated, common property resource) or create a private incentive for the preservation of wild habitat where no incentive existed previously. Rather it is the underlying economic, environmental, and anthropological conditions, together with government regulation, which account for the replacement of an overexploited commons with a commercial structure consistent with conservation.

The tragedy of the commons was circumvented largely because the government moved to exercise control over the wild. The most vulnerable species (the birdwings) were protected. Although other species remained common property and harvesting of them from the wild continued apace, they seemed able to sustain the pressure, and extension of protective controls was unnecessary. While government management of the commons was crucial for the conservation of the butterflies, equally important was the remoteness of the habitat and the absence of any immediate threat from development. PNG's butterfly industry generates too little revenue to stave off deforestation if it were imminent. PNG substituted a butterfly industry in harmony with the environment for other industries that promised not to be. The natives are primarily gatherers not cultivators; hence they put little pressure on the forest. (Slash and burn agriculture is practiced in some parts of PNG.) Much of the country's timber and mineral resources are too remote and inaccessible to make exploitation attractive. Overharvesting is restricted by government prohibitions—prohibitions made easier by natives interested in preserving their traditional life-style and sympathetic with the objectives of the conservation programs. If these conditions could be duplicated elsewhere, there would be little need to establish economic incentives to conserve the wild.

Conceivably, the joint revenues from the *multiple* commercial wildlife programs could be sufficient to make deforestation and traditional development uneconomic, but I doubt it. As we have seen, the wildlife programs are a precarious arrangement, vulnerable to the whims of political change. If traditional development is stalled, it is much more likely to be as a result of the magnitude of the direct costs associated with achieving it than because of the returns from marketing wildlife.

CONCLUSIONS

Some people maintain that the market will solve virtually every allocation problem, even those involving environmental resources. In its purest form, this view is fanciful. Markets function best when values are well captured by market signals; when resources move easily in small increments from one activity to the next; when actions are reversible; and when the consequences of actions are well known and close at hand rather than uncertain and some time distant. These conditions are frequently not met for environmental considerations.

If the concept of property rights is broadened to include the exercise of government authority over the environmental commons, the opportunities for utilizing economic incentives for conservation are greatly expanded. With no claim to either their necessity or sufficiency, the following conditions are listed as favoring the replacement of a commons with commercialization consistent with conservation: (1) habitat has relatively low value in alternative uses; (2) government is stable and committed to conservation and management of the commons; (3) property rights can be established that are not prohibitively expensive to enforce; (4) species and their habitat can be insulated from the effects of contiguous development; and (5) owners of wildlife resources can compete effectively with either poachers or manufacturers of synthetic substitutes.

The analysis of the case studies suggests that although it is often difficult to realize such conditions, there are opportunities for conserving wildlife by substituting property rights structures for uncontrolled commons. But such structures will not necessarily be self-enforcing. They will require varying degrees of government support, management, and regulation to be successful.

Of the cases considered, the management arrangements for the seabirds in Iceland come closest to the pure property rights paradigm. Most of the breeding grounds for eiders are privately owned and farmed. Government protections are extended to eiders on lands that are not farmed. For the other seabirds, local authorities (and, in some cases, private owners) lease the harvesting rights, but the government regulates hunting methods so as to avoid the taking of breeding adults. The birds have fared far better than under a commons.

Although the system for harvesting crocodiles in PNG could have been structured to complement the conservation of crocodiles, there is nothing in the design as implemented that does this. The system remains a commons, and breeding adults continue to be taken. The government did move aggressively to regulate the international skin trade operated by expatriates, reserving the harvesting rights for PNG natives. But there seems to be little significant change in the harvesting system, except for the actors and the distribution of the revenues. If pressure on the crocodile populations has subsided, it is not because management of the commons has improved dramatically, but because of the continuing

262

remoteness and inaccessibility of the habitat and the inefficiency with which the current trade is managed by the government.

The depletion of PNG's threatened butterfly populations has been interrupted, not because private property rights have replaced the commons, but because the government has exerted control over the harvest and protected the most vulnerable species. The farms are satellite entities on the periphery of the commons, and remain dependent on the commons for their breeding stock. The habitat itself remains largely undisturbed because of its remoteness and the absence of pressure for traditional development, not because butterfly farming is sufficiently lucrative to displace such development.

Although there are opportunities to use market mechanisms to conserve wildlife, and to substitute property rights for unmanaged commons, one cannot be too optimistic that this strategy can be widely applied and sustained. It takes careful planning to design the appropriate economic incentives and to gain the support from government authorities to implement them and enforce the property rights. LDCs have immediate and overwhelming economic and political problems, and the likelihood that future ecological consequences will be heavily discounted in favor of short-run benefits from development remains strong.

REFERENCES

Arrow, K. J., and A. C. Fisher. 1974. Environmental preservation, uncertainty, and irreversibility. *Quart. J. Econ.* 88(2):312–319.

Benedict, P., A. H. Ahmed, R. Ehrich, S. F. Lintner, J. Morgan, and M. A. M. Salih. 1982. Sudan: The Rahad Irrigation Project: Project impact evaluation no. 31. U.S. Agency for International Development. Washington, D.C.

Caufield, C. 1985. The rain forests. *The New Yorker* 60(48):41–101.

Clark, C. W. 1974. The economics of overexploitation. *Science* 181:630–634.

Doughty, R. W. 1979. Farming Iceland's seafowl: The eider duck. *Sea Frontiers* 25(6): 342–350.

FAO. 1980. Report on the Study Tour FAO/SIDA/CIDIAT on Incentives for Community Involvement in Forestry and Conservation Programs: Honduras, Jamaica, Colombia, Venezuela. Rome: Food and Agriculture Organization of the United Nations.

Fife, D. 1971. Killing the goose. *Environment* 13(3):20–27.

Haspel, A. E., and F. R. Johnson. 1982. Substitutability, reversibility, and the development-conservation quandary. *J. Envir. Manage.* 15:79–91.

Hyman, R. 1982. Iceland's harvest of plenty. *International Wildlife* 12(3):4–11.

Inskipp, T., and S. Wells. 1979. *International Trade in Wildlife*. London: Earthscan.

Joyce, S., and B. Burwell. 1985. Community-level forestry development: Options and guidelines for collaboration in P.L. 480 Programs. AID/PC Forest Resource Management Project. Washington, D.C.

Myers, N. 1981. A farewell to Africa. *International Wildlife* 11(6):36–47.

———. 1982. Sustainable exploitation of wildlife as a conservation strategy. Lecture presented at The First Annual National Zoological Park Symposium on Animal Extinctions, Washington, D.C.

National Oceanic and Atmospheric Administration. 1984. Survey of natural resource management: Applications to fisheries (draft). Washington, D.C.

National Research Council. 1983a. *Butterfly Farming in Papua New Guinea*. Washington, D.C.: National Academy Press.

———. 1983b. *Crocodiles as a Resource for the Tropics*. Washington, D.C.: National Academy Press.

Rose, M. 1983. Crocodile management and husbandry in Papua New Guinea. Report prepared for the Sixth Working Meeting of the IUCN Crocodile Specialists Group, Konedobu, Papua New Guinea.

Secretary of State. 1984. Humid tropical forests: AID policy and program guidance. Telegram to AID Worldwide, Washington, D.C.

Smith, R. J., J. H. Goldstein, and R. K. Davis. 1983. Economic incentives as a conservation strategy for nongame and endangered species of wildlife. In *Transactions, North American Wildlife and Natural Resources Conference* 48:457–467.

Subcommittee on International Development Institutions, Committee on Banking, Finance and Urban Affairs, House of Representatives. 1984. Draft recommendations regarding environmental concerns associated with multilateral development bank activity. Washington, D.C.

Thomson, J. T. 1981. Public choice analysis of institutional constraints on firewood production strategies in the West African Sahel. In *Public Choice and Rural Development: The Proceedings of a Conference Sponsored by the U.S. Agency for International Development*, ed. C. S. Russell and N. K. Nicholson, 119–152. Washington, D.C.: Resources for the Future.

Wallace, M. B. 1983. Managing resources that are common property: From Kathmandu to Capitol Hill. *J. Policy Anal. Manage.* 2(2):220–237.

Commentary

Ronald N. Johnson

Jon Goldstein's paper is suggestive of the immense problems that need to be overcome if the decline in biological diversity is to be retarded. Rapid population growth in many Third World countries has resulted in the loss of critical wildlife habitat, and pressure has thereby increased on publicly maintained areas. The failure of trade restrictions on endangered species to significantly reduce illegal harvest and trade is noted by Goldstein. In light of these problems, he addresses the question whether allowing trade and the establishment of private property rights in conjunction with traditional regulatory tools can help maintain wildlife and natural habitats.

Goldstein's approach is to examine five cases in which the degree of private property rights and other conditions vary. His conclusions differ for each case, and the tone of his conclusions is not optimistic. Nevertheless, some of the information presented suggests that the development of trade and private property rights can play a positive role.

In the case of eider farming, for example, Goldstein notes that an increase in the degree of exclusivity in terms of private rights has enhanced the stock. He does not, however, detect a positive effect in the other cases involving seabirds. The mere assignment of territorial hunting rights has not been sufficient to foster the enhancement of habitats. Given that individuals of most species of wildlife range over broad areas, the assignment of relatively small individual territorial rights offers only a limited incentive to conserve and invest in the resource. According to Goldstein, there are numerous right holders, and control over harvesting has usually been entrusted to government regulation. The seabird cases are indicative of a general reluctance to grant sufficiently large exclusive rights to a single entity (Johnson and Libecap, 1982). Even where the participants are many, the existence of rights can provide an incentive to conserve the resource. Goldstein is careful to point out that many seabirds have a tendency to return to the same areas each year. With territorial rights in that setting, an incentive to conserve the stock may be present.

In the case of butterfly farming in Papua New Guinea, the rights are again territorial. Farmers own the land but apparently have little direct control over the species in the wild. Goldstein's conclusions here are rather negative: "In and of themselves, however, the farms do nothing to prevent overharvesting of wild butterflies (which except for birdwings remain an unregulated, common property resource) or create a private incentive for the preservation of wild habitat where no incentive existed previously."

The final case that Goldstein examined, crocodile farming in Papua New Guinea, appears to offer the setting with the most exclusive rights structure, but many of the farms are no longer in operation. The difficulties experienced in this case are representative of many investment projects in LDCs. All too often, projects are poorly managed and financed and governmental policies are subject to frequent change. These problems act as a constraint on the use of commercialization in relieving pressure on wildlife and natural habitats. The notion that commercialization could relieve pressure on wildlife stocks is counter to most policies advocated over the last few decades. The groups that sought trade restrictions have a vested interest in the success of those programs and remain highly influential in the policy-making arena. Hence, detailed studies on the role of private property rights in preserving wildlife have not been given much attention. What has been produced is a collection of largely ad hoc stories from which the success or failure of commercialization is to be judged.

The study of property rights involves a close examination of just what rights individuals have and over what they can contract. The extent of a right holder's ability to exclude others, transfer rights, and derive income from the resources is generally the main area of attention. In none of his cases does Goldstein offer such a detailed description of the property rights structure. Furthermore, the cases presented do not provide a basis for simultaneous comparisons of a single species with varying degrees of property rights. Crocodiles and butterflies are each represented by a single case. While the seabird cases provide some opportunity for comparison, there are still significant differences among these species. If single cases are used, then implications about resource use require detailed measurement over time as the property rights structure is altered. The researcher also requires corresponding data on factors such as price. Such information as is provided by Goldstein serves mainly to initiate speculation.

Even though this is an area where facts are in short supply, an assessment of Goldstein's conclusions is in order. While Goldstein apparently does not see much in the way of commonality between the seabird case and the two cases from Papua New Guinea, they all seem to fit the property rights paradigm; hence they may allow more consistent conclusions to be drawn.

The major issue under consideration is whether commercialization can result in an increase of stocks in the wild. To facilitate a comparison between commercialization and a local ban on trade, consider three possible states: (1) a commons or, more appropriately, an open access setting with trading in the species allowed (generally a distinction is made between open access and the commons: the latter applies to a setting where entry is restricted to a certain group and regulations over use are sometimes present); (2) trade but varying degrees of exclusivity (property rights); and (3) a ban on trade. It is generally acknowledged that stocks are lowest and extinction more likely under state 1. The problem arises in making a comparison between states 2 and 3.

Consider a setting in which husbandry and enhancement of the species habitat are not economically practical. Rights exist solely in the form of territorial hunting or limited entry (entry requires a license, and the issuance of licenses is restricted below the open access quantity). Under that arrangement, stocks will be drawn down below their maximum (assuming a steady state solution with its associated potential problems); but just because stocks are reduced by commercial activity does not mean that they would be higher under state 3. What matters is enforcement which Goldstein has argued is often lacking. Accordingly, there may in practice be little difference between states 1 and 3.

Limited entry or territorial rights, on the other hand, provide not only a degree of exclusivity but also a constituency with an interest in policing the illegal harvest of the species. Indeed, much of the game and commercial fishery regulation in the United States was instituted because of the political support of the very groups that were involved in harvesting (Johnson and Libecap, 1982). This is not to say that environmental groups have played no role, but rather that they have not been the only force. Once property rights are in place, individuals will attempt to protect those rights. While heterogeneity among the right holders can make agreement over remaining margins costly to achieve, the potential for the adoption of more stringent regulations remains, provided the new rules can enhance the value of existing rights. The fact that a government is policing the harvest of a species may mean only that the right holders are capable of enforcing their rights.

Goldstein is cognizant of the role the trade constituency can play in enforcement when he notes:

> There is a symbiotic relationship here: protection of eiders encourages the industry, and a healthy industry results in a constituency that supports government protection. Should the industry continue to decline, interest in protection could erode.

Goldstein may well have the sequence wrong. The industry encourages protection rather than protection encouraging the industry. This is important. Treating governmental regulation as exogenous ignores the interplay between the evolution of those rules and trade. Later, Goldstein ignores this interaction when drawing his rather negative conclusions in the butterfly case. Yet the trade and the intensity of regulations seem to have worked well together here.

This discussion has not, however, considered the possibility of habitat enhancement or the establishment of farms. In the former case, the likelihood that stocks in the wild will be greater under state 2 is increased. The establishment of farms, however, adds complexity. Consider that initially farms take their breeding stock from the wild (e.g., crocodile eggs) and over time build up their own breeding stock. In the initial stage the wild habitat provides a resource for the farms and they have an incentive to protect it; but once those farms have their own breeding stock, the wild stocks present a potential source of competition.

There are at least two ways in which farmers may attempt to deal with the problem of competition from the wild. On one side, they may seek stringent regulation and enforcement on harvests in the wild, in which case, wild stocks can increase. On the other hand, if enforcement is impractical (too costly), the incentives of farmers may run counter to the objective of habitat preservation. Alternatively, economies in production may be realized and prices lowered to the extent that taking from the wild becomes unattractive. What is important here is that the incentive for protecting wildlife habitat depends on its use value to the farmers. Contrary to the criteria that Goldstein appears to be using to judge the effects of farming, the taking of breeding stock from the wild provides the incentive to maintain the wild habitat.

What facts Goldstein offers appear to be consistent with the behavior of farmers as described above. Stricter regulation over the exploitation of crocodiles in the wild appears to have coincided with the development of the farms, and the failure of many of these farms saw the decline in protection efforts by the government. Although Goldstein attributes this curtailment of enforcement to exogenous political factors, the pressure for enforcement may have declined with the failure of the farms. He notes no similar decline in the enforcement of rules pertaining to butterfly taking, where farms were apparently more successful. Again, the problem with these case studies is that they do not offer sufficient information to make sound judgments.

Clearly, the idea that private contracting can solve all the problems is precarious at best. It is evident, for example, that the vagaries of the marketplace pose a serious constraint to the reliance we can place on commercialization to preserve wildlife habitat. If wildlife habitat is exclusively owned by private parties, we can expect much of that habitat to be converted to alternative uses whenever returns from those uses are higher. In many settings, the preservation of wildlife and its natural habitat may best be done by the government. Still, commercialization provides a local political constituency who will support enforcement and maintenance of habitat. In the absence of such local support, authorities may find a total ban on harvesting too costly to enforce. This problem is further exacerbated if the pressure for the ban is of foreign origin. It could be argued that those who desire the ban should be prepared to pay for it; but that approach is unlikely to provide the least costly way of achieving the objective. Nor does the electorate in the United States appear willing to pay the necessary fee.

In focusing on the role of trade in maintaining wildlife habitat, Goldstein provides an interesting but perhaps too narrow format. Largely missing from his discussion is the role of private parties in preserving endangered species. Private preserves constitute a means for saving wildlife in the face of a decline in their habitat, although private incentives to preserve genetic stocks are likely to be less than what would prevail in an "ideal state." Arguments showing why and under what conditions the market is likely to fail or succeed are instructive; but, given the current state of affairs, a more appropriate policy-oriented approach would be to compare the real workings of the marketplace with the alternatives at hand. In that regard, much work remains to be done.

REFERENCE

Johnson, R. N., and G. D. Libecap. 1982. Contracting problems and regulation: The case of the fishery. *Amer. Econ. Rev.* 72:1005–1022.

Commentary
Roger G. Noll

The problem of designing effective policies to protect endangered species presents interesting intellectual challenges in several scholarly domains. To an economist, the core issues are inevitably framed in terms of the incentives that alternative policy regimes provide for the individuals whose actions will significantly affect the survivability of a species. The fundamental axiom of economic analysis is that individuals will pursue opportunities for personal economic gain. In using economic analysis to assist in the design of policy institutions, the central question is how to construct a policy so that the incentives acting upon decision makers are compatible with broader social objectives: how do we channel rational self-interest so that the common good is served?

In answering this question, it is useful to divide the problem into two components. Part one consists of constructing a list of alternative policies and institutions for implementing them, and examining each to determine the outcome, driven by self-interested behavior, that they will produce. Part two addresses the question of how we induce the selection of the "best" arrangement, given that this choice, too, will be affected by self-interest? The act of creating a new policy regime will do more than simply alter the survivability of a species; it will affect the distribution of economic activity among various industries, locations, and people, and hence redistribute income and wealth. Because public officials in all political processes, even nondemocratic ones, face constraints on the extent to which they can redistribute wealth and still survive, an important element of institutional design is to identify policies that, comparatively speaking, are compatible with the economic interests of those affected by them.

The range of possible approaches to saving a species is wide. One can write behavioral rules (regulations, criminal laws) and enforce them through policing. One can establish reserves in the public domain, and take measures to protect them. One can impose taxes or bans on economic activities that threaten a species, or one can subsidize activities that are substitutes for its exploitation. And one can redefine property rights in either a species or its habitat in ways that give property owners an incentive to protect it.

Jon Goldstein's contribution to this book examines some examples of programs that attempt to use property rights and commercial exploitation as a means to preserve a species. The idea is to create an institutional setting in which the property rights to a specific population of a species are held by a single decision-making unit. In principle, this can solve the common property resource problem, which arises when a single decision-making unit does not bear the full costs of using a resource. In the case of an endangered species, one among many hunters or land-clearing farmers contributes relatively little to its extinction, and bears only part of the cost if it does become extinct. Hence, each may take actions that, collectively, cause extinction, even if all would prefer a collective harvesting program that would preserve the species (see Levhari and Mirman [1980] for probably the most sophisticated development of this argument). If the farmers and hunters are organized into a collective that owns a population and makes collective decisions about its exploitation, at least some of the external costs of individual exploitation are borne by the collective, and therefore are taken into account. Moreover, the commercialization solution has the additional advantage that it can provide economic benefits to users of the resource—unlike taxes, bans, criminal sanctions, or protected reserves.

In this paper, I set forth a simple economic explanation of the conditions under which species extinction takes place and apply this analysis to characterize the situations in which the "commercialization solution" may or may not work. My objective is to explain the underlying logic of policies that seek to save a species by redefining property rights, and to identify the conditions that must be present for the property rights approach to work.

A useful analytical jumping-off place is provided by two papers from the mid-1970s: Vernon L. Smith's (1975) analysis of the extinction of megafauna in the late Pleistocene, and A. Michael Spence's (1973) analysis of the problem of the blue whales. Smith argued that human hunting and farming could plausibly account for the extinction of large mammals in prehistory. The argument was that very large animals, especially animals that run in herds on the plains, are quite easy to hunt but extremely difficult to domesticate, especially in a primitive society. Hence a tribal group about to drive the last herd of the giant buffalo off a cliff in South Dakota could kill all of the herd at very little additional cost compared to killing only some; but if it spared some animals, it would have little likelihood of being the group that harvested the animals at some future date. Hence, low hunting costs and low incentives for preservation produced extinction.

Spence (1973) analyzed a quite similar situation, the problem of the possible extinction of the blue whale through overharvesting. He argued that competition among whaling nations produced much the same situation as Smith described regarding Pleistocene megafauna. Thus Spence (1973) produced what amounts to the property rights solution: give a single nation the exclusive rights to hunt blue whales, allowing it to decide for itself how to allocate those rights among its own whalers and (presumably through a sale of rights) the whalers of other nations. He presented a simple, empirically based model of the biology and economics of whales which concluded that such a monopolist would, for purely *economic* reasons, stop harvesting the species until its population recovered.

Of course, the monopoly solution may not actually preserve the species. Lewis and Schmalensee (1977) delineated three possible strategies a monopolist might follow to maximize long-run profits from exploiting a species, any one of which

270

might be best depending on costs, demand, and biology. One is to harvest the species continuously at the maximum sustainable economic yield, or some variant thereof that takes account of changes in demand for the species over time. The second is to harvest intermittently, each time taking very large quantities but not risking extinction. The third is to drive the species to extinction over some specified period by overharvesting. In some situations, the last strategy produces the greatest wealth for the monopolist. Generally speaking, the factors favoring the extinction route are (1) a low reproductive rate for the species, (2) a high rate of discount, (3) an elastic demand for harvested members of the species, and (4) a harvesting cost that is insensitive to the size of the current population of animals or that exhibits economies of scale.

As an illustration, consider the following example. Suppose a harvested animal is subject to a perfectly elastic demand at $1 (e.g., a hunter can sell all that is harvested at that price). Suppose also that the species reproduces at a rate of 1 percent per year, that harvesting costs are independent of the number of animals remaining and are very low, and that the interest rate is 10 percent. In this case, the best the monopolist can do is harvest the entire species immediately and collect $1 for every member; anything less leaves a stock that will increase in value by 1 percent per year (the rate of growth of the population), whereas the hunter can kill that stock and earn 10 percent on the sales by putting his revenues in the bank.

Of course, all of the conditions assumed in the illustration are extreme, but the central concept is general. The example illustrates the kind of species that is likely to be endangered regardless of the market institutions and property rights that surround their exploitation: species that reproduce slowly, that are easy to hunt, and that are used in the production of goods for which there are relatively close substitutes (the most likely source of elastic demand), such as a species that is harvested for standard food.

A second cause of species extinction is likely to be more important than excessive harvesting: the destruction of habitat by the conversion of land to uses that are incompatible with the species. Examples are intensive agriculture and man-made water impoundments. The fundamental analytics of this problem are the same as the overharvesting case. Here the focus is the habitat of a population: the area (range) necessary to support a self-sustaining group. If property rights in land are fragmented, so that individual landowners each own a small fraction of the range, a series of small decisions to take away range can lead to extinction. The corresponding property rights solution is to cause all of the range to have a single owner.

Again, whereas this solves a part of the problem, it does not deal with the case in which the wealth-maximizing solution is, in fact, to drive the species to extinction by converting all of the land to a new use. Even a monopolist in range faces the possibility that converting it all to, say, intensive agriculture

271

is more profitable than undertaking a use that also preserves the species while contributing a steady harvest of its members.

In addition, the fact of monopoly does not necessarily mean that the monopolist has full control of the property. To protect against poachers is costly, especially for species with a wide range. After all, African game refuges are, in essence, governmental monopolies in rangeland for protected species, but poaching persists. Were the species harvested, the funds for protecting the stock would increase; however, it remains doubtful whether these funds would be sufficient to cover the costs of enough protection to maintain a viable population.

Still another complication can arise if the species is of little or no commercial value. Obviously, overhunting cannot cause an economically useless species to be driven to extinction, but habitat destruction clearly can. In this case, there is no adequate property rights solution: a group of fragmented landholders will find the species unattractive, and so will a monopolist. Creating a monopoly will not increase the incentive to manage the range in a manner that protects the species. Other than the creation of game refuges or other forms of restrictive controls on land use, an effective preservation strategy here requires providing economically attractive land uses that are compatible with preservation. Two examples are state subsidies for the products of land uses that are compatible, or research to develop more compatible ways to use the range. This is similar in concept to the attempt to create "game farming" as discussed in Goldstein's paper, but deals with the more general case of finding *any* land use that is compatible with the preservation of the range, not just the specific one of organizing and subsidizing exploitation in a way that is consistent with species preservation. The general policy is subsidy of any habitat-preserving land use.

More than likely, there will be many cases in which the alternative uses that protect endangered species will be less attractive economically than more exploitative uses; that is, even the monopoly property right will not work. What this means is a permanent, significant subsidy; indeed, the more economically unattractive the species is as a harvestable commodity, the more expensive will be the preservation program (owing to the absence of any significant revenues from harvest). The rationale for preservation is that a species is valued beyond its usefulness as a commodity: first, in some sense we are all richer if the California condor continues to exist, even those of us who will never see one; second, greater genetic diversity leaves open more options for possible economic uses in the future, as discussed by Brown and Swierzbinski (1985). Because the vast majority of species are confined to a limited geographic area, species preservation for these reasons involves another form of externality. If one nation preserves a species, all other nations benefit. From the preceding analysis, however, the more useless a species is from an economic standpoint, the higher will be the net costs of preserving it; so it is more likely that international transfers of funds will be necessary if preservation is to occur. The conclusion, of course, is that the ultimate commercial potential of a species ought to enter negatively into a nation's calculations about how to allocate a limited budget to preservation programs, all other things being equal.

The discussion so far emphasizes situations in which the creation of undivided property rights in a population is an insufficient measure to ensure its protection. Another possibility is that divided or shared rights will not lead to overharvesting and extinction because the traditional model of the use of a common property resource does not apply. Recent work in game theory has identified conditions under which rational, self-interested behavior leads to cooperative rather than excessively exploitative behavior. These conditions, laid out in greater detail in Lewis (in press), include the following. First, the use of the resource must be expected to continue indefinitely if it is not driven to extinction. Second, each person who is using it must stand a reasonable chance of detecting noncooperative behavior by others. Third, each person has a credible retaliatory threat, should cheating by another be detected. That is, each person must have available a response to cheating that (1) inflicts substantial losses on the cheater and (2) makes economic sense to the person who carries out the threat: if one person cheats, another person's best response is to punish. These are, of course, very stringent requirements, but nonetheless they may apply in certain situations. For example, some instances of cooperative behavior among small, closely knit communities arise in situations that seem to satisfy the conditions of infinite repetition, likely detection, and rational retaliation.

The problem of detecting cheating or poaching is important to most any protection policy. Assume that the situation is one in which adequately policing the wild habitat to protect against poachers is extraordinarily expensive, so that neither the cooperative-competitive nor the monopoly solution (including the government refuge) is viable. Presumably the situation is one in which harvesting from the natural habitat, including the costs of avoiding detection and the risk of being caught poaching, is a cheaper way to deliver harvested animals than to participate in intensive husbandry. If it is not, the easiest solution is to zone certain rangeland for commercial wildlife husbandry, and allow game farms to drive out poachers on game refuges simply by underpricing them. Thus we will assume here that commercial operations cannot compete effectively as long as there is any wild population. This means that in order to use commercialization to preserve a species, one must not only zone land for commercial wildlife husbandry but also subsidize commercial operations (e.g., to use commercialization to protect wild elephants, one must subsidize, on a large scale, elephant farms). The magnitude of the subsidy must be sufficiently large that game farms can underprice poachers, so that it is cheaper to buy an elephant from a game farm than to hunt one.

The proposal to subsidize game farms has two major problems. One is its distastefulness to the most ardent advocates of preservation. The other is that for many species it may be more expensive than effective in protecting game refuges. For megafauna, in particular, the economics seem quite unattractive. Large mammals are very easy to find and hunt, even when their numbers are few; the payoff per unit of harvest (e.g., pounds of food or tusk) is very large; and creating private reserves can require the creation of enormous private ranges (e.g., consider a private range for the blue whale). At the same time, commercialization requires that to some degree the "free" food and range of the natural habitat be replaced by habitat and foodstuffs acquired at some cost by the game farmer.

An upper bound on the necessary subsidy is the total cost of containing, guarding, and feeding the animals. Obviously, if animals are commercially raised at a relatively high population density, harvesting will be relatively inexpensive, compared with the cost of poaching. For species that are endangered, the initial subsidy may be quite a bit lower because their low population density may make poaching costs quite high. But as the program succeeds in making poaching uneconomic and causing the species to become more numerous, poaching costs will fall and the subsidy will have to increase until a viable population (net of poaching) is obtained.

The preceding analysis provides an analytical basis for interpreting the cases discussed in detail by Goldstein. A few examples will serve to illustrate the relation between the overall economic argument and the specific cases.

The keys to the preservation of eiders are (1) the localized habitat and (2) the fact that the economic value of eiders exceeds the costs of preserving them. Because of (1), individual farmers can become "monopolists" of a colony of birds and reap the full benefit of proper husbandry. Because of (2), it makes sense to preserve a colony. But (2) is becoming less true as substitutes are developed. Moreover, (1) is not totally true: apparently the habitat includes the common property of coastal waters, which are becoming increasingly polluted (an alternative use of the habitat that is destructive). Obviously, eider farmers have no credible threat against industrial polluters, so no cooperative, repeated-play solution is possible. Hence, the population is declining. Note that an eider monopoly, extending to coastal waters, is probably no solution. A monopoly in the entire eider habitat, including coastal waters, probably could make more money by leasing water pollution rights to the oil and chemical industries than by farming eiders. Assuming that eiders are, or are about to be, endangered, the solution must involve pollution control activities and perhaps a tax on substitutes for eiders (or, alternatively, a subsidy for eiders), so that eider farmers can beat the price of the substitutes.

The Papua New Guinea experiment in commercial crocodile farming apparently failed because the subsidy was not large enough. Most skins continued to be harvested in the wild, detection of harvesting in the natural habitat was difficult, and conditions in the commercial yards were substandard. This indicates that commercialization could succeed only if the government spent much more on commercial farms—through training as well as capital and operating subsidies—to make them a rational alternative to collecting in the wild. Although Goldstein's analysis does not specifically address the point, apparently the government believed that the program would pay for itself. Judging from the ease with which large, slow reptiles can be captured in the wild and the extraordinary availability of food and range for them in PNG, such an assumption seems wildly unrealistic. To underprice a product (harvested wild crocodile) that is very inexpensive requires a continuing and massive subsidy of commercial operations, or very intensive protection efforts in the natural habitat to raise the cost of poaching.

PNG has also promoted commercial butterfly farming, this time with success. The key here seems to be the fact that harvesting butterflies is quite difficult: one must catch them at exactly the right moment. Hence, a very large population in a specific, cultivated area enhances the likelihood that the "catch" will be valuable. This explains why farmed butterflies produce a substantially larger share of revenues than their quantity share of the market. In addition, the remainder of the commercial farming operation apparently is low in both cost and skill requirements. The supporting plants are verdant and easy to cultivate. The issue of the value of alternative uses of land is not addressed; but it seems likely in the tropics of PNG that these, too, are low if not nonexistent. Thus, with relatively minimum effort, butterfly collectors can increase their productivity by participating in a commercial venture. The only substantial barrier to overcome is to establish a collective property right in the hibiscus field. But because the range of habitat is small, this can be achieved by using village-level enforcement mechanisms. Whereas the last may have noneconomic overtones, the repeated-play, cooperative game seems applicable here because villages can make credible threats against cheaters, and have a good chance of detecting them. But whatever the motivation for cooperative behavior, the key fact seems to be that commercial butterfly farms are profitable because, as a popular playwright observed not so long ago, butterflies are free—or nearly so.

REFERENCES

Brown, G. M., and J. Swierzbinski. 1985. Endangered species, genetic capital and cost-reducing R&D. In *Economics of Ecosystems Management*, ed. D. O. Hall, N. Myers, and N. S. Margaris. Dordrecht, Netherlands: Junk.

Levhari, D., and L. J. Mirman. 1980. The great fish war: An example using a dynamic Cournot-Nash solution. *Bell J. Econ.* 11:322–334.

Lewis, T. R. Cooperation in the commons: An application of repetitious rivalry. *Economica*. In press.

Lewis, T. R., and R. Schmalensee. 1977. Nonconvexity and optimal exhaustion of renewable resources. *Internat. Econ. Rev.* 18:553–552.

Smith, V. L. 1975. The primitive hunter culture, Pleistocene extinction, and the rise of agriculture. *J. Pol. Econ.* 83:727–756.

Spence, A. M. 1973. Blue whales and applied control theory. In *Systems Approaches and Environmental Problems*, ed. H. W. Gottinger. Gottingen, Federal Republic of Germany: Vandenhoeck and Ruprecht.

Summary of the Discussion

William J. Strang

Hal Salwasser began the discussion by pointing out the existence of a fairly extensive literature on the economic incentives for maintaining wild habitats associated with big game hunting on former cattle ranches in Texas, California, Utah, and Montana, among other states. The hunting ranches seem to be fairly successful financially. He wondered whether we could learn anything from these domestic examples about how to design incentive structures that might work in preserving genetic resources in other countries.

Partha Dasgupta felt that we were overemphasizing private property rights as a solution for many open access resource problems. For many resources it is costly to define and enforce property rights. This is probably why they are open access resources in the first place. One example of such a resource is wood in sub-Sahelian Africa, where it takes eight to nine hours to collect enough firewood to cook a meal. In this case, governments can improve the incentives for conserving wood by subsidizing substitutes like kerosene. Similarly, Amazonian forest is being cleared to grow cattle to supply meat for hamburgers. This is causing many extinctions. Creating private property rights won't work, but taxing wood or beef could help solve the problem.

There are special problems in setting up a private property rights structure when externalities are unidirectional rather than when damage is mutually distributed among a number of parties. For example, prevailing winds carry sulfur dioxide released in Great Britain to Sweden, where it falls as acid rain. The solution here would involve a mutually agreed upon *transfer* of property rights between the two countries.

Jon Goldstein responded to Dasgupta's remarks concerning alternatives (like taxes) to setting up property rights schemes by saying he was looking for commercial mechanisms requiring very little regulation and little subsidy, because such mechanisms will be the most attractive to undeveloped countries. These countries have few resources with which to subsidize or regulate. He also felt that property rights schemes would bring about less conservation than trying to internalize externalities caused by multilateral development banks. For example, AID is now trying to avoid projects that may destroy additional rain forest.

Ron Johnson questioned whether any "tremendous externalities" really exist. One can be quite successful in economics by solving mathematical models and making up stories without confronting real cases. He wondered whether we could tell the difference between private forest lands (e.g., Weyerhaeuser and Georgia-Pacific lands) and Forest Service land. Johnson continued to say he felt that the term "multiple use" was invented by the Forest Service to increase its budget, rather than to manage forests so as to solve externality problems. Because the market *always* fails in mathematical models of externalities, the answer we need is empirical. Johnson also felt that the private sector probably does as good a job if not better than the public sector in allocating resources in markets with "externalities."

John Krutilla disagreed with this, saying that the Forest Service was not allowed under law to do some of the things done on private land. Roger Noll and Jon Goldstein agreed, but noted that the incentives could be structured so that Weyerhaeuser would manage its land precisely the same way the Forest Service does.

Kai Lee felt that property rights schemes weren't used more because it is difficult to get people to adopt them, and asked Noll for some principles that could be used in designing such schemes that would facilitate their adoption. Noll responded by saying that when a government attempts to impose a system of property rights, selling them off is a bad idea, because both those who must then buy the rights and those who lose from the implementation of the rights structure may resist the change. Paying those who lose from the scheme is more likely to achieve sufficient support for adoption.

Another procedure, which is becoming more widely used in air pollution regulation, is to grant rights to pollute to firms in proportion to their past output per unit pollution (e.g., emissions banking and offset trading). This procedure, in the context of a fishery, would probably be adopted, because if it is truly efficient, the profits of each fisherman under the new scheme should exceed profits under common access. This is why, in the case of air pollution control, it was easier to convince industry to accept grandfathered tradable emissions permits, rather than emissions taxes or emissions rights that are auctioned off.

Michael Hanemann felt that experimentation with various alternative systems could be useful in providing information on what is likely to be effective and what isn't. Also, he noted that the government's practice of building water resource projects and then making the water available very cheaply has been criticized by economists for a long time. The availability of cheap water encourages its use and economic growth. This results in a perceived shortage of water and discourages new projects from being built. The way to stop this is to charge an economic price for water, thus cutting down on use. Hanemann felt that allowing farmers to resell the cheap water would result in the same efforts toward conservation and efficiency of use as the economic pricing of the water would. He also mentioned Martin Weitzman's analysis of the merits of taxes versus quantity regulation in achieving social objectives under conditions of uncertainty. Taxes are better than regulation to the extent that what gets cleaned up gets cleaned up in an efficient manner; but quantity regulation is preferable in that the regulators can be sure exactly how much gets cleaned up. Given this trade-off, it is possible to derive the *conditions* under which a tax is preferable to quantity regulation and vice versa.

Arthur Weissinger noted that one of the best ways to conserve rare tropical plants (e.g., Brazilian orchids) has been to introduce them into the horticultural trade. Once there, genetic variants can be protected under the Plant Variety Protection Act. Some questioned whether this could protect the wild variety; if not, individuals in the wild would be under collection pressure. It was also mentioned that some species, such as Southeast Asian lady slippers, can't be reproduced *in vitro*, either by seed or tissue culture, so that overcollection can easily lead to extinction.

Bill Kunin noted that if it is cheaper to collect a plant than to grow it *in vitro*, an absence of property rights will lead to extinction, as with some cacti in the Southwest. He also asked how we could structure incentives to help preserve species with no commercial value. Goldstein suggested that we examine the habitat of the species and try to find something else that has commercial value, although management of such an area will be for the commercially viable species.

Noll mentioned that there is still another property rights-market solution to preserving a commercially worthless species. The Department of Interior can issue a request for proposal, which seeks bids (e.g., from seed companies) on how much it would cost to keep a certain number of individuals of an endangered species alive in a natural habitat. Although it may be easier to save the species by including its habitat in, say, a national park, the principle here is that things can usually be accomplished more cheaply through competition than monopoly.

Margery Oldfield mentioned that preserving species by domestication often drives out wild relatives. This can occur via hybridization or displacement. Wheat and cattle are two examples of this. She also agreed that in many cases captive breeding could reduce harvest pressure on wild populations, but if captive stocks can't meet the demand, extinction in the wild is likely unless regulation of the type mentioned by Dasgupta is used. She gave the example of birds of paradise, and other birds producing marketable feathers, which were endangered around the turn of the century because of overcollection. The birds were saved as a result of public outcry and consumer boycotts against feather producers. Although such a last-ditch effort may not prevent extinction in general,

278

it worked in this case, for many of the birds have recovered and are no longer endangered.

Oldfield then summarized some aspects of the Cayman turtle case from which we might learn how better to conserve wild species. First, the turtles are not captive bred. Eggs are collected from beaches and then reared *ex situ* for market, so that the wild stock is the sole source of eggs. Pressure on the wild egg stocks is somewhat balanced by laws that require half of the reared turtles to be returned to the wild. One problem with this practice is that the sex of the turtles is dependent on temperature during incubation (which depends on how deep they are buried in the wild); but the sex of young turtles cannot be ascertained without killing them, so the correct incubation temperature is not known. As a result, large numbers of turtles being returned to the wild may be of the same sex. This could devastate the wild population. Perhaps we should know more about such species before we allow captive breeding.

Lawrence Riggs was concerned with saving the genetic diversity of species as well as species themselves. From his experience, he felt that the private sector was doing more than the public sector to preserve the natural diversity of various tree species. For example, much of Redwood National Park in northern California had been logged before it was set aside as a park. The seedlings used to replant logged areas of the park were from outside the park, however, so that the natural diversity of genotypes native to the park was reduced. Private timber company officials are aware that efforts on public lands are not necessarily going to deliver what is important to private companies: having well-adapted native stocks from which to collect seed in the future. As a result, private companies are becoming convinced that they must undertake these efforts themselves. There is at least one company in California that has designated, on paper, areas to be considered as special managed stands for this purpose.

Miguel Altieri felt that commercialization of the market for genetic information may lead to dependence of the LDCs on firms in developed countries for such information. He also asked who should compensate those who bear the external costs of such agricultural practices as DDT use.

Joe Swierzbinski noted that what had been occupying most of the previous two days' discussion was wild habitat destruction in the Third World. This was in contrast to the present discussion, which focused on institutions for preserving harvested species. He felt that some of the same principles might be used to design the best institutions for habitat preservation, but that this needed to be addressed.

Krutilla then turned the discussion to political aspects of genetic resource conservation. Salwasser began the discussion with a review of the politics of the spotted owl case. The spotted owl is not listed as a threatened or endangered species. The concern over it came about as a result of the National Forest Management Act of 1976, which mandated the maintenance of diversity on the national forests, leaving it to the secretary of agriculture to decide what "diversity" meant. By 1979 the secretary had decided that in federal regulations (over which the secretary has total control) diversity meant, among other things, viable vertebrate populations. He left it to the Forest Service to decide what "viable" means.

The timber industry formally protested to the secretary and to the chief of the Forest Service in that year that the spotted owl was stopping timber sales. In 1981, the Forest Service instructed the foresters to plan for maintaining the numbers and range of reproductive-age individuals of vertebrate populations to ensure that all species were well distributed, at least in their current range. Clearly, the planning process gave consideration to unlisted species that were in danger because of impending changes in their habitats or human uses.

In the spotted owl case, an interagency group in Oregon and Washington composed of the Fish and Wildlife Service, the Oregon and Washington fish and game departments, the Bureau of Land Management, the Forest Service, and experts from Oregon State University developed recommendations that the participants felt would maintain the spotted owl. The participants submitted the recommendations to their respective line officers and received endorsements from all of the agencies except the Bureau of Land Management (which was not operating under the same legal requirements).

The guidelines were adopted and then immediately appealed by the timber industry as being overly restrictive of timber harvesting. They were nevertheless incorporated into the instructions to all national forests in Oregon and Washington. The guidelines called for 1,000 acres of old growth for each pair of owls, the patches not to be more than 6 to 12 miles apart, with the collective set of National Forest and BLM lands to provide for at least 500 pairs of owls in Oregon and Washington. Similar guidelines were developed in California instructing the Forest Service to preserve another 400 to 500 owls.

Experts involved in developing the guidelines included natural historians, population demographers, and geneticists. In the spring of 1982, the Forest Service convened a small workshop of scientists to help them develop a general process for planning to maintain wild populations. Geneticists were especially prominent in this group. In October 1984, the Forest Service reconvened a group of experts in Ann Arbor, Michigan, because a number of things had been learned in the two-year interval, specifically that genetics wasn't the only important consideration. The protocol was revised to include, in a more thorough fashion, demographics, biogeography, and decision analysis, along with genetics.

In December 1984, the National Wildlife Federation formally appealed the decision to manage habitats according to interagency guidelines. The secretary of agriculture, who had also been under pressure from timber interests, saw an opportunity to reassess the original decision. The appeal was granted, and a new environmental impact statement was required. This time it was made clear that an analysis of economic effects was to be included in the decision process.

In May 1985, Oregon Governor Victor Atiyeh appealed to the secretary of agriculture to relieve the Northwest timber industry from spotted owl constraints. Senator Mark Hatfield asked for a halt in planning because the forest plans were reducing timber harvests. The deputy assistant secretary of agriculture made clear, informally, that he is considering removing the requirement that viable populations be maintained on each national forest.

Meanwhile, the National Wildlife Federation has released the findings of a new demographic and biogeographic analysis, done by R. Lande of the University of Chicago, which concludes that the spotted owl is doomed to extinction within 100 years under the planned management scheme. They will request that the secretary of agriculture call for an immediate halt to all harvesting of mature and old-growth timber in the national forests of Oregon and Washington for at least one year. If the secretary refuses, they will file suit to enjoin the Forest Service from that harvesting.

It was asked why the owl is not listed as an endangered species, and Salwasser replied that the Fish and Wildlife Service has decided that it is not in danger of extinction under current planned management. They have reviewed the case as recently as 1984. The secretary of agriculture is considering pulling the viability requirement out of the regulations, which would defuse the issue, at least until the owl is formally listed as an endangered species.

Gardner Brown felt that if conservationists push too hard on trying to save an individual species like the snail darter, they may end up losing lots of species in the legislative backlash. Margery Oldfield felt that saying that every species must be saved at all costs was a mistake, and that there was good justification for establishing priority rankings of species at least partly based on the cost of saving them from extinction. For example, it would take only $10,000 to save the Texas wild rose, while we have spent $2 million on the spotted owl (although this may simply reflect a preference for vertebrates over plants).

Salwasser mentioned that $400,000 will be spent on an environmental impact statement to reanalyze the spotted owl case. Clearly, a trade-off is being made between preservation and doing research to better understand how to preserve other species, or genetic resources in general. Noll concluded the session by suggesting that when the participants of the workshop are acting as advocates of conservation, they may learn from this experience not to overthrow a regulation they can live with for five years in the middle of an administration that would like to cut back on preservation.

7

Synthesis and Recommendations

Gordon H. Orians, Gardner M. Brown, Jr.,
William E. Kunin, and Joseph E. Swierzbinski

Papers and discussions at Lake Wilderness were organized around three major problem areas: (1) *Valuation:* what values can be ascribed to biological resources, and which resources are likely to be of value? (2) *Technology:* how can we best preserve biological resources that we deem valuable? (3) *Incentives:* what institutional structures best ensure that conservation efforts will be carried out in appropriate ways? As Figure 13 indicates, these issues represent areas where the concerns of biology, economics and management overlap. The geometric analogy is useful in revealing the issues in which the different perspectives most frequently find a common focus. For example, the concerns of economists and biologists often converge when it comes to matters of valuation and the choice of what to preserve. These areas of mutual interest generate both challenges and opportunities. They provide challenges because each perspective is likely to need input from another without necessarily being aware of this need. They also represent opportunities for valuable interdisciplinary communication and research.

To bring together the three themes of the workshop in a synthetic perspective that highlights research needs, the final session of the workshop was devoted to meetings of three working groups, one representing each of the three perspectives. To continue the cross-fertilization evident in the discussions throughout the workshop, each group included biologists, economists, and managers. Although we review the recommendations sequentially in terms of the three perspectives, readers should keep in mind the areas of overlap indicated in Figure 13. The recommendations highlight both the overlap in substance and the differences in viewpoints involved in the three perspectives.

THE BIOLOGICAL PERSPECTIVE

The discipline of conservation biology is at a turning point. Until quite recently its efforts have concentrated on a few conceptual models, particularly the theory of island biogeography as originally formulated by MacArthur and Wilson (1967). The first and second conferences on conservation biology (Soulé and Wilcox, 1980; Soulé, 1986) helped to promote consideration of new ideas and new ways of addressing the issues. Both the biological and management aspects of the present workshop have served to open new areas of scientific interest, raising in the process at least as many questions as they answered.

282

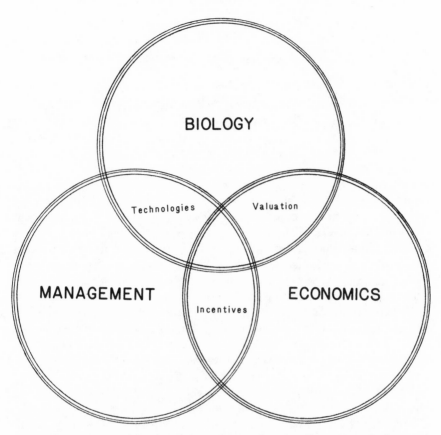

FIGURE 13. A conceptual framework for explaining the roles of economics, biology, and management in the conservation of genetic resources.

The research topics raised in this working group fall into four general areas: refining and expanding deletion sensitivity models, making optimal use of limited information, understanding the role of captive breeding, and improving the theoretical underpinnings of conservation biology.

Refining and Expanding Deletion Sensitivity Models

The intellectual framework of deletion sensitivity, discussed in Chapter 4, requires considerable refinement before it can be applied with any confidence. For example, how are communities affected by the simultaneous loss of several species? Are deletion effects additive? If not, does the order of species losses matter? Do reductions in population size create qualitatively similar effects to those caused by extinctions? How are ecosystem processes such as energy flow, nutrient cycling, and hydrology affected by species losses? A great deal of theoretical and empirical work needs to be done before the answers to questions of this sort can be determined. Only then will it be possible to assess the consequences of species deletions and to use this knowledge in directing conservation efforts.

Making Optimal Use of Limited Information

The body of biological theory potentially useful to managers and conservationists has grown steadily in recent years and is expected to continue to do so in the future. As models grow more sophisticated, however, their input data requirements become more exacting and extensive. In most management situations such technical information is lacking and would be expensive to procure, using resources that could otherwise be devoted to conservation efforts. The marginal value of additional data generally drops as information is accumulated, but the cost of procuring them may remain constant or even rise. Consequently, it is usually preferable, indeed often necessary, to make decisions using incomplete data rather than delaying them while gathering more data. A major challenge is to develop better techniques for using limited information. Several areas where such methods would be especially useful are discussed below.

Keystone Species. If deletion impact models are to prove useful in conservation, there needs to be a practical method for identifying keystone species (those whose loss would result in major changes in the functioning of the systems of which they are a part) in advance. Expanding the system of ecological equations used by Pimm in his simulations is probably not workable, but a number of generalizations emerging from Pimm's work may have broad applicability. Connor's suggestion that keystone species are often those that modify habitats or otherwise create structure provides a profitable avenue of research.

Extinction Proneness. A number of relatively simple predictors of extinction proneness have been proposed in recent years, based on theoretical considerations and inventories of past extinctions in certain well-studied areas. Experimental work is needed to confirm the predictive power of such generalizations and their applicability to geographically and taxonomically distant cases.

Minimum Viable Population Size. Soulé's (1980) original discussion of the minimum threshold for viability of a population was crude, and it predicted the same minimum effective population size for all species. More recent models (e.g., Gilpin and Soulé, 1986) are much more complex and require detailed ecological information before predictions can be made. What are most needed now are models of intermediate levels of complexity that capture some of the important dynamics of the complex models without requiring unattainable data bases.

Species Inventories. The uncatalogued diversity remaining to be described is such that, were we to stop conservation efforts to inventory it, all decisions would have to be postponed for decades. Choices about areas to be preserved must be made before the biological contents of those areas are known in detail. We need to develop ways to predict biological diversity on local scales to assist in the proper location of reserves. We also need to determine if areas of high species richness and endemism of such well-studied groups as birds and mammals correspond to those for other taxa. Geographic and climatic factors that best predict species richness need to be identified.

Understanding the Role of Captive Breeding

In recent years, population genetics has assumed an increasingly important place in the intellectual arsenal of conservation biologists. The applicability of such "pure" research to the management of small populations *in situ* and *ex situ* has often been remarkably direct, and the potential for further progress seems quite good. Members of the working group highlighted the following areas as being of particular interest in the future development of conservation genetics.

Inbreeding and Outbreeding Depression. Conservation efforts often necessitate inbreeding (mating of very close relatives) or severe outbreeding to the degree that losses of fitness can result. Species differ greatly in their susceptibility to such deleterious genetic effects. In the case of inbreeding depression, recent work (Lande and Schemske, 1985; Schemske and Lande, 1985) suggests that naturally inbred (selfing) populations have lower genetic loads, and consequently lower susceptibility. Remarkably little work has been done on the inbreeding levels of natural populations, and seldom is information on dispersal patterns and breeding systems incorporated into those studies.

The Role of Genetic Intervention In Situ. In situ and *ex situ* conservation strategies are alternative but complementary pathways to the preservation of genetic resources. More work needs to be done to integrate them better. For example, under what circumstances is it most useful to store genetic material *ex situ* for future introductions in the event that variability in the wild declines? How does the social organization of the species involved affect the choice of the ages and sexes of individuals to release?

Optimal Breeding Strategies. A breeding population will experience a certain degree of both genetic impoverishment and inbreeding. The former is minimized by subdividing the gene pool into smaller, tightly inbred units, each of which preserves different components of the original genetic mixture. The latter is best avoided by keeping genetic interchanges as free as possible. Managing the trade-offs between these two antagonistic goals is a challenging problem. The best strategies may differ for populations intended for release as opposed to those intended for long-term captive propagation. Also the optimal breeding system depends on the biology of the species involved.

The Role of Selection in Managed Systems. The maintenance of captive or other closely managed populations inevitably brings with it some degree of artificial selection. We need to determine to what degree such selection jeopardizes re-introduction efforts, its role in the maintenance of inbreeding depression, and how it can be minimized, and at what cost.

285

Improving the Theoretical Underpinnings of Conservation Biology

Whereas conservation biology theory has made important advances in the last decade, there remain several areas where additional modeling is badly needed. In addition, much of existing theory is as yet untested. Among these important research needs are the following.

Applying Island Biogeographic Theory. The widely used theory of island biogeography relating equilibrium species richness to island size and degree of isolation is in need of empirical review to determine the current data base supporting it and the geographic and taxonomic generality of its findings. The various phenomena contributing to species-area effects need to be teased apart statistically and experimentally so that the differences among them in terms of their consequences for conservation strategies can be clarified.

The Intermediate Disturbance Hypothesis. A growing body of evidence supports the generalization that habitats with moderate levels of disturbance or stress support more varied biological communities than areas where disturbance is either very high or very low. The theoretical basis for this effect remains in contention. Therefore, conditions under which the introduction of disturbance to managed systems is appropriate are not clear. We cannot yet distinguish between a site that would be enriched by further stress and one that would be impoverished by it.

Donor Control. Keystone predators are unlikely to be found in systems that are "donor controlled" (where the rate of prey supply is largely independent of predator density). Such is clearly the case, for instance, among detritivores and sessile planktivores. As research results accumulate on the importance of places where prey can escape from their predators, it appears that many other systems are, at least in part, donor controlled. More work is needed to explore these cases empirically and theoretically to determine where they are likely to evolve and to what extent partial refugia are similar in dynamics to pure donor controlled cases.

Evolution, Community Structure, and Deletion. Rosenzweig (Chapter 4) proposed an evolutionary model for the development of deletion sensitivity. The generalizations he proposes concerning the strength of deletion impacts as a function of community structure are intriguing but speculative. Further work is needed to test these and similar conclusions theoretically and empirically.

Provincialization and Insularization. The effects of insularization on the evolutionary potential of organisms has been the subject of much recent discussion. Division of continuous habitats into small parcels can terminate speciation processes or stimulate them, depending on the type of process operative. The net effect of such antagonistic pressures and how a reserve can best be managed to promote evolutionary potential when it is desired are in need of intensive study.

Population Replenishment. Relatively little work has been done on restoration of populations and communities after serious perturbations (but see Bradshaw and Chadwick, 1980). The potential usefulness of such work to conservation efforts is clear. Where protected species have failed to flourish, we need to determine to what extent negative density dependence, alternative stable states, or other effects can be demonstrated or inferred.

Overlap and Faunal Collapse. A series of small reserves can protect more species than a single large reserve of similar total size, as long as the various small parks differ sufficiently in what they save. In managing a system of reserves, we need techniques to encourage different species to survive on different, but similar islands. We do not know if removing one species from an island will improve the survival chances of another species or to what extent species are lost to small islands due to competitive interactions, as opposed to such effects as large range requirements and dependence on immigration.

THE ECONOMIC PERSPECTIVE

The discussions during the third day of the workshop led to a tripartite taxonomy of genetic resource values, each associated with a particular kind of investment in conservation: (1) values associated with the phenotype (characteristic) being conserved, (2) values associated with a total stock of genotypes or alleles, and (3) values associated with the knowledge needed to better invest in the conservation of the resources identified in the first two types.

1. There is a broad range of values associated with the phenotype being conserved, including the direct consumptive value obtained when organisms are used for food, clothing, or the production of drugs and other commodities; amenity values obtained by viewing the organisms; ecological values obtained when the organism contributes to the maintenance of a valued ecosystem; and the option values associated with the uncertain future enjoyment of the above values. In addition, there is the possibly important category of existence value.

2. A stock of alleles can have an important insurance value by providing a pool of genetic variability that can be used to facilitate the production of valuable phenotypes at some future date. This category of value can be thought of as a special case of the first category, since the insurance value of the genetic stock is closely related to the option value of the phenotypes that can be produced from it. On the other hand, the distinction is worth making, since the most likely value of the genetic stock may involve future phenotypes that are significantly different from any current ones.

3. Investment in knowledge to improve future choices is a distinct and important part of the total investment process. The working group focused mainly on incentive issues, because economic techniques for valuing phenotypes are well understood and because the unresolved issues concerning economic valuation

287

are not specific to genetic resources. Yet two important concerns emerged and warrant mention.

Techniques of Economic Evaluation

Refining Existence Value Studies. The existence value of a resource (willingness to pay for a resource that one will never use) is considered by some analysts to be a significant component of the value of many species, such as whales. Surveys that necessarily have used hypothetical questions to estimate existence value typically find large dollar values. It is, however, difficult to believe these estimates, which are as high or higher per capita than the values reported by those who directly enjoy the resource. Because the problems usually associated with hypothetical valuation are especially acute for existence issues, it would be useful for economists, psychologists, and other expert questionnaire designers to develop better types of questions. The use of hypothetical valuation to measure existence value may also be a fruitful area of research for experimental economists.

Amalgamating Finance Literature and Uncertain, Future Valuation Questions. A second issue concerns the insurance value of a genetic stock. A large amount of work has already been done on the valuation of risky assets (Chapter 5). Nevertheless, the subject remains an area of active research, and interesting questions remain unanswered. Ongoing research in this area may have important impacts on the methods used to value genetic resources.

Economic Incentives for Preservation

The discussions regarding incentives and institutions can be summarized in terms of three further recommendations for research management.

Institutional Case Studies of Common Property Management. Commercialization is an important issue because of two characteristics of many endangered species. First, the habitat of many species often has a high alternative value (e.g., as farmland) for uses that may be incompatible with species preservation. Second, many species are fugitive in nature, making it difficult for individuals to capture the gains from increasing their numbers or otherwise conserving the species. When both of these factors are present (often one is sufficient), there is no financial incentive to pay the costs of preserving the habitat of the species. The combination of the common property problem and the high opportunity costs of land makes commercialization both attractive and problematic.

A large body of economic theory and empirical study deals with the common property problem and how to solve it. Although the theory typically involves strong assumptions, it has proved to be a powerful tool for predicting the success or failure of particular policies. Hence, the theory deserves to be taken seriously by policy makers. On the other hand, empirical studies indicate a wide variation in the effectiveness of even those policies emerging from the theory.

288

Therefore, we recommend that a number of field studies, primarily in the Third World, analyze past attempts to develop ranching and/or farming of wild stocks of plants and animals. The studies would discover the economic, cultural, and institutional reasons for the success or failure of particular efforts. The goal is to determine how well a richer research framework can extend the predictive power of current economic theory. Such a framework should include, for example, a concern with historical and cultural settings and a detailed description of the actual property rights that were established. A major drawback of previous studies is their lack of grounding in economics, without which the wrong questions are often asked and crucial data are not collected. Avoiding these errors is imperative.

Two types of research benefits should result. In the long run, the descriptive power of the common property paradigm will be enhanced. In the short run, additional economic data for answering current policy questions will be provided. A number of commercialization efforts currently being funded by the United States, by multinational agencies, and by private sources could be enhanced by a richer economic perspective. Using these existing efforts as the basis for case studies makes it possible to combine valuable research with applied policy analysis at low cost.

International Policies for Paying the Opportunity Cost of Maintaining Wilderness. This next recommendation concerns the role of developed countries in encouraging the conservation of wild tropical habitat, located primarily in the Third World. Tropical ecosystems represent a particularly valuable storehouse of genetic resources for research, for use in agricultural breeding, and for other purposes. However, many of those ecological systems are being converted into farmland. Maintaining wilderness involves paying the opportunity cost associated with forgoing those uses. Developing ways for advanced countries to help pay this opportunity cost is essential. Possible policy instruments include purchase or lease of land by private or multinational agencies and other forms of cooperation and subsidy.

There are several reasons why developed countries should be interested in this issue. First, the developed countries are likely to reap a large share of the benefits associated with the preservation of wild tropical habitats. Second, it is likely that the "owners" of the wilderness land will be unable to control access to many of the benefits accruing from the land, such as knowledge from research or stocks of seeds. Lacking a method for obtaining an "economic rent" by selling wilderness benefits, the current owners of the land have no incentive to pay the opportunity cost of preserving it. Even when some of the benefits can be bought and sold, differences in the discount rate between countries, and other imperfections in the capital market, suggest that the developed countries may wish to preserve relatively more wilderness than the developing countries do.

We also recommend that two ongoing efforts to enhance conservation, one

289

in Costa Rica and the other in Kenya, be given special attention and study. The Costa Rican case involves the purchase of large tracts of land through funds raised from many sources external to the country. The Kenyan case involves the design and management of game reserves such that local people, the ones paying the largest opportunity costs from reserve establishment, can reap a larger fraction of the benefits.

We further recommend the study of the conceptual issues involved in the design of stable agreements to enhance wilderness preservation. In the absence of any means for enforcing international contracts, any agreement to preserve wilderness areas must be self-enforcing: it must be in the self-interest of all parties to keep the agreement, not only at the time it is made but at all future times. The economic theory of self-enforcing agreements may offer valuable conceptual insights into the design of stable, long-term preservation agreements.

Although these recommendations focus on the economics of institutional design, "intervention" by developed countries in the Third World involves a number of very sensitive political and cultural issues. For example, the large-scale purchase of land in the Third World could easily raise the specter of imperialism. Therefore, such preservation efforts may be best coordinated by international agencies. In addition, the educational and cultural base of the country in question has much to do with the outcome of conservation efforts. Hence the form and content of education may be vitally important to the success of any conservation plan. Clearly, the design of long-term agreements to preserve tropical wilderness involves a number of important research topics for political scientists and sociologists, as well as for economists.

Evaluation of Alternative Forms of Agriculture in Tropical Countries. Older forms of "subsistence" agriculture, based on native wild crops, may provide attractive alternatives for modern agriculture in tropical countries when the social value of conserving native "land races" of crops is incorporated. These land races are a reservoir of genetic variability valuable for agriculture in both developed and developing countries. This recommendation is a special case of the commercialization and subsidy schemes discussed above. We list it here as a separate category to emphasize its importance.

THE MANAGEMENT PERSPECTIVE

The job of a resource manager begins where that of the policy maker leaves off. Given that a decision has been made to preserve certain genetic resources (however determined), and given that appropriate political and social institutions exist, how do we decide on the best course of action to meet those goals? Several issues were identified by the working group as especially important.

Making Technology Effective

To maintain biological diversity in a world of increasing and changing pressures, new techniques for conservation continually need to be created, developed, and refined. In the process, attention needs to be directed toward the purposes of a particular conservation effort (whether it is for, say, crop germ plasm improvement or aesthetic enrichment) so that the technology chosen is most appropriate to the needs. How flexible are existing institutions to changes in conservation technology? Alternatively, how can technologies be designed to fit within current institutional structures? How promising are the emerging new techniques for *ex situ* preservation?

Choosing from Existing Technologies

Given the technological options available, ranging from wilderness areas to gene sequencing, how can we determine which are most appropriate for a particular conservation effort? What biological data are most relevant for each type of problem, and how should the relative importance of problems be assessed? What social, economic, and institutional considerations must be factored into the management decisions? How do our reasons for conserving influence our choice of technologies? What will be the effects of different preservation techniques on nontarget concerns (other species, people nearby, political constituencies, etc.)?

Communicating Conservation

Information can be one of the most powerful tools for managing genetic resources. Adequately informing practitioners, conservation leaders, and the public about efforts planned or in progress can be at least as important to the success of a program as any of the technical issues that dominated discussions during the workshop. Educating people about the strategies, methods, and reasons for conserving particular genetic resources is a critical aspect of managerial success, and it is too often neglected. Many educational needs and methods are obvious, but the complexity of the subject of genetic resource preservation places heavy demands on educators to explain clearly, but in simple terms, the complicated issues of population genetics, community ecology, and economics. Research on how to communicate these technical matters is needed.

Coping with Uncertainty

Most conservation decisions cannot and should not be delayed until all relevant information can be obtained. Little is known about the biology of many of the organisms with which we must work. The bases of conservation priorities currently held by the general public are both difficult to determine and changeable. The cost and effectiveness of many of the tools at our disposal are

291

poorly known. How should conservation decisions cope with such inherent uncertainty? How can they be made sufficiently flexible to adapt to new information as it becomes available? One approach to this problem, known as adaptive environmental management (Holling, 1978), has been developed and used in several cases. Its main feature is that the uncertainty surrounding the situation and the probable consequences of any actions that might be taken are acknowledged at the beginning. Decisions about which actions to undertake are influenced by expectations about which ones will best help gain valuable information while at the same time not irrevocably committing the full course of action of the project. This allows modifications in the action plan as new information becomes available (the *adaptive* part). Attractive as this approach may be, there are great difficulties in implementing it. Results may be too slow in coming and candid acknowledgment of uncertainty may make the project politically vulnerable. Research on the efficacy of adaptive environmental management in the context of preservation of genetic resources would be especially valuable.

Taken together, the issues discussed above represent a broad spectrum of time frames. The development and application of new technologies are important long-range goals, but it is still necessary to make short- and medium-range choices using the imperfect implements that are at hand. Research on the issues outlined above should help managers make better choices even though great uncertainty will always remain.

EPILOGUE

Improving the foundation from which societies choose those genetic resources to preserve and decide how best to preserve them requires advances in pure and applied biology (especially population genetics and ecology), in theoretical and experimental economics, and in the theory and practice of public institutions. The field is another example in the growing list of technical issues that pose serious problems for the functioning of democratic institutions in a framework of increasing technical, political, and demographic complexity. Embedded in matters of preservation of genetic resources are both the traditional issues of distribution of societal benefits and difficult questions concerning what those benefits are likely to be and how they can be realized. Moreover, important problems of international relations are central to many components of actions undertaken to help preserve those resources. Therefore, in many of its aspects, preservation of genetic resources is typical of problems of democratic institutions in highly technological societies.

Yet many issues concerning the preservation of genetic resources have features unique to themselves. Among the most important are the extensive land requirements for species preservation, the noncoincidence of the political-geographical location of the genetic resource and its direct beneficiaries, the difficulties of charging adequate economic rents for access to the benefits of preservation, the complexities of ecological interactions which mean that any conservation action is likely to have unforeseen side effects, and the fact that all preservation techniques are likely to affect the genetic structure of the populations being preserved, hence their evolutionary potential. Gaining better understanding of these problems will require the attention of the best minds in the appropriate disciplines and, perhaps even more important, close cooperation among people from disciplines that usually pursue their activities independently. The exciting dialogues between geneticists, ecologists, economists, and managers that characterized the Lake Wilderness workshop were perhaps the greatest benefits realized by the participants. Our report is intended as a stimulus to help institutionalize such interchanges and to lead to the accelerated growth of the field of genetic resource preservation. The individual disciplines themselves may reap unexpected benefits in the process.

REFERENCES

Bradshaw, A. D., and M. J. Chadwick. 1980. *The Restoration of Land*. Oxford: Blackwell.

Gilpin, M. E., and M. E. Soulé. 1986. Minimum viable populations: Processes of species extinction. In *Conservation Biology: The Science of Scarcity and Diversity*, ed. M. E. Soulé, 19–34. Sunderland, Mass.: Sinauer.

Holling, C. S. (ed.) 1978. *Adaptive Environmental Assessment and Management*. New York: John Wiley and Sons.

Lande, R., and D. W. Schemske. 1985. The evolution of self-fertilization and inbreeding depression in plants. I. Genetic Models. *Evolution* 39:24–40.

MacArthur, R. H., and E. O. Wilson. 1967. *The Theory of Island Biogeography*. Princeton, N.J.: Princeton University Press.

Schemske, D. W., and R. Lande. 1985. The evolution of self-fertilization and inbreeding depression in plants. II. Empirical observations. *Evolution* 39:41–52.

Soulé, M. E. 1980. Thresholds for survival: Maintaining fitness and evolutionary potential. In *Conservation Biology: An Evolutionary-Ecological Perspective*, ed. M. E. Soulé and B. A. Wilcox, 151–169. Sunderland, Mass.: Sinauer.

Soulé, M. E. (ed.) 1986. *Conservation Biology: The Science of Scarcity and Diversity*. Sunderland, Mass.: Sinauer.

Soulé, M. E., and B. A. Wilcox (eds.). 1980. *Conservation Biology: An Evolutionary-Ecological Perspective*. Sunderland, Mass.: Sinauer.

Contributors

PETER ASHTON, Harvard University Herbaria, Harvard University, Cambridge, Massachusetts

MIGUEL ALTIERI, State of California Division of Biological Control, Albany, California

JANIS ANTONOVICS, Department of Botany, Duke University, Durham, North Carolina

GARDNER M. BROWN, JR., Department of Economics, University of Washington, Seattle

ERIC CHARNOV, Department of Biology, University of Utah, Salt Lake City

EDWARD F. CONNOR, Department of Environmental Science, University of Virginia, Charlottesville

PARTHA DASGUPTA, Faculty of Economics, University of Cambridge, Cambridge, England; Department of Economics, Stanford University, Stanford, California

DOUGLAS E. GILL, Department of Zoology, University of Maryland, College Park

JON GOLDSTEIN, Office of Policy Analysis, Department of the Interior, Washington, D.C.

MARK HAFNER, Museum of Natural Science, Louisiana State University, Baton Rouge

MICHAEL HANEMANN, Department of Natural Resource Economics, University of California, Berkeley

DREW HARVELL, Department of Ecology and Systematics, Corson Hall, Cornell University, Ithaca, New York

BRIAN HARVEY, Department of Biology, University of Victoria, Victoria, B.C., Canada

295

RONALD JOHNSON, Department of Agricultural Economics and Economics, Montana State University, Bozeman

JOHN KRUTILLA, McLean, Virginia (formerly at Resources for the Future)

WILLIAM E. KUNIN, Department of Zoology, University of Washington, Seattle

ROGER NOLL, Department of Economics, Stanford University, Stanford, California

MARGERY L. OLDFIELD, Division of Biological Science, University of Texas, Austin

GORDON H. ORIANS, Institute for Environmental Studies, University of Washington, Seattle

DEBORAH RABINOWITZ (deceased), formerly at the Department of Ecology and Systematics, Cornell University, Ithaca, New York

ROBERT REPETTO, World Resources Institute, Washington, D.C.

LAWRENCE RIGGS, Genrec, Berkeley, California

MICHAEL ROSENZWEIG, Department of Ecology and Evolutionary Biology, University of Arizona, Tucson

OLIVER RYDER, Zoological Society of San Diego, San Diego, California

HAL SALWASSER, U.S. Forest Service, Washington, D.C.

CHRISTINE M. SCHONEWALD-COX, Division of Environmental Studies, University of California, Davis

RUTH SHAW, Department of Botany and Plant Science, University of California, Riverside

WILLIAM STRANG, The Treasury, Wellington, New Zealand

MICHAEL S. STRAUSS, Congress of the United States, Office of Technology Assessment, Washington, D.C.

JOSEPH SWIERZBINSKI, Department of Economics and the Institute for Environmental Studies, University of Washington, Seattle

ARTHUR WEISSINGER, Pioneer Hi-bred International, Inc., Johnston, Iowa

BRUCE WILCOX, ISD, Palo Alto, California

DAVID WOODRUFF, Department of Biology, University of California, San Diego, La Jolla, California

Index

The Preservation and Valuation of Biological Resources

Edited by Gordon H. Orians, Gardner M. Brown, Jr., William E. Kunin, and Joseph E. Swierzbinski

Human activities are threatening the extinction of a substantial number of species living on earth. The immense scale of the impending "extinction crisis" means that hard choices must be made about what to conserve with limited resources. How can we estimate the value of species in natural ecosystems, their value in terms of the products and services they provide, and their role in enhancing the quality of human life? Answers to these questions are necessary for the development of wise policies concerning the allocation of societal resources for the preservation and management of the planet's living resources.

In this book, a distinguished group of biologists, economists, and resource managers presents a multidisciplinary approach to the importance of the earth's biological richness. The six major chapters focus on methods of preserving genetic resources, both in the laboratory and in the field; the biological value of species richness, both genetic and ecological; management of biological resources; and economic methods of assessing the value of species. The essays are accompanied by commentaries and summaries written by specialists from a variety of disciplines. A final chapter provides an overview, culminating in suggestions for needed research.

Gordon H. Orians is director of the Institute for Environmental Studies and professor of zoology; *Gardner M. Brown, Jr.,* is chairman of the Department of Economics, and *William E. Kunin* is a graduate student in zoology, all at the University of Washington. *Joseph E. Swierzbinski* is associate professor of economics, University of Michigan.